AF360904

Intelligent Sensors for Positioning, Tracking, Monitoring, Navigation and Smart Sensing in Smart Cities

Intelligent Sensors for Positioning, Tracking, Monitoring, Navigation and Smart Sensing in Smart Cities

Editors

Tiancheng Li
Junkun Yan
Yue Cao
Javier Bajo

MDPI • Basel • Beijing • Wuhan • Barcelona • Belgrade • Manchester • Tokyo • Cluj • Tianjin

Editors
Tiancheng Li
Northwestern Polytechnical
University
China

Junkun Yan
Xidian University
China

Yue Cao
Wuhan University
China

Javier Bajo
Universidad Politécnica de
Madrid
Spain

Editorial Office
MDPI
St. Alban-Anlage 66
4052 Basel, Switzerland

This is a reprint of articles from the Special Issue published online in the open access journal *Sensors* (ISSN 1424-8220) (available at: https://www.mdpi.com/journal/sensors/special_issues/intelligent_smart_cities).

For citation purposes, cite each article independently as indicated on the article page online and as indicated below:

LastName, A.A.; LastName, B.B.; LastName, C.C. Article Title. *Journal Name* **Year**, *Volume Number*, Page Range.

ISBN 978-3-0365-0122-2 (Hbk)
ISBN 978-3-0365-0123-9 (PDF)

Contents

About the Editors

Tiancheng Li (Professor) is currently a Professor at the School of Automation, Northwestern Polytechnical University (NPU), Xi'an, China. He received two bachelor's degrees from Harbin Engineering University, Harbin, China (2008), his first Ph.D. degree from London South Bank University, London, U.K. (2013), and his second doctoral degree from NPU (2015). He received the Excellent Doctoral Thesis Award of Shaanxi Province in 2017 and the Marie Sklodowska-Curie Individual Fellowship from the European Commission in 2016. He was a Postdoctoral Researcher with the University of Salamanca, Spain, from June 2014 to the Fall of 2018, and a Visiting Scholar with the Vienna University of Technology, Austria, in the Summer of 2017 and Fall of 2018. His research is focused on collaborative mobile robots, distributed-linear information fusion, and intelligent data-driven algorithms for target detection, tracking, and trajectory forecasting. He is an Associate Editor for three peer-reviewed journals *Frontiers of Information Technology and Electronic Engineering*, *Advances in Distributed Computing and Artificial Intelligence Journal*, and *Aero Weaponry*.

Junkun Yan (Professor, IEEE Senior Member) was born in Sichuan, China, in 1987. He received his B.S. and Ph.D. degrees in electronics engineering from Xidian University, Xi'an, China, in 2009 and 2015, respectively. He is currently an associate professor of the National Laboratory of Radar Signal Processing, Xidian University. His research interests include adaptive signal processing, target tracking, and cognitive radar. He has published more than 50 scientific articles in refereed journals, such as the *IEEE Transactions on Signal Processing*, *Information Fusion*, *IEEE Transactions on Vehicular Technology*, *Signal Processing*, *IEEE Transactions on Aerospace and Electronic Systems*, and *IEEE Sensors Journal*. He received the Excellent Doctoral thesis Award from the SHANNXI Institute of Electronics in 2017. He also received the Young Talent fund of the China Association for Science and Technology in 2020.

Yue Cao (Professor) received his Ph.D. degree from the Institute for Communication Systems (ICS) formerly known as the Centre for Communication Systems Research, at the University of Surrey, Guildford, UK in 2013. Further to his Ph.D. study, he was a research fellow at the University of Surrey, and academic faculty at Northumbria University, Lancaster University, UK, and Beihang University, China; he is currently a Professor at the School of Cyber Science and Engineering, Wuhan University, China. His research interests focus on Intelligent Transport Systems, including E-Mobility, V2X, Edge Computing.

Javier Bajo (Profesor) is a Full Profesor at the ETS Ingenieros Informáticos—Universidad Politécnica de Madrid (UPM). He is currently the Coordinator of the Research Master in Artificial Intelligence at UPM. Previously, he held the position of Associate Professor at the Universidad Pontificia de Salamanca (2003 to 2012) and Director of the Data Center in the same University (2010–2012). He obtained a Ph.D. in computer sciences from the Universidad de Salamanca (with honors) (2007) and a Master in E-Commerce from the same University (2006). He obtained a bachelor's in computer sciences at the Universidad de Valladolid (2001) and an MSc in computer sciences at the Universidad Pontificia de Salamanca (2003). His research efforts are focused on multi-agent systems, social computing, and ambient intelligence. He has participated in more than 50 research projects (funded by European Commission, National or Regional entities) and contracts with private companies, acting as the principal researcher in 11 of these projects. He is the co-author of more than 300 scientific publications (books, journal papers, and conference papers). Of these publications, 49 have been published in international journals that are indexed in the ISI JCR reference index. He has been the co-chairman of the organizing committee of more than 30 recognized international conferences (ACM SAC, IEEE FUSION, PAAMS, etc.).

Preface to "Intelligent Sensors for Positioning, Tracking, Monitoring, Navigation and Smart Sensing in Smart Cities"

We are in a new era of intelligence and data. The rapid development of advanced sensors and their massive deployment provide a foundation for new paradigms to combat the challenges that arise in significant tasks such as positioning, tracking, and smart sensing in harsh environments with poor a priori information. Relevant advances in artificial intelligence and machine learning are also rapidly adopted by industry and further fan the fire. Together with the classic least squares and linear fusion paradigms, these advances provide another unlimited means to utilize the pattern of data and probabilistic models for positioning and tracking, showing great promise as a means of restoring sensor capability over a range of challenging operating conditions. Consequently, research on advanced sensing and estimation approaches has burgeoned into two promising and intertwined directions. The first is intelligent systems and data-driven algorithms for information fusion, which have led to a variety of effective applications related to intelligent transportation, autonomous vehicles/robots, wearable computing, smart sensing, and the internet of things. For example, the least squares estimator plus a clustering algorithm based on appropriate target trajectory modeling can deal with the complicated problem of joint target detection and tracking in clutter with little a priori knowledge. The second is distributed systems and multi-agent frameworks, which are gaining considerable popularity due to their low power consumption and simple installation and high performance and strong reliability, compared with a centralized setting. To this end, arithmetic average fusion of multi-target densities is a notable yet simple approach, which demonstrates outstanding robustness and tolerance to local failures. The advent in both directions provides rich observation at high frequencies but low financial costs, which facilitates novel perspectives based on data clustering, fusion, and learning to deal with noise, false alarms, and misdetection, given little a priori knowledge. As such, the sensors community has a great interest in novel information fusion, resource optimization, and data mining methods coupled with intelligent algorithms for substantial performance enhancement, especially for challenging scenarios that make traditional approaches inappropriate.

This book is a reprint of the Special Issue *"Intelligent Sensors for Positioning, Tracking, Monitoring, Navigation and Smart Sensing in Smart Cities"* which was published in the journal *Sensors*. This book consists of 14 papers contributed by scientists and technicians and aims to provide a cutting-edge coverage of recent advances in sensor signal and data mining techniques, algorithms, and approaches, particularly applied for positioning, tracking, navigation, and smart sensing.

Tiancheng Li, Junkun Yan, Yue Cao, Javier Bajo
Editors

Article

MIMU/Odometer Fusion with State Constraints for Vehicle Positioning during BeiDou Signal Outage: Testing and Results

Kai Zhu [1,2]**, Xuan Guo** [1]**, Changhui Jiang** [3,*]**, Yujingyang Xue** [1]**, Yuanjun Li** [1]**, Lin Han** [3] **and Yuwei Chen** [4,*]

[1] School of Automobile and Traffic Engineering, Jiangsu University of Technology, Changzhou 213001, China; fatkyo@jsut.edu.cn (K.Z.); gx1836226090@163.com (X.G.); 2019500013@jsut.edu.cn (Y.X.); 2019560099@jsut.edu.cn (Y.L.)

[2] Zhenjiang Zhongao AI Institute, Zhenjiang 212001, China

[3] School of Automation, Nanjing University of Science and Technology, Nanjing 210094, China; jiajialage@163.com

[4] Department of Photogrammetry and Remote Sensing, Finnish Geospatial Research Institute, Masala, FI-0245 Espoo, Finland

[*] Correspondence: Changhui.jiang@njust.edu.cn (C.J.); Yuwei.chen@nls.fi (Y.C.)

Received: 29 February 2020; Accepted: 16 April 2020; Published: 17 April 2020

Abstract: With the rapid development of autonomous vehicles, the demand for reliable positioning results is urgent. Currently, the ground vehicles heavily depend on the Global Navigation Satellite System (GNSS) and the Inertial Navigation System (INS) providing reliable and continuous navigation solutions. In dense urban areas, especially narrow streets with tall buildings, the GNSS signals are possibly blocked by the surrounding tall buildings, and under this condition, the geometry distribution of the in-view satellites is very poor, and the None-Line-Of-Sight (NLOS) and Multipath (MP) heavily affects the positioning accuracy. Further, the INS positioning errors will quickly diverge over time without the GNSS correction. Aiming at improving the position accuracy under signal challenging environment, in this paper, we developed an MIMU(Micro Inertial Measurement Unit)/Odometer integration system with vehicle state constraints (MO-C) for improving the vehicle positioning accuracy without GNSS. MIMU/Odometer integration model and the constrained measurements are given in detail. Several field tests were carried out for evaluating and assessing the MO-C system. The experiments were divided into two parts, firstly, field testing with data post-processing and real-time processing was carried out for fully assessing the performance of the MO-C system. Secondly, the MO-C was implemented in the BeiDou Satellite Navigation System (BDS)/integrated navigation system (INS) for evaluating the MO-C performance during the BDS signal outage. The MIMU standalone positioning accuracy was compared with that from the MIMU/Odometer integration (MO), MO-C and MIMU with constraints (M-C) for assessing the Odometer, and the influence of the constraint on the positioning errors reduction. The results showed that the latitude and longitude errors could be suppressed with Odometer assisting, and the height errors were suppressed while the state constraints were included.

Keywords: GNSS; MIMU; odometer; state constraints

---

1. Introduction

Unmanned Ground Vehicles (UGV) with complete automatic operation are regarded as the most promising technology to be available in the near future [1,2]. Precise and reliable position and navigation information are fundamental for the autonomous driving vehicles [3]. Currently, the Global Navigation Satellite System (GNSS) and the Inertial Navigation System (INS) are the most popular solutions for providing comparatively reliable positioning information [4]. GNSS is usually integrated with INS since they are highly complementary. The GNSS works by relying on the geometry distribution of the in-view satellites and signal quality, however, if the satellite signals are blocked by the surrendered buildings or obstacles, the GNSS will fail to generate precise positioning results [5]. Under this condition, the INS could provide moderate navigation solutions in a short time. However, due to the complex noises contained in the raw measurements from the gyroscope and accelerometer, the INS errors will accumulate quickly over time [4–6].

In the past decade, there has been a lot of literature focusing on improving the positioning accuracy under GNSS signal-challenging environments. These methods could be divided into two categories. The first solution is to suppress the INS noise and compensate for its positioning errors. The INS generates navigation solutions through processing the raw accelerator and gyroscope outputs. Limited by the manufacturing technology, there are complex noises contained in the raw measurements. Sheimy employed the Allan Variance method to characterize and quantify the noise [7,8]. Grip proposed an exponentially stable attitude and gyroscope bias estimation method in GNSS/INS integration [9]. Machine learning (LS-SVM, LSTM-RNN) methods were employed for modeling the errors [10–13]. Some calibration methods were also proposed to improve positioning accuracy [14–18]. Wu investigated the self-calibration of the Inertial Measurement Unit (IMU)/odometer integrated system for land vehicle navigation [14]. In addition, in the GNSS/INS integrated navigation system, some machine learning methods were employed and investigated to compensate for the INS errors during the GNSS signal outage [15–18]. These machine learning methods were well trained while the GNSS signal was normal.

The second solution is to employ more sensors in the GNSS/INS integration system and construct a multi-sensor fusion system. Among these sensors, LiDAR, vision cameras, altitude barometers, Chip Scale Atomic Clock (CSAC), and the odometer are the most popular sensors [19–24]. LiDAR is a sensor collecting the point cloud of the surrounding environment. With the continuous matching of the point cloud sequences, LiDAR can generate relative displacements and attitudes [19–25]. In aspects of the vision sensors, with the matching of the image's sequences, attitude changes could also be extracted. With two well-calibrated vision cameras or depth cameras, this method could also provide positioning information [23,24]. An altitude barometer and odometer could provide height and odometer information, respectively. GNSS/LiDAR/HD-Map/INS integration system is a popular solution for autonomous driving vehicle positioning and navigation [25]. In addition, with the size and accuracy improvement, the CSAC is employed for augmenting the GNSS accuracy by providing a more precise frequency base [21].

In general, ground vehicles are usually equipped with an odometer for measuring the traveled distance. Therefore, it is convenient to develop a GNSS/MEMS-INS/odometer fusion system. Some works have revealed and demonstrated its effectiveness in improving positioning accuracy [26–29]. Georgy investigated the stochastic drift model of a MEMS (Micro-Electro-Mechanical System) gyroscope in a gyroscope/odometer/GPS integrated navigation system [26]. A mixture of particle filter and fuzzy neural network was employed for enhancing the MEMS-IMU/odometer/GPS integration for land vehicle applications [27,28]. An odometer and MEMS IMU were also employed for enhancing Precise Point Positioning (PPP) under weak satellite observability environments [29]. However, these studies were conducted while the GNSS was available, and the influence of the constraints and odometer on the positioning errors were not presented respectively and clearly.

Scientists have explored and investigated this issue using the vehicle trajectory constraints to reduce the INS errors while the GNSS is unavailable [30–33]. Non-Holonomic Constraints (NHC) were employed as the measurements for suppressing the INS errors while GNSS was unavailable. While the vehicle is driving on the road, the velocity of both the up direction and perpendicular to the direction of vehicle traveling are almost zero [33]. The observability of the NHC was analyzed for demonstrating its effectiveness in land vehicle navigation systems [33]. However, NHC could not suppress the positioning errors in the forward direction. Therefore, in this paper, apart from the NHC, an odometer was also employed to suppress the positioning errors in the forward direction. We developed an

MEMS-IMU/odometer integration navigation system considering the vehicle state constraints (MO-C) for ground vehicle positioning without GNSS. Both data post-processing and real-time processing experiments were carried out for assessing the navigation solution accuracy. Comparisons between MO-C, MO, and M-C were presented for evaluating and validating the odometer and constraints influence on the navigation solution accuracy. The contribution and innovation of this paper are summarized as follows:

(1) The model of the IMU/odometer with constraints is comprehensively given, detailed equations are listed and analyzed, and the influence of the odometer and constraints on the positioning errors were numerically compared and evaluated, which might be a reference for implementing these algorithms for different conditions.

(2) The odometer and constraints were firstly implemented in a BeiDou Satellite Navigation System (BDS)/MIMU loosely-integrated navigation system for evaluating its performance and effectiveness in reducing and suppressing INS positioning errors while GNSS was unavailable, and positioning errors were presented for assessing these methods' feasibility in a GNSS/INS integration framework.

(3) In the experiments, both post-processing and real-time filed tests were carried out for assessing the odometer and NHC performance in improving the positioning accuracy, respectively, and the NHC and odometer were employed in the BDS/MIMU integrated navigation system, which was of great significance for improving the positioning accuracy during the BDS signal outage.

The rest of the paper is organized as follows: Section 2 introduces the model of the MO, MO-C (state measurement equations), the integration filter, and the vehicle state detection method. Section 3 presents the results and numerical analysis of the field tests. Then, we discuss and conclude the paper, and the limitations and the future work are detailed.

2. Model

2.1. GNSS/MIMU Loose Integration Model

The state vector of the GNSS/MIMU loose integration model contains 15 states, and the state vector X_I is defined as:

$$X_I = [\delta\phi, \delta v, \delta p, \delta\varepsilon, \delta V]^T \tag{1}$$

where $\delta\phi = [\alpha, \beta, \gamma]$ denotes the three-axis attitude errors (pitch, roll, and yaw angle errors), $\delta v = [\delta v_e, \delta v_n, \delta v_u]$ denotes the three-axis velocity errors (east, north, and up velocity errors) in the ENU navigation frames, $\delta p = [\delta L, \delta\lambda, \delta h]$ denotes the three-axis positioning errors (latitude, longitude, and height errors), $\delta\varepsilon = [\varepsilon_x, \varepsilon_y, \varepsilon_z]$ denotes the bias errors of the three-axis gyroscopes in body frame, and $\delta V = [\nabla_x, \nabla_y, \nabla_z]$ denotes the bias errors of the three-axis accelerometers.

The state equation GNSS/MIMU loose integration can be written as:

$$\dot{X}_I(t) = F_I(t) \cdot X_I(t) + G_I(t) W_I(t) \tag{2}$$

where $F_I(t)$ is the state transferring matrix; W_I denotes the state model noise matrix [18–21]. Specifically, the detailed description of the state equation is as:

$$
\begin{bmatrix}
\delta\dot{p}_{3\times1} \\
\delta\dot{v}_{3\times1} \\
\delta\dot{\phi}_{3\times1} \\
\dot{\varepsilon}_{3\times1} \\
\dot{\nabla}_{3\times1}
\end{bmatrix}
=
\begin{bmatrix}
F_{pp} & F_{pv} & 0_{3\times3} & 0_{3\times3} & 0_{3\times3} \\
F_{vp} & F_{vv} & F_{v\phi} & 0_{3\times3} & C_b^n \\
0_{3\times3} & 0_{3\times3} & F_{\phi\phi} & C_b^n & 0_{3\times3} \\
0_{3\times3} & 0_{3\times3} & 0_{3\times3} & F_{\varepsilon\varepsilon} & 0_{3\times3} \\
0_{3\times3} & 0_{3\times3} & 0_{3\times3} & 0_{3\times3} & F_{\nabla\nabla}
\end{bmatrix}
\begin{bmatrix}
\delta p_{3\times1} \\
\delta v_{3\times1} \\
\delta\phi_{3\times1} \\
\varepsilon_{3\times1} \\
\nabla_{3\times1}
\end{bmatrix}
+
\begin{bmatrix}
0_{3\times3} & 0_{3\times3} & 0_{3\times3} & 0_{3\times3} \\
C_b^n & 0_{3\times3} & 0_{3\times3} & 0_{3\times3} \\
0_{3\times3} & C_b^n & 0_{3\times3} & 0_{3\times3} \\
0_{3\times3} & 0_{3\times3} & I_{3\times3} & 0_{3\times3} \\
0_{3\times3} & 0_{3\times3} & 0_{3\times3} & I_{3\times3}
\end{bmatrix}
\begin{bmatrix}
w^v_{3\times1} \\
w^\phi_{3\times1} \\
w^\varepsilon_{3\times1} \\
w^\nabla_{3\times1}
\end{bmatrix}
\tag{3}
$$

Further, the first-order discrete form of the state equation is as:

$$\dot{X}_I(k+1) = (I + F_I \cdot T) \cdot X_I(k+1) + G_I \cdot T \cdot W_I(k+1) \tag{4}$$

$$
\begin{bmatrix} \delta\dot{p}_{3\times1} \\ \delta\dot{v}_{3\times1} \\ \delta\dot{\phi}_{3\times1} \\ \dot{\varepsilon}_{3\times1} \\ \dot{\nabla}_{3\times1} \end{bmatrix} =
\begin{bmatrix}
I_{3\times3} + F_{pp} & F_{pv}T & 0_{3\times3} & 0_{3\times3} & 0_{3\times3} \\
F_{vp}T & I_{3\times3} + F_{vv}T & F_{v\phi}T & 0_{3\times3} & C_b^n T \\
0_{3\times3} & 0_{3\times3} & I_{3\times3} + F_{\phi\phi}T & C_b^n T & 0_{3\times3} \\
0_{3\times3} & 0_{3\times3} & 0_{3\times3} & I_{3\times3} + F_{\varepsilon\varepsilon}T & 0_{3\times3} \\
0_{3\times3} & 0_{3\times3} & 0_{3\times3} & 0_{3\times3} & I_{3\times3} + F_{\nabla\nabla}T
\end{bmatrix}
\begin{bmatrix} \delta p_{3\times1} \\ \delta v_{3\times1} \\ \delta\phi_{3\times1} \\ \varepsilon_{3\times1} \\ \nabla_{3\times1} \end{bmatrix}
$$

$$
+ \begin{bmatrix}
0_{3\times3} & 0_{3\times3} & 0_{3\times3} & 0_{3\times3} \\
C_b^n T & 0_{3\times3} & 0_{3\times3} & 0_{3\times3} \\
0_{3\times3} & C_b^n T & 0_{3\times3} & 0_{3\times3} \\
0_{3\times3} & 0_{3\times3} & I_{3\times3}T & 0_{3\times3} \\
0_{3\times3} & 0_{3\times3} & 0_{3\times3} & I_{3\times3}T
\end{bmatrix}
\begin{bmatrix} w^v_{3\times1} \\ w^\phi_{3\times1} \\ w^\varepsilon_{3\times1} \\ w^\nabla_{3\times1} \end{bmatrix} \tag{5}
$$

where T denotes the integration filter updating interval.

Then, the measurement model of the GNSS/MIMU integration is expressed as:

$$Z_{k+1} = H_{k+1}X_{k+1} + \mu_{k+1} \tag{6}$$

where Z_{k+1} denotes the measurement matrix, H_{k+1} denotes the observation matrix, and μ_{k+1} denotes the measurement noise. In the loose integration model, the measurement vector is composed of GNSS and MIMU position and velocity difference, and the detailed description is as:

$$
Z_{k+1} = \begin{bmatrix}
\left(L^{MIMU}_{k+1} - L^{GNSS}_{k+1}\right) \cdot (R_M + h) \\
\left(\lambda^{MIMU}_{k+1} - \lambda^{GNSS}_{k+1}\right) \cdot (R_N + h)cos(L) \\
h^{MIMU}_{k+1} - h^{GNSS}_{k+1} \\
v^{MIMU}_{e,k+1} - v^{GNSS}_{e,k+1} \\
v^{MIMU}_{n,k+1} - v^{GNSS}_{n,k+1} \\
v^{MIMU}_{u,k+1} - v^{GNSS}_{u,k+1}
\end{bmatrix} \tag{7}
$$

A detailed description of the observation matrix H_{k+1} is:

$$
Z_{k+1} = \begin{bmatrix} diag[R_M + h(R_N + h)\cos(L)\ 1] & 0_{3\times3}\ 0_{3\times3} \\ 0_{3\times3} & 0_{3\times3} \end{bmatrix} X_{k+1} + \mu_{k+1} \tag{8}
$$

2.2. State Constraints

The definition of the vehicle body coordinates is presented in Figure 1. In the coordinates, the origin is the center of gravity of the vehicle body, the **Y** axis points to the direction of the vehicle traveling, the **X** axis points to the right side of the vehicle body, and the **Z** axis points to the up direction of the vehicle.

According to the driving characteristics of the vehicle on the road, while the vehicle is running normally on the road without sideslip or jump, e.g., the vehicle is driving on an expressway, the X-axis and Z-axis speeds of the vehicle in the defined vehicle frame are approximately zero. The characteristic is modeled as:

$$
\begin{cases} V^v_x \approx 0 \\ V^v_z \approx 0 \end{cases} \tag{9}
$$

In the dynamic trajectory, vehicle kinematic constraints are employed for suppressing the diverging positioning errors under the GNSS-denied environment. Vehicle kinematic constraints are constructed in the vehicle body coordinates. Although the Inertial Measurement Unit (IMU) is installed on the vehicle body, there is usually a misalignment angle between the IMU body coordinates and the vehicle body coordinates. The misalignment angles will affect the velocity constraints listed in Equation (1).

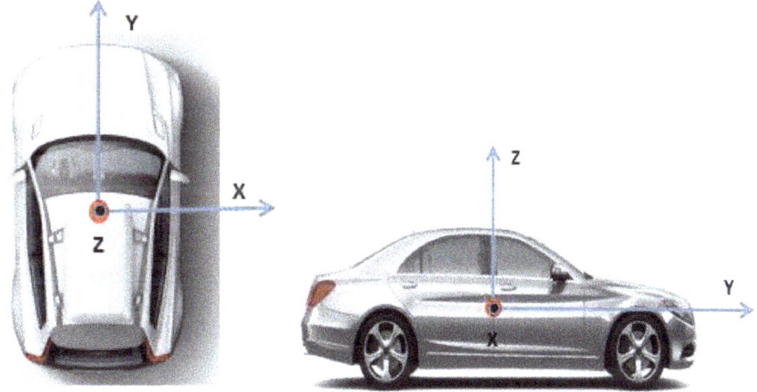

Figure 1. Illustration of the vehicle body coordinates.

Assuming the conversion matrix between the IMU body coordinates and the vehicle body system is C_v^b. The heading misalignment angle is α_ψ, the pitch misalignment angle is α_θ, and the roll misalignment angle is α_γ. Converting the velocity in the IMU body system to the vehicle body system can be modeled as:

$$\mathbf{V}^b = C_v^b \mathbf{V}^v \tag{10}$$

where $\mathbf{V}^v$ is the velocity vector in the IMU body coordinates, and $\mathbf{V}^b$ is the velocity vector in vehicle body coordinates.

Specifically,

$$\mathbf{V}^b = \begin{bmatrix} V_x^b \\ V_y^b \\ V_z^b \end{bmatrix} = C_v^b \begin{bmatrix} V_x^v \\ V_y^v \\ V_z^v \end{bmatrix} = \begin{bmatrix} \sin\alpha_\psi \cos\alpha_\theta \\ \cos\alpha_\psi \cos\alpha_\theta \\ \sin\alpha_\theta \end{bmatrix} V_y^v \tag{11}$$

While the vehicle is moving, the V_y^v is not zero. The V_y^v will be projected to the V_x^b and V_z^b through the heading and pitch misalignment angle. The roll angle does not influence the V_x^b and V_z^b. The influence of the misalignment angle on the velocity V_x^b and V_z^b can be described as

$$\delta C_b^v = -\begin{bmatrix} \delta\alpha_\theta \\ 0 \\ \delta\alpha_\psi \end{bmatrix} \times C_b^v = -\delta\alpha \times C_b^v \tag{12}$$

While employing the constraints in the GNSS/MIMU loose integration, the misalignment angle between the MIMU body frame and the vehicle body frame should be considered and added to the state vector. The new state vector is as:

$$\mathbf{X} = \begin{bmatrix} \mathbf{X}_I & \mathbf{X}_\alpha \end{bmatrix}^T \tag{13}$$

where $\mathbf{X}_I$ is the same as that in Equation (1), $\mathbf{X}_\alpha = \begin{bmatrix} \delta\alpha_\theta & \delta\alpha_\psi \end{bmatrix}^T$, $\delta\alpha_\theta$ is the misalignment heading angle, and the $\delta\alpha_\psi$ is the misalignment pitch angle.

Once the IMU is fixed on the vehicle, the misalignment angles can be regarded as constant values. Therefore, the first-order derivative of the misalignment angles is zero.

$$
\begin{cases}
\delta\dot{\alpha}_\theta = 0 \\
\delta\dot{\alpha}_\psi = 0
\end{cases}
\tag{14}
$$

Then, the state equation of the MIMU can be as:

$$
\begin{aligned}
\dot{\mathbf{X}}(t) &= \begin{bmatrix} \dot{\mathbf{X}}_I(t) \\ \dot{\mathbf{X}}_\alpha(t) \end{bmatrix} = \begin{bmatrix} \mathbf{F}_I(t) & 0 \\ 0 & \mathbf{F}_\alpha(t) \end{bmatrix} \begin{bmatrix} \mathbf{X}_I(t) \\ \mathbf{X}_\alpha(t) \end{bmatrix} + \begin{bmatrix} \mathbf{G}_I(t) & 0 \\ 0 & \mathbf{G}_\alpha(t) \end{bmatrix} \begin{bmatrix} \mathbf{W}_I(t) \\ \mathbf{W}_\alpha(t) \end{bmatrix} \\
&= \mathbf{F}(t) \cdot \mathbf{X}(t) + \mathbf{G}(t) \cdot \mathbf{W}(t)
\end{aligned}
\tag{15}
$$

where $\mathbf{F}_I(t)$ is the state transformation matrix of the IMU's states, $\mathbf{F}_\alpha(t)$ is the state transformation matrix of the misalignment angles, $\mathbf{W}_I(t)$ is the IMU state noise matrix, and $\mathbf{W}_\alpha(t)$ is the misalignment angle state noise matrix.

Commonly, the MIMU and GNSS loose integration model is constructed in East–North–Up (ENU) coordinates. Positioning and velocity information from the GNSS and INS are subtracted and employed as the measurement information. Converting the SINS velocity from the ENU coordinates to the vehicle body frame.

$$
\mathbf{V}^v = C_b^v C_n^b \mathbf{V}^n
\tag{16}
$$

where C_b^v means the velocity conversion from the vehicle body frame from the IMU body coordinates, C_n^b is the velocity transformation matrix from the ENU navigation frame to the vehicle body coordinates, and $\mathbf{V}^n$ is the velocity vector in the ENU navigation frame.

Combining Equations (11)–(16), the differential equation is

$$
\begin{aligned}
\delta\mathbf{V}^v &= C_b^v\left(C_n^b\varphi \times \mathbf{V}^n + C_n^b\delta\mathbf{V}^n\right) - \delta\alpha \times C_b^v C_n^b \mathbf{V}^n \\
&= \left(-C_b^v C_n^b(\mathbf{V}^n)\times\right)\varphi + C_b^v C_n^b\delta\mathbf{V}^n + \left(\left(C_b^v C_n^b \mathbf{V}^n\right)\times\right)\delta\alpha \\
&= M_{3\times3}^1\varphi + M_{3\times3}^2\delta\mathbf{V}^n + M_{3\times3}^3\delta\alpha
\end{aligned}
\tag{17}
$$

The measurement equation is

$$
\mathbf{Z}_V = \begin{bmatrix} V_x^v - 0 \\ V_z^v - 0 \end{bmatrix} = \mathbf{H}_V\mathbf{X} + \mathbf{V}_V
\tag{18}
$$

where $\mathbf{H}_V$ is the measurement matrix, and the $\mathbf{V}_V$ is the noise matrix.

$$
\mathbf{H}_V = \begin{bmatrix} M_{3\times3}^1(1,\times) & M_{3\times3}^2(1,\times) & 0_{1\times11} & 0 & M_{3\times3}^3(1,3) \\ M_{3\times3}^1(3,\times) & M_{3\times3}^2(3,\times) & 0_{1\times11} & M_{3\times3}^3(3,1) & 0 \end{bmatrix}
\tag{19}
$$

where $M_{3\times3}^1(1,\times)$ is the first row of the matrix $M_{3\times3}^1$, $M_{3\times3}^3(1,3)$ is the element in the first row and third column, $M_{3\times3}^1(3,\times)$ is the third row of the matrix $M_{3\times3}^1$, and $M_{3\times3}^3(3,1)$ is the element in the third row and first column.

2.3. MIMU/Odometer Measurement Model

The state vector of the MIMU/Odometer integration model is the same as Equations (13)–(15), however, the measurement equation is different. The odometer output is listed as:

$$
\hat{\mathbf{V}}_{odo}^b = \begin{bmatrix} 0 & \hat{V}_{odoy}^b & 0 \end{bmatrix}^{\mathbf{T}}
\tag{20}
$$

We then convert the odometer velocity from the IMU body frame to the ENU navigation frame, then subtracting them with the velocity from the MIMU. The equation is listed as:

$$\mathbf{Z}_O = \mathbf{V}_I^n - C_b^n \hat{\mathbf{V}}_{odo}^b = \begin{bmatrix} V_{IE} - V_{odoE}^n \\ V_{IN} - V_{odoN}^n \\ V_{IU} - V_{odoU}^n \end{bmatrix} = \mathbf{H}_O \mathbf{X} + \mathbf{V}_O \tag{21}$$

where $\mathbf{V}_I^n$ is the velocity from the MIMU, and $\hat{\mathbf{V}}_{odo}^b$ is the velocity from the odometer, $\mathbf{H}_O$ is the measurement matrix, and $\mathbf{V}_O$ is the noise matrix.

$$\mathbf{H}_O = \begin{bmatrix} 0_{1\times3} & 1 & 0 & 0 & 0_{1\times11} \\ 0_{1\times3} & 0 & 1 & 0 & 0_{1\times11} \\ 0_{1\times3} & 0 & 0 & 1 & 0_{1\times11} \end{bmatrix} \tag{22}$$

2.4. MIMU/Odometer Measurement Model with Constraints

Combining Equations (8)–(14), the MIMU/odometer measurement model with constraints is as:

$$\mathbf{Z}_{OV} = \begin{bmatrix} V_x^v - 0 \\ V_{IN} - V_{odoN}^n \\ V_z^v - 0 \end{bmatrix} = \mathbf{H}_{OV} \mathbf{X} + \mathbf{V}_{OV} \tag{23}$$

where V_x^v is the X-axis velocity in the vehicle body coordinates, V_z^v is the Z-axis velocity in the vehicle body coordinates, $\mathbf{H}_{OV}$ is the measurement matrix, and $\mathbf{V}_{OV}$ is the measurement noise matrix.

$$\mathbf{H}_{OV} = \begin{bmatrix} M_{3\times3}^1(1,\times) & M_{3\times3}^2(1,1) & M_{3\times3}^2(1,2) & M_{3\times3}^2(1,3) & 0_{1\times11} & 0 & M_{3\times3}^3(1,3) \\ 0_{1\times3} & 0 & 1 & 0 & 0_{1\times11} & 0 & 0 \\ M_{3\times3}^1(3,\times) & M_{3\times3}^2(3,1) & M_{3\times3}^2(3,2) & M_{3\times3}^2(3,3) & 0_{1\times11} & M_{3\times3}^3(3,1) & 0 \end{bmatrix} \tag{24}$$

2.5. Integration Method

The state model is listed in Equations (5)–(7), and the measurement models under different conditions are given in Equations (10), (14), and (15). Here, a Kalman filter is employed for carrying out the integration. The Kalman filter is as:

The Kalman filter state vector and state covariance prediction are as:

$$\hat{X}_k^- = \Phi_{k|k-1} \hat{X}_{k-1} \tag{25}$$

$$P_k^- = \Phi_{k|k-1} P_{k-1} \Phi_{k|k-1}^T + Q_{k-1} \tag{26}$$

The updating of the gain matrix, state vector, and the covariance are as follows:

$$K_k = P_k^- H_k^T (H_k P_k^- H_k^T + R_k)^{-1} \tag{27}$$

$$\hat{X}_k = \hat{X}_k^- + K_k(Z_k - H_k \hat{X}_k^-) \tag{28}$$

$$P_k = (I - K_k H_k) P_K^- \tag{29}$$

where $\Phi_{k|k-1}$ is the state transformation matrix; $\hat{X}_k^-$ is the predicted state vector through the state transformation matrix and the state vector at previous epoch; P_k^- is the covariance matrix; K_k is the gain matrix at the k_{th} epoch, which decides the updating weight between the predicted state vector and the new measurements; $\hat{X}_k$ is the estimated state vector at the k_{th} epoch; P_k is the covariance matrix.

Based on the above model, Figure 2 shows the structure of the integration system with constraints. When the GNSS is available, the GNSS/MIMU integration system can provide satisfying navigation solutions. While GNSS is unavailable, constraints are employed in the integration system for estimating the IMU state errors and compensating them.

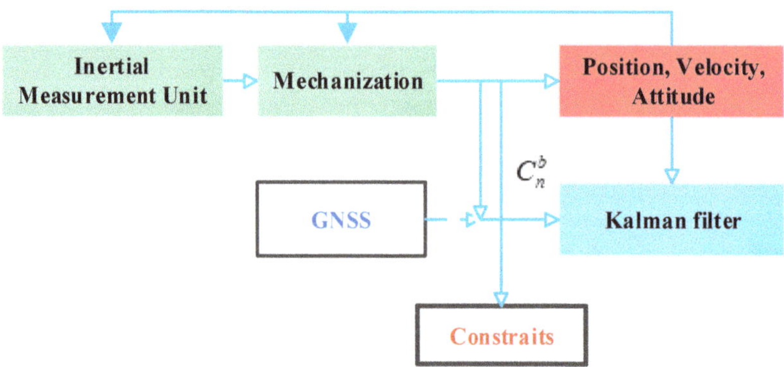

Figure 2. Structure of the Global Navigation Satellite System (GNSS)/Inertial Navigation System (INS)/odometer/Non-Holonomic Constraints (NHC) integrated system.

3. Results

For fully testing and assessing the performance of the MO-C system, two different field tests were carried out. The equipment employed in the field testing are given in Figure 3, the employed MIMU is presented in Figure 3a, and the vehicle is shown in Figure 3b,c. In the first field test, the MIMU and Odometer dataset was collected and post-processed through the software implemented in Matlab [34]. In the second field test, the algorithm was implemented using the hardware platform DSP+FPGA (Digital Signal Processor, DSP; Field Programmable Gate Array, FPGA), and the results were obtained from real-time processing of the data. The MO-C was also implemented in the BDS/MIMU integrated navigation system for improving the effectiveness during BDS signal outage.

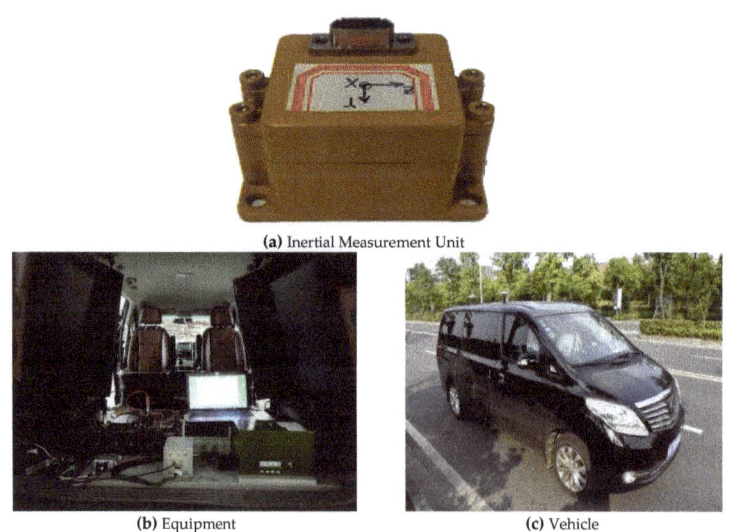

(a) Inertial Measurement Unit

(b) Equipment   **(c)** Vehicle

Figure 3. Field testing equipment.

Parameters of the employed MIMU were given in Table 1. The gyroscope bias stability was 3 degrees/h, and the accelerometer bias stability was 0.1 mg. The MIMU sampling frequency was 400 Hz, and the odometer data output frequency was 20 Hz. The integration filter operation period was one second.

Table 1. Inertial Measurement Unit (IMU) specifications.

Gyroscope	Bias stability (degree/h)	≤3 degree/h
	Scale factor nonlinearity (ppm)	≤200 ppm
	White noise (degree/h)	0.1 degree/h
Accelerometer	Bias stability (mg)	0.1 mg
	Scale factor nonlinearity (ppm)	≤150 ppm
	White noise (mg)	0.05 mg

3.1. Field Testing with Data Post-Processing

MIMU and Odometer data were collected with the equipment presented in Figure 3. Following positioning values comparisons (GNSS, MIMU, MIMU/odometer (MO), and MIMU/Odometer with constraints (MO-C) are presented in Figures 4–6. Trajectories obtained from different methods were presented in Figure 4. The MO-C and MO results were similar in this trajectory. The positioning curves in Figure 5 are plotted as the trajectory. Velocity results comparisons are presented in Figure 6. Positioning and velocity error analysis including the Root Mean Square Error (RMSE) and 90-s errors without GNSS are listed in Table 2. Through a comparison of the results of the GNSS, MIMU, MO, and MO-C, it could be seen that:

(1) Compared with the MIMU standalone, the positioning errors were suppressed with the odometer and constraints included, the latitude and longitude curves were almost consistent with the GNSS curves. The east and north velocity were also consistent with the GNSS results. The height and up velocity were also converging over time.

(2) Compared with the MIMU standalone, within 90 s, the MO and MO-C latitude errors reduced by 98.8% and 98.9%; the MO and MO-C longitude errors reduced by 95.1% and 95.5%; the MO and MO-C height errors decreased by 81.1% and 95.9%. In aspects of the velocity errors, both MO and MO-C east velocity errors reduced by 87.2%; the MO and MO-C north velocity errors decreased by 96.8% and 96.9%; the MO and MO-C up velocity obtained a 63.4% and 99.2% improvement. Among them, the up direction position and velocity errors obtained the largest reduction compared with that of other directions.

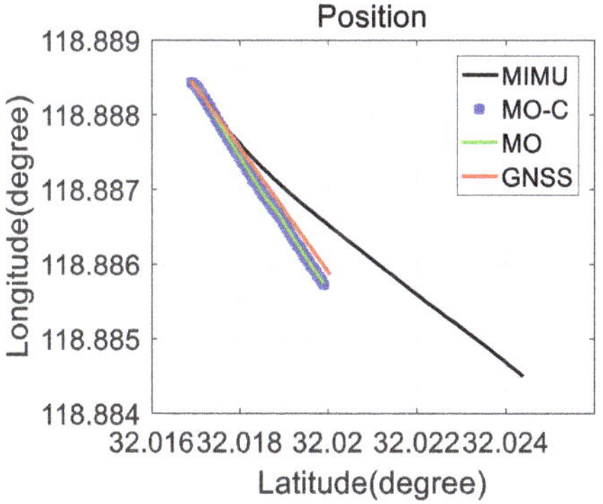

Figure 4. The trajectory from different methods.

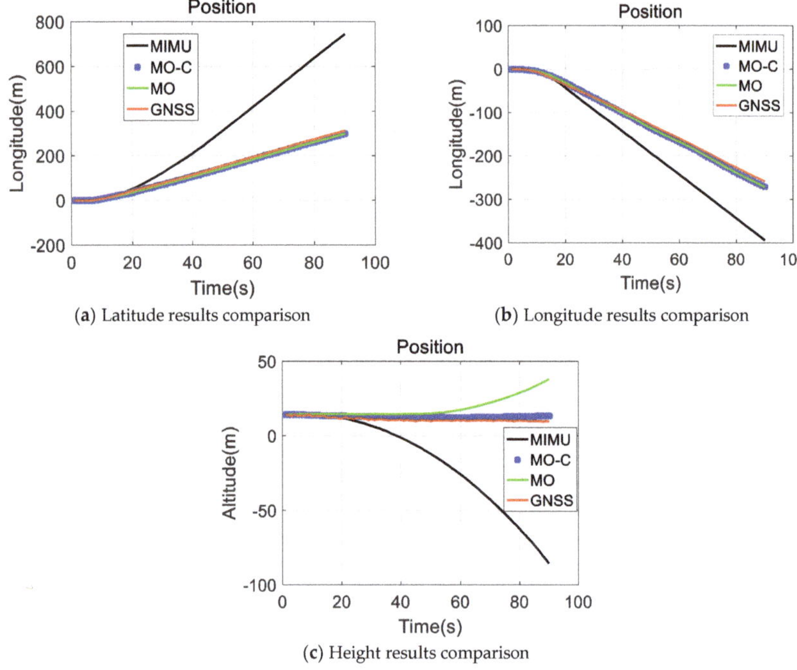

(**a**) Latitude results comparison

(**b**) Longitude results comparison

(**c**) Height results comparison

Figure 5. Positioning results comparison.

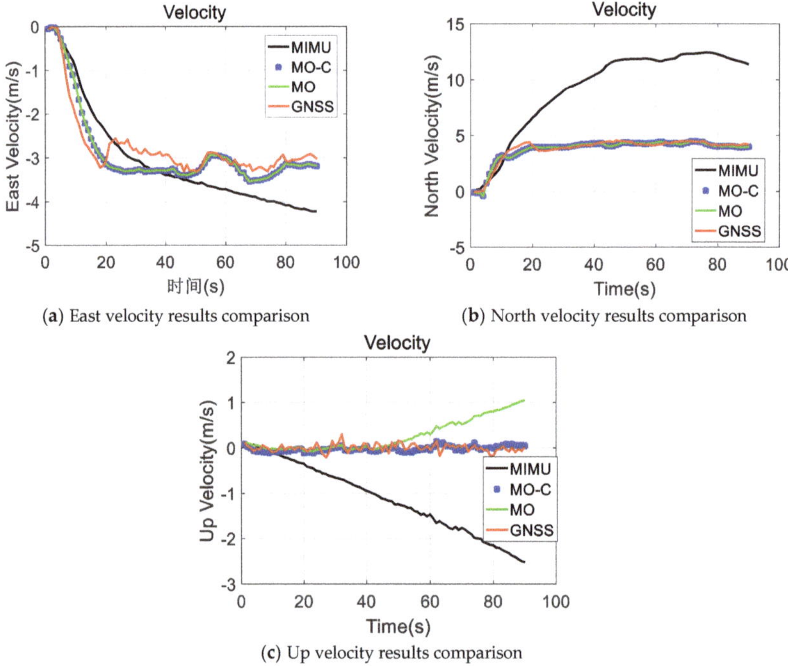

(**a**) East velocity results comparison

(**b**) North velocity results comparison

(**c**) Up velocity results comparison

Figure 6. Velocity results comparison.

Table 2. Positioning and velocity error comparison.

		Latitude (m)	Longitude (m)	Height (m)	East Velocity (m/s)	North Velocity (m/s)	Up Velocity (m/s)
MIMU	90s error	480.1	−150.6	−95.37	−1.189	7.144	−2.544
	RMSE	234.56	81.31	40.36	0.715	6.121	1.365
MO	90s error	−5.55	−7.30	18.06	−0.152	−0.225	0.931
	RMSE	8.59	7.01	10.34	0.330	0.286	0.416
MO-C	90s error	−5.28	−6.74	3.90	−0.152	−0.223	0.020
	RMSE	8.54	6.93	2.08	0.304	0.286	0.101

3.2. Field Testing with Real-Time Data Processing

After the post-processing field testing, we carried out real-time data processing-based field testing for further evaluation of the performance of the method. The algorithm was implemented using the DSP+FPGA hardware platform with real-time processing data fusion. This sub-section is divided into four parts. In the first part, we evaluate the MIMU with constraints; in the second part, the MIMU/odometer integration is assessed; in the third part, the MO-C results are presented and analyzed; in the last part, the MO-C was integrated into GNSS/MIMU integrated navigation system, and the navigation solutions are presented and compared during a signal outage.

3.2.1. MIMU with Constraints

The field-testing trajectory was presented in Figure 7, and the positioning errors and velocity errors were presented in Figures 8 and 9. The latitude and longitude errors of M-C gradually accumulated, but gradually tended to be flat, the altitude errors gradually stabilized, and the errors were small; although the three-dimensional speed errors were stable, the east and north speed errors were relatively large, and the up speed errors were small. Without GNSS, the position and velocity errors within 90 s were as follows: latitude error was −25.85 m, longitude error was −28.80 m, altitude error was −3.06 m, east velocity error was −0.91 m/s, north velocity error was 0.27 m/s, up velocity error was −0.38 m/s. The results showed that the constraints were effective for suppressing the errors of the MIMU in the dynamic trajectory. However, the M-C 90-s error values were still not ideal, which was also affected by the heading angle. The heading angle was presented in Figure 10, and it varied between 40° and 42°. The vehicle kinematics constraints could only suppress the X-axis and Z-axis position and velocity errors.

Figure 7. Field testing trajectory.

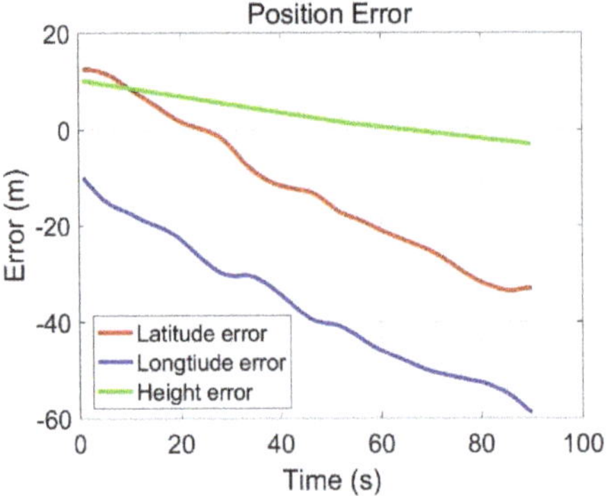

Figure 8. Three-axis positioning errors.

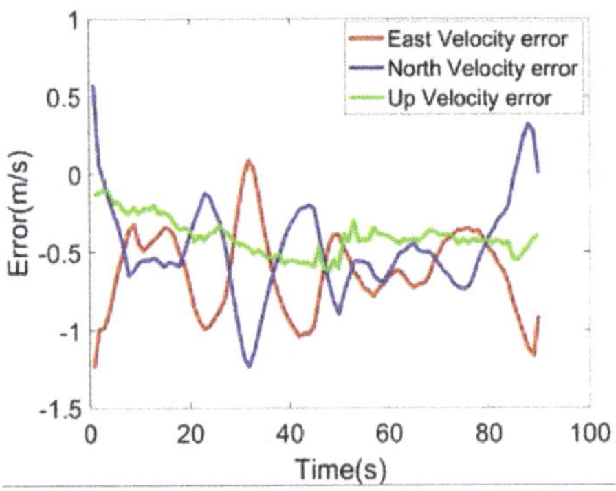

Figure 9. Three-axis velocity errors.

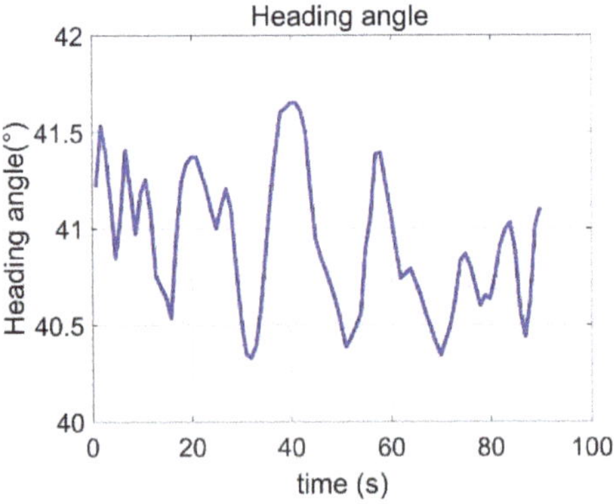

Figure 10. Heading angle.

3.2.2. MIMU/Odometer Integration

Three-axis position errors and velocity errors were presented in Figures 11 and 12, and it could be seen that the MO latitude and longitude errors were stable and small, the altitude errors firstly decreased and then increased; the east and north speed errors were stable, and kept within 0.2 m/s, the up velocity errors gradually increased, and it trended to diverge. The 90-s position and velocity errors were as follows: latitude error was 1.74 m, longitude error was 4.73 m, altitude error was −13.35 m, east velocity error was 0.03 m/s, north velocity error was −0.06 m/s, and the up velocity error was −0.85 m/s.

The results showed that the MO was effective for reducing the MIMU positioning errors. It was worth of noting that the error of up velocity increased gradually without good constraints, which was the same as the results in Section 3.1. If we wanted to obtain high-precision three-dimensional positioning and speed measurement under GNSS-denied environments, additional sensors or methods were necessary to suppress the height and up-direction velocity errors.

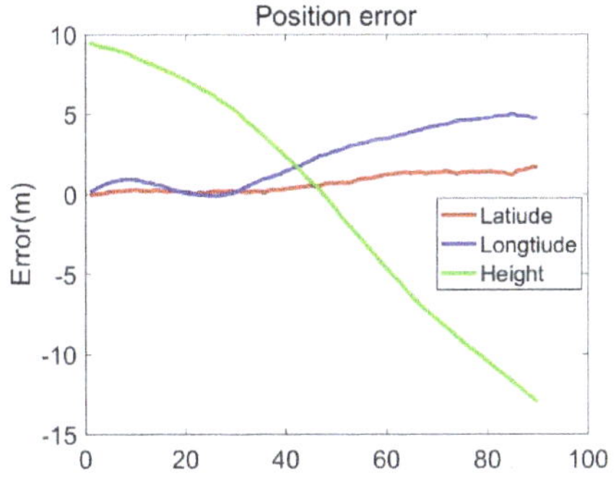

Figure 11. Positioning errors.

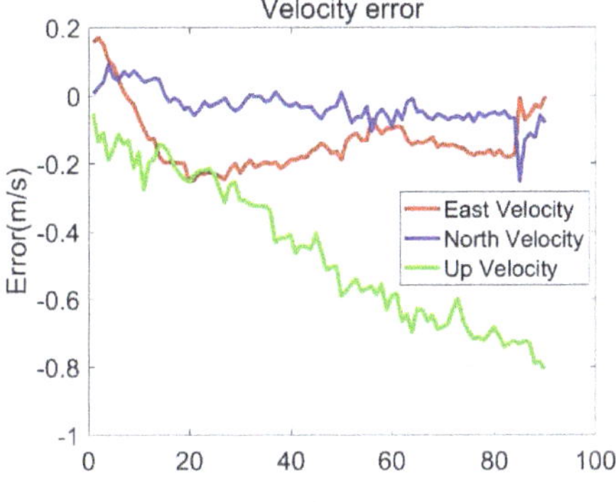

Figure 12. Velocity errors.

3.2.3. MIMU/Odometer Integration with Constraints

In Figures 13 and 14, the three-dimensional position and velocity errors of MO-C were stable and small. The height errors and the up velocity errors were significantly smaller than that of the MC. The added kinematic constraints information performed well in suppressing the up velocity errors and height errors. The 90-s position and velocity errors were as follows: latitude error was 1.72 m, longitude error was −1.36 m, altitude error was −4.38 m; east velocity error was 0.02 m/s, north velocity error was −0.04 m/s, and up velocity error was −0.23 m/s. The position and velocity errors comparison for MINS/BDS, M-C, MO, and MO-C are listed in Table 3. It could be seen that the latitude and longitude errors were suppressed while the odometer was included in the system. With the odometer assisting, the MO-C 90-s latitude and longitude errors decreased over 90% compared with that of M-C. While adding the constraints to the MO, the MO-C longitude and height errors performed with a 71.2% and 67.2% decrease, and the north and up velocity decreased by 33.3% and 72.9%. These results demonstrated the effectiveness of the odometer and constraints in position and velocity error suppression.

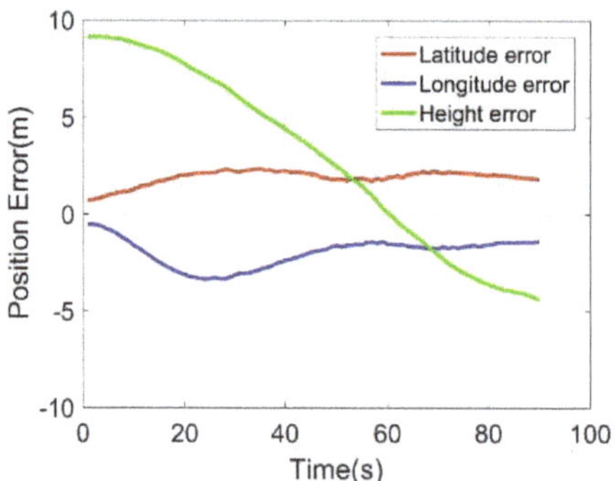

Figure 13. Positioning errors.

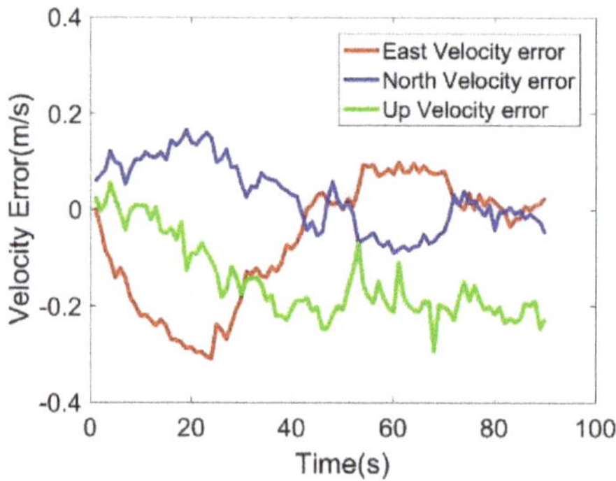

Figure 14. Velocity errors.

Table 3. Position and velocity error comparison.

		Latitude (m)	Longitude (m)	Height (m)	East Velocity (m/s)	North Velocity (m/s)	Up Velocity (m/s)
M-C	90s	−25.85	−28.80	−3.06	−0.91	0.27	−0.38
	RMSE	16.662	18.761	4.931	0.686	0.567	0.367
MO	90s	1.74	4.73	−13.35	0.03	−0.06	−0.85
	RMSE	0.876	2.910	7.301	0.162	0.060	0.503
MO-C	90s	1.72	−1.36	−4.38	0.02	−0.04	−0.23
	RMSE	1.911	2.112	5.487	0.142	0.077	0.168

3.2.4. Implementation of MC-O in MIMU/BDS during Signal Outage

In this part, we integrated into the BDS/MINS coupled navigation system for improving its positioning accuracy during a signal outage. Figure 15 presented the field-testing trajectory in Google maps. The BDS satellite amount of change was presented in Figure 16. The red line represented the in-view satellite amount of GPS and BDS, which was employed as the reference. The blue line represented the BDS satellite amount employed in the experiment. At 75 s, the antenna was removed to simulate the signal outage. Therefore, the satellite amount was zero after 75 s.

Position and velocity errors were presented in Figures 17 and 18. During 0–75 s, the system worked on BDS/MINS integration mode, and the position and velocity errors were within the normal range. The system worked on GNSS/MIMU/Odometer mode during 0–75 s, the benefits could be summarized as follows: firstly, the GNSS provided the initial position and velocity information for the MIMU, and the GNSS velocity was helpful for the attitude estimation; secondly, the odometer could also help the navigation solutions estimation, and in this mode, some parameters of the odometer could be estimated from the reliable navigation solutions.

Figure 15. Trajectory.

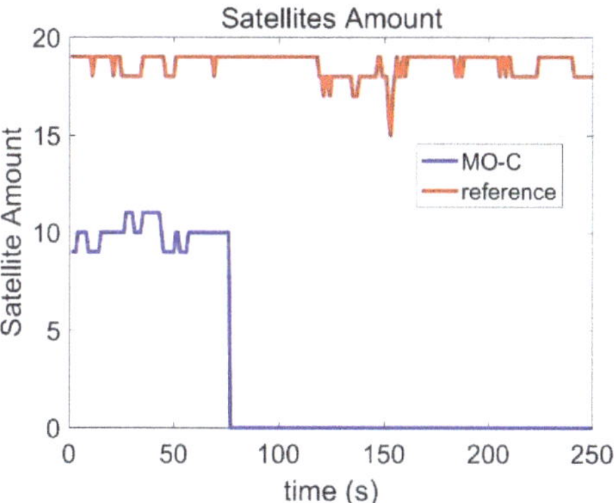

Figure 16. Satellites amount.

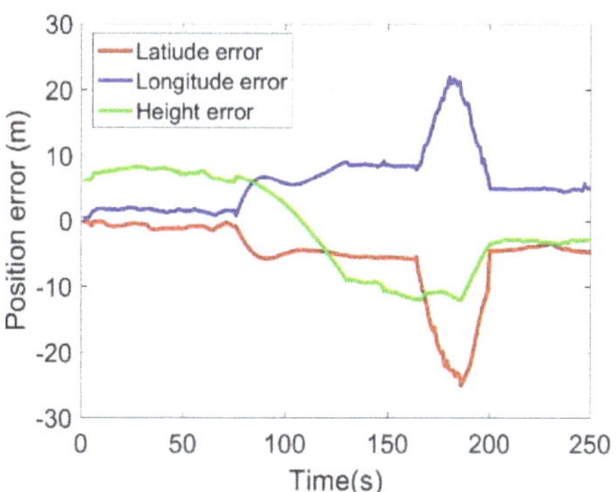

Figure 17. Position errors.

After the 75 s, the BDS antenna was removed to simulate the signal outage and to assess the performance of the MC-O. During 75 s and 165 s, the position errors experienced a minor decrease; however, the position errors still kept within 10 m. After 165 s, the odometer was disconnected for the assessment of the M-C performance. The position and velocity errors obtained a dramatic decrease. The latitude and longitude errors were over 20 m. However, the height errors still kept within 10 m, which demonstrated the effectiveness of the up velocity constraint. After 180 s, the odometer was re-connected to the system, and the positioning and velocity errors converged quickly to normal range.

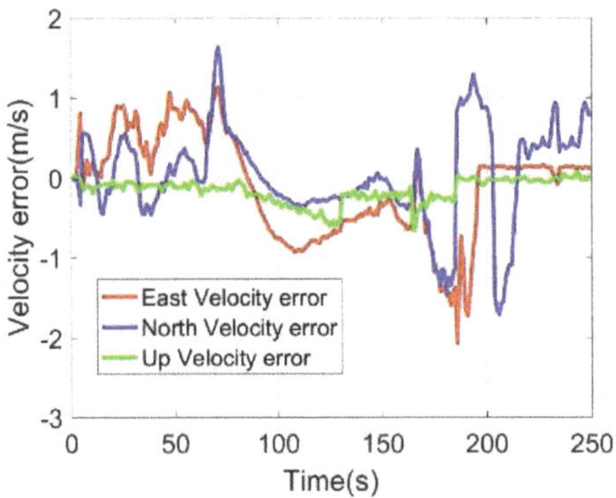

Figure 18. Velocity errors.

4. Discussion

Our experimental results demonstrated that the odometer and the state constraints were effective for suppressing the positioning errors of the MIMU while GNSS was unavailable. The odometer was effective for reducing the errors of the vehicle moving direction, and the constraints also performed well in reducing the height errors. During the 90 s testing time, the MO-C three-dimensional position errors could keep within five meters above IMU. However, we thought the following work was worthy of further investigation:

(1) In the above experiments, the testing time was 90 s, and the position errors would diverge due to the odometer errors and the heading angle errors. In the MO-C, there were no constraints for the heading angle. Other sensors or methods, providing better heading angles, could certainly improve the MO-C position accuracy while GNSS was unavailable for a long time.

(2) In the experiments, we removed the GNSS antenna for simulating the signal outage for assessing the MO-C, in fact, in urban areas, although part of the GNSS satellites were blocked by the surrounding buildings, there were still a few satellites in view. However, there were not enough for generating precise three-dimensional navigation solutions, the remaining satellites might be helpful in the MO-C for aiding the navigation solutions estimation.

(3) Although the NHC and odometer were effective during the GNSS signals outage, it was still necessary for GNSS/MIMU/odometer integration system, while the GNSS was normal, some MIMU and odometer parameters could be estimated and calibrated, which could help reduce the positioning errors during GNSS signal outage.

5. Conclusions

In this paper, we present a comprehensive investigation of the MIMU/odometer integrated navigation system with vehicle state constraints. The algorithm is described and listed in detail. Abundant experiments were conducted for evaluating and comparing the performance of the MO, M-C, and MO-C methods. We could conclude that:

(1) Odometer was effective for reducing the latitude and longitude errors, however, it has almost no influence on height accuracy.

(2) These constraints were effective for the height error reduction, but its influence on the latitude and longitude errors were related to the moving direction of the vehicle.

(3) With the odometer and constraints aiding, the heading angle heavily affects the accuracy of the navigation solutions. If the heading angle could be determined precisely, the multi-sensor fusion method could provide long-time three-dimensional navigation solutions without GNSS.

(4) This paper firstly presented the implementation and evaluation of these methods in the BDS/MIMU loose integration system, and the satisfying results could support the BDS for vehicles in urban areas.

(5) The methods discussed in the paper could also be implemented in a BDS chip receiver, and MIMU could be connected to the BDS chip-scale receiver for improving the reliability and robustness of the navigation solutions.

Author Contributions: K.Z. wrote the first draft of this paper, X.G. conducted the field tests, C.J. proposed the method, Y.X. and Y.L. reviewed and revised the paper, L.H. developed the software, and Y.C. guided the paper writing and discussed the method. All authors have read and agreed to the published version of the manuscript.

Funding: The authors acknowledge the support of the National Natural Science Foundation of China (Grant No. 61601225).

Conflicts of Interest: The authors declare no conflict of interest.

References

1. Imparato, D.; El-Mowafy, A.; Rizos, C.; Wang, J. A review of SBAS and RTK vulnerabilities in Intelligent Transport Systems applications. In Proceedings of the IGNSS Symposium 2018, Sydney, Australia, 7–9 February 2018; pp. 1–18.
2. Chen, Y.; Tang, J.; Jiang, C.; Zhu, L.; Lehtomäki, M.; Kaartinen, H.; Kaijaluoto, R.; Wang, Y.; Hyyppä, J.; Hyyppä, H.; et al. The accuracy comparison of three simultaneous localization and mapping (SLAM)-based indoor mapping technologies. *Sensors* **2018**, *18*, 3228. [CrossRef] [PubMed]
3. Quack, T.M.; Reiter, M.; Abel, D. Digital map generation and localization for vehicles in urban intersections using LiDAR and GNSS data. *IFAC-PapersOnLine* **2017**, *50*, 251–257. [CrossRef]
4. Jiang, C.; Chen, S.; Chen, Y.; Bo, Y. Research on a chip scale atomic clock driven GNSS/SINS deeply coupled navigation system for augmented performance. *IET Radar Sonar Navig.* **2019**, *13*, 326–331. [CrossRef]
5. Fernández, A.; Wis, M.; Silva, P.F.; Colomina, I.; Pares, E.; Dovis, F.; Ali, K.; Friess, P.; Lindenberger, J. GNSS/INS/LiDAR integration in urban environment: Algorithm description and results from ATENEA test campaign. In Proceedings of the 2012 6th ESA Workshop on Satellite Navigation Technologies (Navitec 2012) & European Workshop on GNSS Signals and Signal, Noordwijk, The Netherland, 5–7 December 2012; pp. 1–8. [CrossRef]
6. Jiang, C.; Chen, S.; Chen, Y.; Bo, Y.; Han, L.; Guo, J.; Feng, Z.; Zhou, H. Performance analysis of a deep simple recurrent unit recurrent neural network (SRU-RNN) in MEMS gyroscope de-noising. *Sensors* **2018**, *18*, 4471. [CrossRef]
7. El-Sheimy, N.; Hou, H.; Niu, X. Analysis and modeling of inertial sensors using Allan variance. *IEEE Trans. Instrum. Meas.* **2007**, *57*, 140–149. [CrossRef]
8. Syed, Z.F.; Aggarwal, P.; Goodall, C.; Niu, X.; El-Sheimy, N. A new multi-position calibration method for MEMS inertial navigation systems. *Meas. Sci. Technol.* **2007**, *18*, 1897–1907. [CrossRef]
9. Grip, H.F.; Fossen, T.I.; Johansen, T.A.; Saberi, A. Globally exponentially stable attitude and gyro bias estimation with application to GNSS/INS integration. *Automatica* **2015**, *51*, 158–166. [CrossRef]
10. Jiang, C.; Chen, Y.; Chen, S.; Bo, Y.; Li, W.; Tian, W.; Guo, J. A mixed deep recurrent neural network for MEMS gyroscope noise suppressing. *Electronics* **2019**, *8*, 181. [CrossRef]
11. Ismail, M.; Abdelkawy, E. A hybrid error modeling for MEMS IMU in integrated GPS/INS navigation system. *J. Glob. Position. Syst.* **2018**, *16*, 6. [CrossRef]
12. Xing, H.; Hou, B.; Lin, Z.; Guo, M. Modeling and compensation of random drift of MEMS gyroscopes based on least squares support vector machine optimized by chaotic particle swarm optimization. *Sensors* **2017**, *17*, 2335. [CrossRef]
13. Jiang, C.; Chen, S.; Chen, Y.; Zhang, B.; Feng, Z.; Zhou, H.; Bo, Y. A MEMS IMU de-noising method using long short term memory recurrent neural networks (LSTM-RNN). *Sensors* **2018**, *18*, 3470. [CrossRef]
14. Wu, Y.; Goodall, C.; El-Sheimy, N. Self-calibration for IMU/odometer land navigation: Simulation and test results. In Proceedings of the ION International Technical Meeting, San Diego, CA, USA, 25–27 January 2010.

15. Nieminen, T.; Kangas, J.J.J.; Suuriniemi, S.; Kettunen, L. An enhanced multi-position calibration method for consumer-grade inertial measurement units applied and tested. *Meas. Sci. Technol.* **2010**, *21*, 105204. [CrossRef]

16. Qin, H.; Cong, L.; Sun, X. Accuracy improvement of GPS/MEMS-INS integrated navigation system during GPS signal outage for land vehicle navigation. *J. Syst. Eng. Electron.* **2012**, *23*, 256–264. [CrossRef]

17. Nassar, S.; Niu, X.; El-Sheimy, N. Land-vehicle INS/GPS accurate positioning during GPS signal blockage periods. *J. Surv. Eng.* **2007**, *133*, 134–143. [CrossRef]

18. Chen, L.; Fang, J. A hybrid prediction method for bridging GPS outages in high-precision POS application. *IEEE Trans. Instrum. Meas.* **2014**, *63*, 1656–1665. [CrossRef]

19. Davidson, P.; Hautamäki, J.; Collin, J. Using low-cost MEMS 3D accelerometer and one gyro to assist GPS based car navigation system. In Proceedings of the 15th Saint Petersburg International Conference on Integrated Navigation Systems, Saint Petersburg, Russia, 26–28 May 2008.

20. Yang, L.; Li, Y.; Wu, Y.; Rizos, C. An enhanced MEMS-INS/GNSS integrated system with fault detection and exclusion capability for land vehicle navigation in urban areas. *GPS Solut.* **2013**, *18*, 593–603. [CrossRef]

21. Jiang, C.; Chen, S.; Bo, Y.; Sun, Z.; Lu, Q. Implementation and performance evaluation of a fast relocation method in a GPS/SINS/CSAC integrated navigation system hardware prototype. *IEICE Electron. Express* **2017**, *14*, 20170121. [CrossRef]

22. Soloviev, A.; Venable, D. Integration of GPS and vision measurements for navigation in GPS challenged environments. In Proceedings of the IEEE/ION Position, Location and Navigation Symposium, Indian Wells, CA, USA, 4–6 May 2010; pp. 826–833. [CrossRef]

23. Wang, J.; Garratt, M.; Lambert, A.; Wang, J.J.; Han, S.; Sinclair, D. Integration of GPS/INS/vision sensors to navigate unmanned aerial vehicles. *Int. Arch. Photogramm. Remote Sens. Spat. Inf. Sci.* **2008**, *37*, 963–969.

24. Yun, S.; Lee, Y.J.; Sung, S. IMU/Vision/Lidar Integrated Navigation System in GNSS Denied Environments. In Proceedings of the 2013 IEEE Aerospace Conference. Available online: citeseerx.ist.psu.edu/viewdoc/download?doi=10.1.1.436.2328&rep=rep1&type=pdf (accessed on 16 April 2020).

25. Cai, H.; Hu, Z.; Huang, G.; Zhu, D.; Su, X. Integration of GPS, monocular vision, and high definition (HD) map for accurate vehicle localization. *Sensors* **2018**, *18*, 3270. [CrossRef]

26. Georgy, J.; Noureldin, A.; Korenberg, M.J.; Bayoumi, M.M. Modeling the stochastic drift of a MEMS-based gyroscope in Gyro/Odometer/GPS integrated navigation. *IEEE Trans. Intell. Transp. Syst.* **2010**, *11*, 856–872. [CrossRef]

27. Georgy, J.; Karamat, T.; Iqbal, U.; Noureldin, A. Enhanced MEMS-IMU/odometer/GPS integration using mixture particle filter. *GPS Solut.* **2010**, *15*, 239–252. [CrossRef]

28. Li, Z.; Wang, J.; Li, B.; Gao, J.; Tan, X. GPS/INS/Odometer integrated system using fuzzy neural network for land vehicle navigation applications. *J. Navig.* **2014**, *67*, 967–983. [CrossRef]

29. Gao, Z.; Ge, M. Odometer and MEMS IMU enhancing PPP under weak satellite observability environments. *Adv. Space Res.* **2018**, *62*, 2494–2508. [CrossRef]

30. Simon, D. Kalman filtering with state constraints: A survey of linear and nonlinear algorithms. *IET Control. Theory Appl.* **2010**, *4*, 1303–1318. [CrossRef]

31. Zhou, G.; Li, K.; Kirubarajan, T.; Xu, L. State estimation with trajectory shape constraints using pseudomeasurements. *IEEE Trans. Aerosp. Electron. Syst.* **2019**, *55*, 2395–2407. [CrossRef]

32. Zhou, G.; Li, K.; Kirubarajan, T. Constrained state estimation using noisy destination information. *Signal. Process.* **2020**, *166*, 107226. [CrossRef]

33. Niu, X.; Li, Y.; Zhang, Q.; Cheng, Y.; Shi, C. Observability analysis of non-holonomic constraints for land-vehicle navigation systems. *J. Glob. Position. Syst.* **2012**, *11*, 80–88. [CrossRef]

34. MathWorks. Available online: https://www.mathworks.com/products/matlab.html (accessed on 16 April 2020).

Article

Autonomous Road Roundabout Detection and Navigation System for Smart Vehicles and Cities Using Laser Simulator–Fuzzy Logic Algorithms and Sensor Fusion

Mohammed A. H. Ali [1,*]**, Musa Mailah** [2]**, Waheb A. Jabbar** [3]**, Khaja Moiduddin** [4]**, Wadea Ameen** [4] **and Hisham Alkhalefah** [4]

[1]   Faculty of Manufacturing Engineering, Universiti Malaysia Pahang (UMP), Pekan 26600, Malaysia
[2]   School of Mechanical Engineering, Faculty of Engineering, Universiti Teknologi Malaysia, UTM Johor Bahru 81310, Malaysia; musa@fkm.utm.my
[3]   Faculty of Electrical and Electronic Engineering Technology, Universiti Malaysia Pahang (UMP), Pekan 26600, Malaysia; waheb@ump.edu.my
[4]   Advanced Manufacturing Institute, King Saud University, Riyadh 11451, Saudi Arabia; khussain1@ksu.edu.sa (K.M.); wqaid@ksu.edu.sa (W.A.); halkhalefah@ksu.edu.sa (H.A.)
*   Correspondence: hashem@ump.edu.my; Tel.: +60-942-459-12

Received: 30 April 2020; Accepted: 24 June 2020; Published: 1 July 2020

Abstract: A real-time roundabout detection and navigation system for smart vehicles and cities using laser simulator–fuzzy logic algorithms and sensor fusion in a road environment is presented in this paper. A wheeled mobile robot (WMR) is supposed to navigate autonomously on the road in real-time and reach a predefined goal while discovering and detecting the road roundabout. A complete modeling and path planning of the road's roundabout intersection was derived to enable the WMR to navigate autonomously in indoor and outdoor terrains. A new algorithm, called Laser Simulator, has been introduced to detect various entities in a road roundabout setting, which is later integrated with fuzzy logic algorithm for making the right decision about the existence of the roundabout. The sensor fusion process involving the use of a Wi-Fi camera, laser range finder, and odometry was implemented to generate the robot's path planning and localization within the road environment. The local maps were built using the extracted data from the camera and laser range finder to estimate the road parameters such as road width, side curbs, and roundabout center, all in two-dimensional space. The path generation algorithm was fully derived within the local maps and tested with a WMR platform in real-time.

Keywords: wheeled mobile robot; path panning; laser simulator; fuzzy logic; laser range finder; Wi-Fi camera; sensor fusion; local map; odometry

---

1. Introduction

The navigation of autonomous vehicle in urban-building and roads is still a tricky topic in robotics research worldwide, due to many uncertain cases that prevail while navigating on roads. The autonomous vehicle is currently not just used as a transportation medium but can be also utilized for performing some road services like road marks painting, grass cutting, and side-road cleaning [1–3].

A complete navigation algorithm that can consider and deal with all the conditions encountered in the road environment has yet to be thoroughly constructed and developed [4]. One such case is navigation and path determination in the open space area, e.g., in a roundabout and T, Y, and cross junctions. Three main challenges are likely to be encountered during navigation in such areas: firstly, the capability to detect the printed marks during weather changes; secondly is the ability to discover and analyze the traffic light signals; and finally, to detect various intersections or junctions using the natural landmarks like borders and edges in such areas, when the printed road marks are missed.

The open space road areas include branched/unbranched roads [5,6] and road junctions/intersections [7–13]. It has been remarked that a limited number of studies have addressed the autonomous navigation of mobile robot in open-space areas of the roads such as cross, roundabout, T, and Y intersections. Another situation for open space area on the roads can be seen when there is a sudden change of the road width along the span of roads [13].

Several works for navigation on the road intersections (e.g., T and Y branches and cross areas) and road following, have been reported [7–18] through building a suitable local map of terrains using sensors fusion technique. In fact, a handful of the above-mentioned reported works have dealt with a roundabout environment where the surroundings and rules of the road are significantly different from other kinds of junctions or intersections. The roundabout intersection typically consists of a circular area in the middle and several inlet/outlet branches on the sides, which enables the vehicles to rotate around until finding the suitable exit branch without a need for traffic light signals, such as in the cross junction, where the vehicle selects accordingly its exit branch based on appropriate traffic light signals.

2. Related Works

The open space areas such as T, Y, cross, and roundabouts intersections are considered as the main challenge during autonomous vehicle navigation on the roads, due to sudden changes of the road's direction, losses of sensors' signals, and difficulty to make the right decision in such environments. Other open space area challenges can be found during autonomous navigation and right maneuver searching in underwater vehicles [19–21].

This paper focuses on the road roundabout environments navigation, since it needs a further development and consideration as stated in Section 1; however, the other types of junctions such as T/Y and cross intersections have been already studied well in the literature [7–18].

Most of the current methods used for roundabout navigation and detection depend on the offline information coming from GPS and maps to recognize the roundabout and find the proper exit of roundabout. Jorge et al. [22] have developed an algorithm to recognize the roundabout setting and find the proper maneuvers using so-called open street map and GPS, which help to find the appropriate direction of vehicle to be undertaken by the driver either in left, right, and straight manners. In fact, this is a manual and offline navigation system that utilizes a yaw-rate profile to determine the exit of roundabout. A 3D simulation of CyberCar to detect a roundabout setting from well-known digital maps is developed by Rastelli et al. [23]. In this simulator, once the car arrives at the entrance of the roundabout setting, it generates a circular path in the map with a radius slightly larger than the roundabout radius.

A steering system to maintain the vehicle on a circular path during navigation in roundabout using a fusion of fuzzy-logic controller with distance curve gain and angular-error strategies has been proposed by Katrakazas et al. [24] and Rastelli et al. [25]. In addition, a control system using inertial measurement unit (IMU) and GPS is used for roundabout trajectory tracking. In fact, the features of a roundabout are completely known, such as center, entrance, and exit of the roundabout from GPS and the maps data; however, the path generation is determined from inlet to outlet of roundabout using Bezier curves strategy.

A guidance system that allows the vehicle to travel along an optimum traffic lane in a roundabout setting and passes the information to the driver has been developed by Okusa et al. [26]. In this system, the maps and GPS are used to localize the car in the roundabout setting; however, the traffic signals and vehicle turning in roundabouts are determined using a group of sensor data that can be generally classified into two subgroups, namely; traveling distance and direction sensors.

Laura et al. [27] has presented a machine learning algorithm-based predictive model to estimate the vehicle speed and steering angles in a roundabout intersection. In fact, this model depends on Open-Streets-Maps and recorded video to identify the roundabout geometry and dimensions, which is unsafe for driving in such a dangerous area and is subject to errors; thus there is a need for online estimation of roundabout parameters.

A sensor fusion model has been used to predict 3D dimensions of surrounding environments during autonomous driving on roundabouts using LIDAR point's cloud and camera [28]. It builds a 3D objects map by integrating the data of the images with LIDAR using 3D bounding boxes. This work is just focused on the roundabout environment detection; however, it does not show how to generate the path within the roundabout environment. A 2D image has been combined with 3D point cloud to recognize the road's traffic signs for the purpose of autonomous transportation systems [29]. It uses the bag of visual phrases and Gaussian Bernoulli deep Boltzmann algorithm to extract the features of the traffic signs of the road environment, which is useful for detection of traffic signs; however, it is not able to generate the path within the road environments.

Tesla and Google autonomous cars have used a sensor fusion technique for building a 3D map of a vehicle's surrounding environments and finding the cars located at the back or side of the autonomous vehicle [30]. It uses velodyne laser range finder (LRF) for construction of an online map, camera for estimating the obstacles, car on the side or back of the autonomous car, and GPS with Google maps to localize the vehicle within the road environments from the start to goal positions. Such autonomous vehicles depend mainly on GPS and Google maps to recognize the roundabout geometry, which might cause damage to pedestrians and road infrastructure when there is a recent change on the roundabout that has not been yet registered on the Google maps or there are signal losses in GPS sensor.

Owing to the accuracy and safety issues such as incapability to identify a roundabout in a specific location using GPS, GPS signal losses, and issues in recognizing the unexpected changes of the road roundabout terrain, it may result in particular damage and unwarranted harm to the road infrastructure and pedestrians. Hence, there is a necessity to utilize onboard sensors such as camera, odometry, and LRF for online detection, path determination, and decision making during navigation of roundabout environments. It is deemed to be amongst the pioneering researches that advocates the implantation of real-time navigation in a road roundabout environment using cameras, odometry, and LRF sensors. This paper presents an onboard navigation system in a roundabout using a novel algorithm, called Laser Simulator (LS), which is integrated with a fuzzy logic (FL) algorithm and sensor fusion for autonomous navigation in roundabout environments. The combination between LS and FL algorithms was introduced to make the right decision on the existence of the roundabout, with the sensor fusion implemented to determine the appropriate robot path in a roundabout environment.

In this work, the robot is navigating autonomously in the road following and later moves effectively in a circular path of a standard roundabout. It finds the path starting from the predefined initial position until it reaches the predefined goal using the LS–FL algorithms and sensor fusion. This approach was implemented for recognizing the roundabout within the local map with sensor fusion involving the LRF and odometry measurements to determine the robot path. The predefined initial and goal positions were determined using a DGPS system and the robot uses such information to detect the direction of the roundabout exit. Because the robot is navigating in a relatively small area (within 10–500 m diameter), it is not useful to use GPS, and instead, the robot could be informed about the roundabout exit direction and goal position even prior to its activation. The goal position can be determined in terms of the distance the robot should travel after passing the roundabout, e.g., robot to proceed to the east direction of the roundabout (270°) for about 20 m distance to a goal position.

In this work, two tasks were performed to enable the mobile robot to pass autonomously through the roundabout:

1. Autonomous road roundabout detection
2. Autonomous road roundabout navigation

The details of each task are described in the subsequent sections.

3. Autonomous Road Roundabout Detection

It is not an easy task to decide if the circular/elliptical object located ahead of the robot during navigation is indeed a road roundabout or otherwise. Thus, a decision making procedure is needed to distinguish clearly the roundabout and other similar circular/elliptical shaped objects, such as obstacles, vehicles, etc. It is worthwhile to note that the Google maps and GPS are also currently and typically used in conjunction with many navigation systems to identify a roundabout location [23]. However, these methods have some major limitations on the accuracy and safety issues previously mentioned that may contribute to adverse effects and undesired consequences to road users, pedestrians and road curbs. Thus, the proposed study is regarded as an attempt to implement an online detection and navigation system in a roundabout setting using LRF, camera, and odometry sensors.

Three roundabout conditions were considered to detect the presence of a roundabout ahead of the robot during autonomous navigation on the road, as shown in Figure 1, assuming a right hand drive convention:

1. The right curb of the road is suddenly faded
2. The left curb of the road is slightly faded
3. There is a circular/elliptical curve located in front of the robot

All these conditions should occur simultaneously.

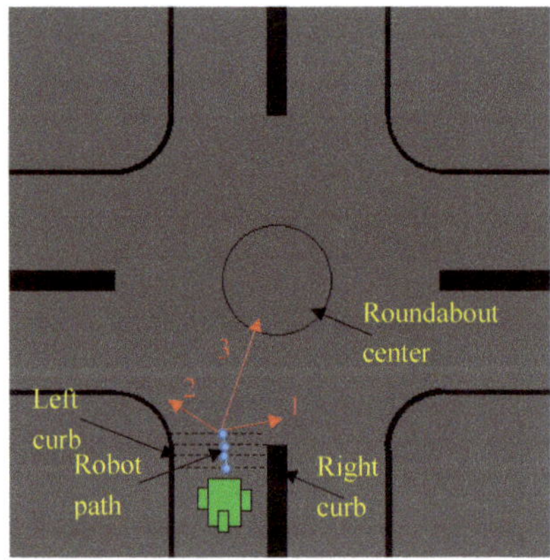

Figure 1. Conditions of the roundabout used for roundabout detection: (1) right curb is faded, (2) left curb is slightly faded, and (3) circular path is detected.

Prior to encountering the above conditions, a local map has to be created for the road environment using a camera and an image processing algorithm as follows:

3.1. Developing of a Local Map for the Road Environment

This was first done by performing video streaming. It was captured by a camera with suitable resolution, and then the captured image was processed using MATLAB software involving suitable image acquisition and processing toolboxes to perform an online image capturing and processing procedure. The image processing algorithm has been used to extract the road environment features and build the local map using laser simulator algorithm as shown in Figure 2.

The brightness of the image sequences was adjusted, since the setting of the camera aperture is not automatic. The following operations were applied selectively for edge detection and noise filtering: 2D Gaussian filters (fspecial), multidimensional images (imfilter), canny edge algorithm, morphological structuring element (strel), dilating image (imdilate), 2D order-statistic filtering (ordfil2), removing small objects from binary image (bwareaopen), and filling image regions and holes (imfill).

The results of the image processing algorithm and local mapping of the road environment will be shown in Section 3.2.

After building the local map of the road, the subsequent image post processing procedure was implemented to check whether the object located in front of the robot is a roundabout or otherwise as follows.

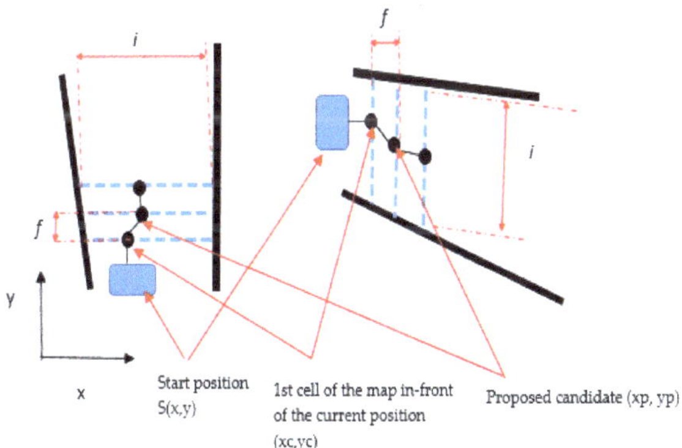

Figure 2. Basics of the Laser Simulator (LS) principle.

3.2. Laser Simulator-Based Roundabout Detection

A new algorithm for observing the road curbs and roundabout in the developed local maps using the so-called LS algorithm integrated with FL algorithm has been introduced in this section. It is actually emulating the laser beams by sending a series of points as horizontal or vertical lines in the local map that has been prepared in the previous step as shown in Figure 2, to detect either the faded road curbs or the location of a roundabout center. This will be followed by the application of fuzzy logic algorithm to make the right decision in regard to roundabout detection.

3.2.1. Right and Left Side Detection

The subsequent algorithm has been used to detect the road curbs in the camera's local map:

Generation of Points' Center Reference in the Image

Because the robot with its camera was assumed to be placed between the road sides as depicted in Figures 3–5, the position of the camera has been located in the middle of the image's frame sequence, exactly in bottom horizontal area of each image's frame. This is the 1st point's reference center c(x,y). It is followed by next points' reference centers which are produced by series points as horizontal line as in Equation (1):

$$y = x_c + i \tag{1}$$

where i is an incremental value located between 1 and the image's height (image's resolution in pixels in y-direction). The horizontal lines lengths are chosen between the subsequent limits as expressed in Equation (2):

$$y_{right} = y_c + R_p; \, y_{left} = y_c - L_p \tag{2}$$

where y_{right} is the limit of the lines on the right side. y_{left} is the lines' limit on the left-side. R_p is the pixels' number starting from the center to the curb of right side. L_p is pixels' number starting from the center to the curb of left side. The reference center points are then described as in Equation (3):

$$x_{cnew} = x_c - i; \, y_{cnew} = y_{left} + (R_p + L_p)/2 \tag{3}$$

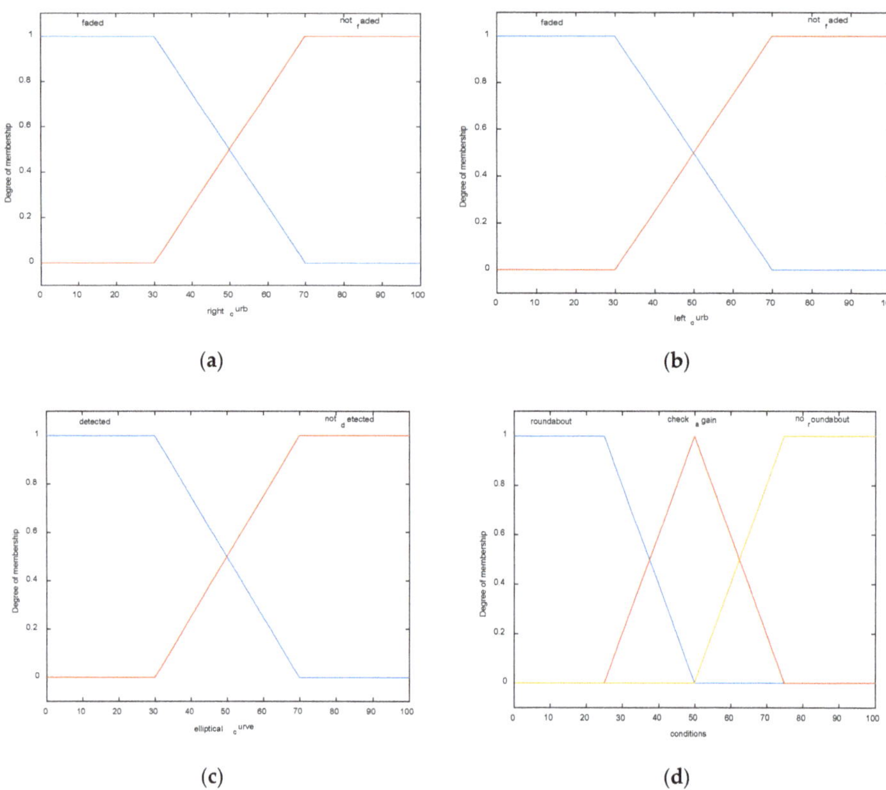

Figure 3. Input/output fuzzy membership functions: (**a**) input: right curb, (**b**) input: left curb, (**c**) input: elliptical curve, and (**d**) output: conditions.

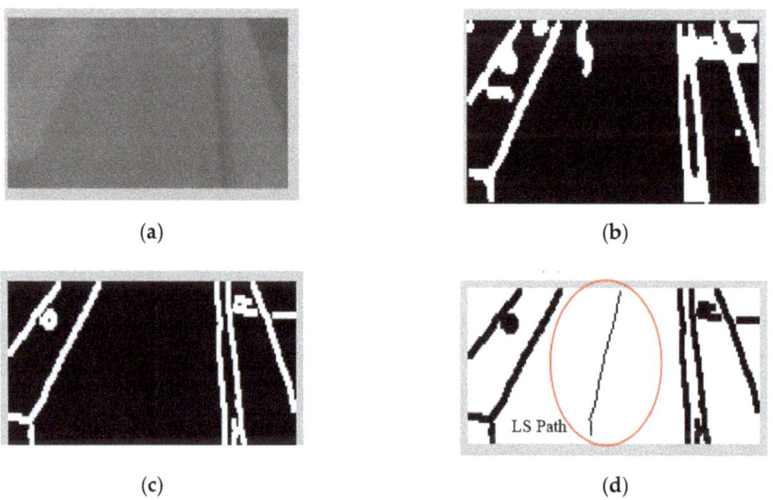

Figure 4. Image sequences processing where no roundabout can be detected using LS: (**a**) original image gray scale, (**b**) image after applying curbs detection, (**c**) image after removing the noise, and (**d**) image after applying the LS (continuous line in the middle).

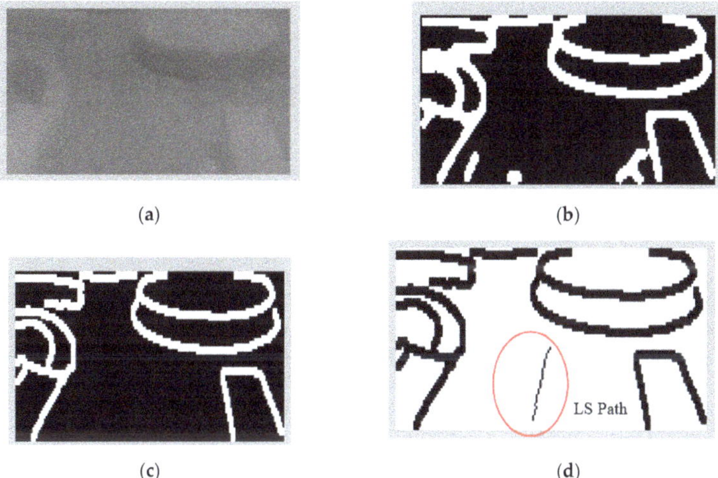

(a)

(b)

(c)

(d)

Figure 5. Image sequences processing where roundabout is detected using Laser Simulator: (**a**) original image in gray scale, (**b**) image after applying curbs and roundabout detection, (**c**) image after removing the noise, and (**d**) image after applying the LS (discontinuous line in the middle).

Detecting of the Curbs in the Right and Left Sides

To detect the road curbs, a series of points as inclined lines with slight changes of angles were produced in the image's frame in the video, initiating from the points centers x_{cnew} and y_{cnew} that have been calculated in Equation (3) and shown in Figure 2. The equations that are used to determine these lines can be written as follows:

With the left curb of the points' centers x_{cnew} and y_{cnew}, these lines were generated by:

$$x_L = x_{cnew} - f; y_L = y_{cnew} - (x_{cnew} - x_L)\tan(\delta) \tag{4}$$

And they are generated for the right curb of the points' centers x_{cnew} and y_{cnew} by:

$$x_r = x_{cnew} + k_r; y_r = y_{cnew} - (x_r - x_{cnew})\tan(\delta) \tag{5}$$

where $f = 1{:}L_p$. $k_r = 1{:}R_p$. L_p and R_p are the pixels' number which are existed between the points' centers x_{cnew} and y_{cnew} and right or left side, in respectively. δ is the slope of inclined lines to the ground. The number of pixels (P) for Equation (4) can be expressed as:

$$P_L = \sum_{j=1}^{L_P} P_j \tag{6}$$

The number of pixels (P) for Equation (5) can be expressed as:

$$P_r = \sum_{k=1}^{L_P} P_k \tag{7}$$

Because the dimension between the road sides as in actual case is almost equal; the pixels in each two successive inclined lines are compared with each other as follows:

$$Al = pl_i - pl_{i-1} Ar = pl_i - pl_{i-1} \tag{8}$$

If *Al* and *Ar* override the threshold as calculated in Equation (9), it indicates that such new line is not belonged to the road side or curb:

$$Al < A_{l1} + d Ar < A_{r1} + d \tag{9}$$

where A_{r1} and A_{l1} are the 1st tangent line's pixels in Equations (4) and (5). d is a value measure as a ratio of image resolution. Here in this work, this ratio was considered as 10% in both x and y directions.

3.2.2. Roundabout Center Detection:

The road roundabout presents in a camera's local map as an ellipse. If the curbs of road could not be determined in the former step, the roundabout detection algorithm will start to be executed. From the reference center point, the algorithm generates multi-tangent lines on the right curb with an angle ranging between 0 until 90 till they intersect with edges. The intersection points between these tangent lines and roundabout center that is represented as an ellipse were calculated. If the intersected points can verify the equation of ellipse as illustrated in Equation (10), it indicates that the center of roundabout center has been detected.

$$\frac{(x - x_0)^2}{a^2} + \frac{(y - y_0)^2}{b^2} = 1 \tag{10}$$

One can choose four points from intersection points, the first two points could be used to calculate the constants a and b in Equation (11) and the rest two points for the verification if the shape is an ellipse, which must be achieved based on the following conditions:

$$comp = \frac{x^2}{a^2} + \frac{y^2}{b^2} - 1 < ad \tag{11}$$

where *comp* = 0, *ad* represents the allowed deviation from zero (in the program, *ad* = *10*).

3.3. Fuzzy Logic-Based Decision Making

Once the three features of the roundabout are determined using the LS (faded right and left curbs and roundabout center detection), the fuzzy logic algorithm is applied subsequently to make the right decision on the roundabout detection. Three fuzzy sets are used as input sets, namely; right_curb, left_curb, and elliptical curve with one as output set called conditions. The input sets have been fuzzified into the linguistic variables as follows: right-curb = {faded (less or equal to *d* as in Equation (9), not_faded (bigger than *d* as in Equation (9))}, left-curb = {not faded (less or equal to *d* as in Equation (9), faded (bigger than *d* as in Equation (9))}, and elliptical curve = {not detected (*comp* > *ad* as in Equation (11)), detected (*comp* < *ad* as in Equation (11))}. The output fuzzy set has been fuzzified into the following linguistic variables: conditions = {roundabout (all condition have been verified), check again (only one condition hasn't been verified, no roundabout (two conditions out of three haven't been verified)}. The membership functions of the input and output sets are illustrated in Figure 3.

Four rules have been used in fuzzy logic to identify the existence of the roundabout as follows:

IF (the right curb is faded) and (elliptical curve is detected) and (left curb is faded) THEN (there is a roundabout in front.)

IF (the right curb is faded) and (elliptical curve is detected) and (left curb is not faded) THEN (check again in the next laser simulator lines)

IF (the right curb is not faded) and (elliptical curve is detected) and (left curb is faded) THEN (check again in the next laser simulator lines).

IF (the right curb is not faded) and (elliptical curve is not detected) and (left curb is faded) THEN (this is not a roundabout).

The results of the LS with rules decision are shown in Figures 4d, 5d, 6, 7, 8 and 9c. The continuous lines in the middle of the LS images as shown in Figure 4d denote that the roundabout has not occurred yet; however, the discontinuous line in Figure 5d indicates the detection of roundabout.

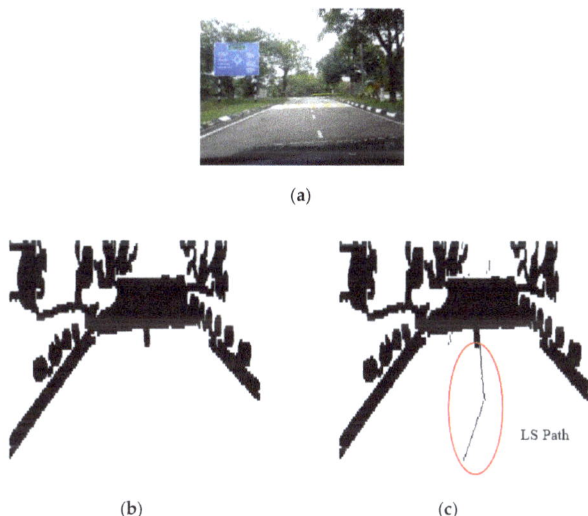

Figure 6. Image sequence with applying LS for roundabout determination at 100 m from the roundabout: (**a**) original image, (**b**) processing image, and (**c**) implementation of LS (continuous dotted line at the middle).

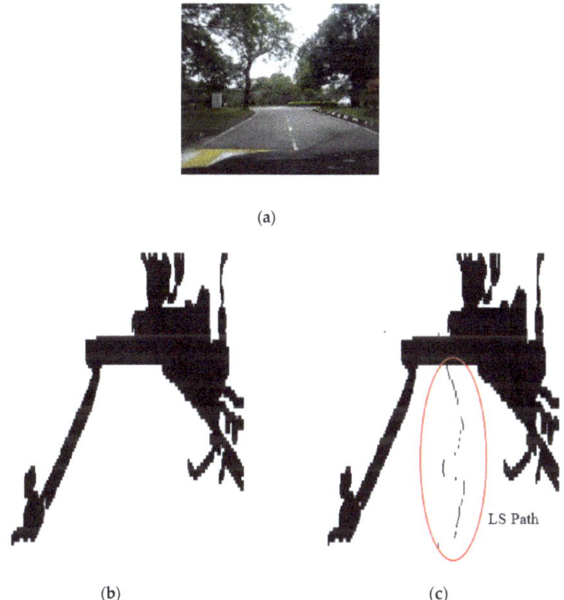

Figure 7. Image sequence with applying LS for roundabout determination at 50 m from the roundabout: (**a**) original image, (**b**) processing image, and (**c**) implementation of LS (continuous dotted line at the middle).

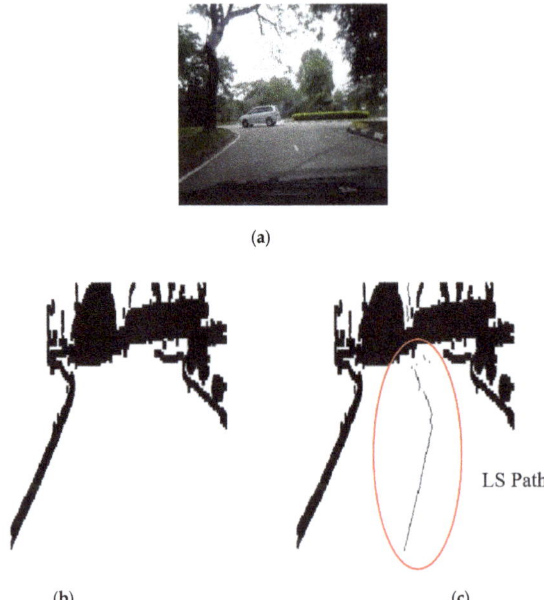

Figure 8. Image sequence with applying LS for roundabout determination at 10 m from the roundabout: (**a**) original image, (**b**) processing image, and (**c**) implementation of LS (continuous dotted line at the middle).

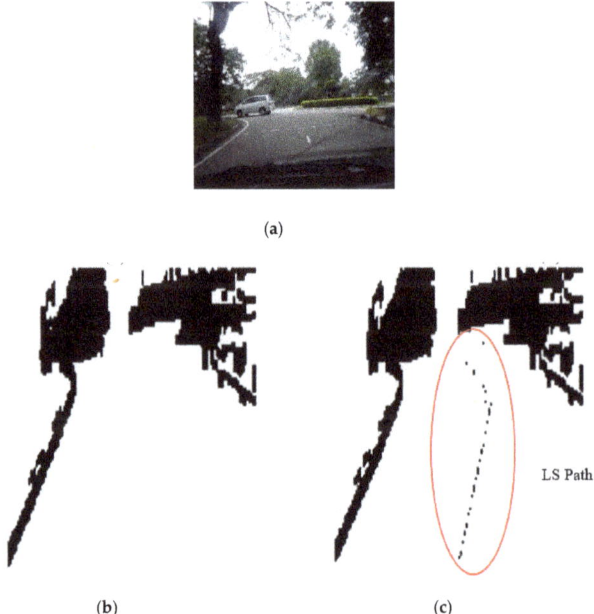

Figure 9. Image sequence with applying LS for roundabout determination at close distance to the roundabout: (**a**) original image, (**b**) processing image, and (**c**) implementation of (continuous dotted line at the middle).

The LS algorithm for the roundabout detection has been applied to the real roundabout as shown in Figures 6–9. It was observed that the algorithm can effectively detect the existence of the roundabout from the sequences of the images as shown in Figures 6–9.

As previously mentioned, the continuous lines in the middle of the image after applying the LS algorithm as shown in the results of Figures 6, 7 and 8c denote that the roundabout has yet to exist; however, the discontinuous line in Figure 9c indicates the early detection of the roundabout.

4. Sensor Fusion for Path Planning and Roundabout Navigation

A wheeled mobile robot (WMR) platform, equipped with LRF, camera, and odometry sensors, has been developed in the laboratory as shown in Figure 10 to accomplish the autonomous navigation in the roundabout intersection.

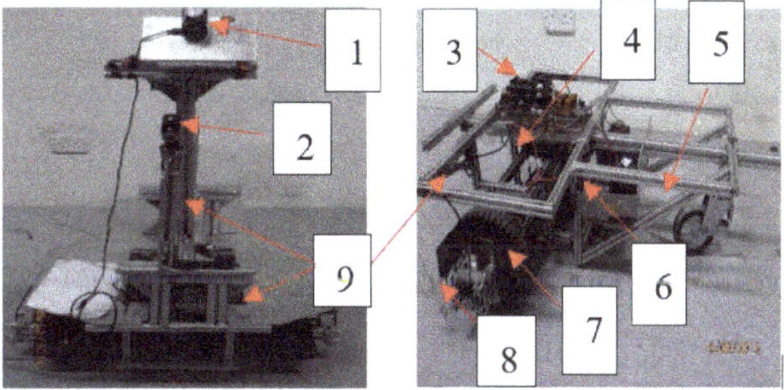

Figure 10. Developed wheeled mobile robot (WMR) platform in this research: (1) LRF, (2) Wi-Fi camera, (3) interface free controller cards, (4) DC-motors driver card, (5) castor wheel, (6) battery, (7) differential drive wheels, (8) rotary encoder, and (9) aluminum profiles and plates.

A sensor fusion technique involving the use of LRF, camera, and odometry has been used to determine the robot path starting from a specific entrance to an appropriate exit of the roundabout. As mentioned in Section 2–A–1, the camera's image processing algorithm has been used to extract some features of the road curbs and roundabout, e.g., the road roundabout center and borders. Other features will be extracted using the LRF and odometry sensors.

4.1. Sensor Fusion Modeling

In this section, the road features extracted by the LRF and odometry sensor will be explained in detail, as follows:

4.1.1. Odometry-Based Measurements

Two encoders were utilized to determine the robot position. They have been linked to the two differential wheels through the encoder's pins in dual brush card motor. The complete rotation of this rotary encoder is around 500 pulses/rotation; thus the linear displacement can be computed as follows:

$$C = \frac{2\pi \, r P_{cur}}{P_{fr}} \tag{12}$$

where P_{fr} and P_{cur} are the pulses of the complete and current rotation, respectively.

4.1.2. LRF-Based Measurements

The LRF measurements were used for building a 2D local map as shown in Figure 11.

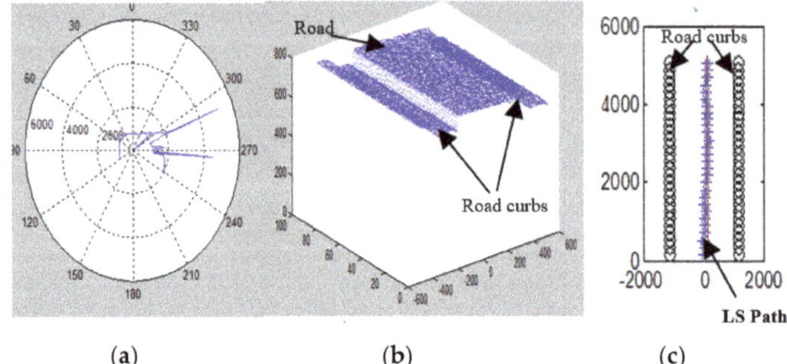

Figure 11. Principle of LRF measurement and calculation: (**a**) one scan measurement (mm), (**b**) road with curbs in 3D (mm), and (**c**) LS path generation (mm).

Two parameters can be extracted from the LRF measurements as follows:
The road fluctuations (height of objects with respect to the laser device) can be expressed as:

$$rf_n = r_n \cos\theta \tag{13}$$

The road width (side distance measurement with respect to the laser device) is presented as:

$$rw_n = r_n \sin\theta \tag{14}$$

where r_n is the length of the LRF signal for n-th LRF measurement, θ ($-120°$, 0, $120°$) is the angle of the laser beam deviation from extreme right at $120°$ to extreme left at $-120°$.

If rf_0 is the road fluctuation at $\theta = 0°$, which is used as a reference point, rf_i is the other fluctuation measurements located on the left and right side of this point are being compared, and d is the threshold of the curb detection, which was set to 10 cm in this work, then:

$$rf_i - rf_0 \geq \pm d \tag{15}$$

This implies that if the deviation between the reference point and other measured values exceeds the predefined threshold value, then this point is considered as the road curb or obstacles as in Figure 11b. This operation is repeated with all measurements as in Equation (15).

4.2. Road Roundabout Navigation

The sensor fusion data were used to enable the robot to navigate autonomously in the roundabout setting starting from the entrance to the exit sections of the roundabout. The algorithm used for driving the robot on the path approaching the entrance or after the exit areas of the roundabout is called road following, whereas the road roundabout center algorithm was used to navigate about the center of the roundabout. This is described as follows:

4.2.1. Navigation in the Road Following

The sensor fusion including camera, LRF and encoders were used effectively to determine the collision-free path in the road following areas. The camera was used to recognize the roundabout area when it exists by LS algorithm as expressed in Section 3.2. The LRF was used for finding the curbs of roads and determine the location of the robot within its environment. The measurement of encoder was utilized to estimate the location of robot during navigation in the given environment.

The path determination of the robot using LRF and odometry is discussed as follows:

If the robot's start position is (x_1, y_1) as depicted in Figure 12, the LS algorithm will determine the planned pose x_2 as a center of the LRF measurements within the two curbs of the road as presented in Equation (15). y_2 is computed from the measurement of encoder as shown in Equation (18).

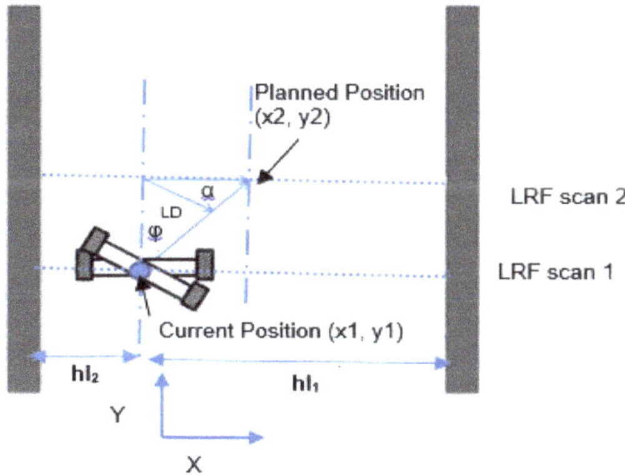

Figure 12. Robot path planning calculation for road following section.

In Figure 12, hl_1 and hl_2 are the distances between the current position and road curbs in x. The differential wheels of WMR should be moved by an angular velocity as shown in Equation (16):

$$\dot{\varphi} = \frac{V_r - V_l}{b} \qquad \dot{\varphi}b = V_r - V_l \tag{16}$$

where V_l and V_r are the velocities of the wheels in the left and right sides, respectively, and b is the dimension between the right and left wheels in the differential drive mechanism. The left side of Equation (16) can be written as in Equation (17):

$$\dot{\varphi}b = b\frac{\varphi_{max} - \varphi_0}{T} = \frac{b\varphi_{max}}{T} = \frac{LD}{T} \tag{17}$$

where LD is the displacement required to be shifted by WMR to reach the required position starting from the current measurement of LRF as illustrated in Figure 12. T is the periodic time of the acuring the measurement, φ_0 is current position heading angle (set at the beginning as 0), and φ_{max} is the heading angle's new location.

The planned position (x_2, y_2) can be calculated as follows:

$$x_2 = x_1 + |hl_2 + hl_1|Ra \qquad y_2 = y_1 + \Delta T. V \quad \text{where} \quad V = \frac{V_r + V_l}{2} \tag{18}$$

where x_1 and y_1 indicate the current position of the robot, and Ra indicates the place where WMR is planning to drive. If Ra is setup to 0.5, that means the robot will drive in the mid-distance between the road curbs.

φ_{max} in Equation (17) can be further calculated as:

$$\varphi = \tan^{-1}\left(\frac{x_2}{y_2}\right) \tag{19}$$

The rotational distance LD can then be calculated as follows:

$$LD = (x_2 - x_1)\frac{\sin(\alpha)}{\sin\left(\frac{\pi + \varphi}{2}\right)} \quad \text{where} \quad \alpha = \frac{\pi}{2} - \varphi \tag{20}$$

4.2.2. Navigation in Road Roundabout Center

When the LS integrated with FL algorithm indicates that the roundabout is located in the camera's local map, the camera will be shut-off and the algorithm for approaching the roundabout entrance starts to operate. The algorithm for the roundabout entrance is operating based on a mixed information coming from the last camera local map calculation and the last measurements from the LRF.

In Figure 13, *s-ang* is the slip angle of the robot, *r–ang* is the rotation angle, and *n-ang* is navigation angle. All these angles can be calculated using Equation (21) from the camera local map, since the coordinates of (x_1, y_1), (x_2, y_2), and (x_3, y_3) are known from the LS algorithm.

$$s\_ang = \arctan\frac{x_2-x_1}{y_1-y_2}$$
$$n\_ang = \arctan\frac{x_3-x_1}{y_1-y_3} \tag{21}$$
$$r\_ang = \frac{\pi}{2} - s\_ang$$

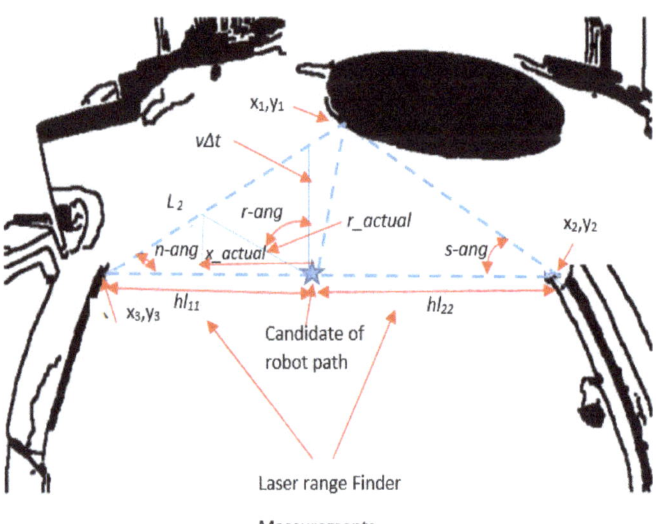

Figure 13. Entrance parameters and path determination of roundabout.

The robot position, x can be computed based on the angles that have been calculated from the camera local map and the last LRF measurements in the road following:

$$x\_actual\,\tan(s\_ang) = (hl_{11} - x\_actual)\,\tan(n\_ang) \tag{22}$$

where hl_{11} is the left measurements of the robot in the road following part as defined in Figure 13.

The x and y of the planned position for the robot can be then calculated as follows:

$$x\_actual = hl_{11}\frac{\tan(n\_ang)}{\tan(s\_ang)+\tan(n\_ang)}$$
$$Y = \Delta T.\,V \quad \text{where} \quad V = \frac{V_r+V_L}{2} \tag{23}$$

The angles of the roundabout entrance were calculated from the camera local map as shown in Figure 13; however, the dimensions were calculated from the last measurements of the LRF. The entrance parameters are illustrated in Figure 13.

The robot angular velocities to reach the planned path can be calculated as:

$$V_l - V_r = \dot{\varphi}b = b\frac{\varphi_{max} - \varphi_0}{t_{max} - t_0} = \frac{b\varphi_{max}}{T} = \frac{V \, \sin(r\_ang)}{\sin(s\_ang + n\_ang)} \tag{24}$$

The planned path of the mobile robot must be in the direction of r_actual, which can be calculated as:

$$r\_actual = \frac{x\_actual}{\cos(s\_ang)} \tag{25}$$

If the robot can move by an angle r_ang and reach the r_actual, this means that the robot has reached and passed the roundabout entrance.

The calculation of the exit parameters is almost similar to that of the entrance counterpart. It is in fact exactly an inverse calculation of the entrance if the roundabout is assumed to be standard.

In the roundabout center as shown in Figure 14, a combination between the LRF and encoders were used for rotating the robot in the correct path. Because the robot will move in a circular path, two coordinate systems can be used for describing the robot path; one is moving with the robot while the other is fixed as shown in Figure 14. The rotation will be clockwise and thus, the transformation matrix may be presented as:

$$\begin{bmatrix} X_{las-fix} \\ Y_{las-fix} \end{bmatrix} = \begin{bmatrix} \cos(rd - rd_0) & \sin(rd - rd_0) \\ \sin(rd - rd_0) & \cos(rd - rd_{0)} \end{bmatrix} \times \begin{bmatrix} X_{las-rot} \\ Y_{las-rot} \end{bmatrix} \tag{26}$$

where $X_{las\text{-}fix}$, $Y_{las\text{-}fix}$ is the fixed coordinate system and $X_{las\text{-}rot}$, $Y_{las\text{-}rot}$ is the robot coordinate system. $rd - rd_0$ is the total angle that robot should rotate in the roundabout.

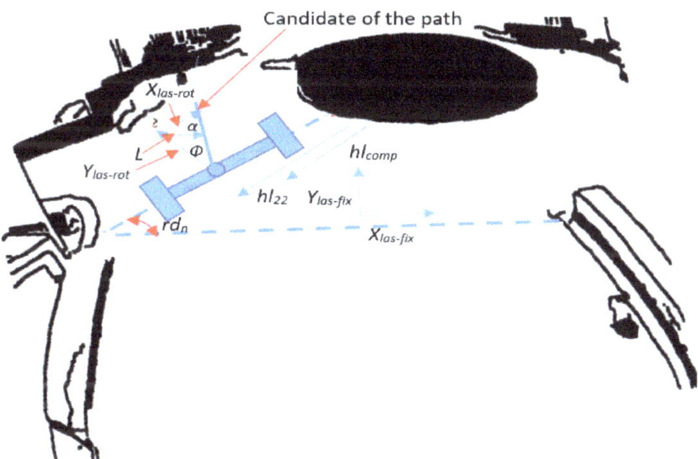

Figure 14. Roundabout center parameters and path determination.

The first measurement of LRF at the right side of robot after passing the entrance area (hl_{comp}) was used as a reference for the robot when rotating around the roundabout, and the other LRF measurements were continuously compared to the reference to determine the ($x_{las\text{-}rot}$, $y_{las\text{-}rot}$) position of the robot as expressed in Equation (27):

$$X_{las-rot} = hl_{22} - hl_{comp}$$
$$Y_{las-rot} = \Delta T. V \quad \text{where} \quad V = \frac{V_r + V_L}{2} \tag{27}$$

where *hl22* is the current measurement of the LRF at the right side of the robot. The rotation angle φ can be calculated as follows:

$$\varphi = \arctan\left(\frac{X_{las-rot}}{Y_{las-rot}}\right) \tag{28}$$

The rotation distance of robot *L* can be calculated as:

$$L = X_{las-rot}\frac{\sin(\alpha)}{\sin\left(\frac{\pi+\varphi}{2}\right)} \tag{29}$$

where:

$$\alpha = \frac{\pi}{2} - \varphi$$

The angular velocity can be calculated as follows:

$$V_l - V_r = \dot{\varphi}b = b\frac{\varphi_{max} - \varphi_0}{t_{max} - t_0} = \frac{b\varphi_{max}}{T} = \frac{L}{T} = \frac{X_{laser}\frac{\sin(\alpha)}{\sin\left(\frac{\pi+\varphi}{2}\right)}}{T} \tag{30}$$

The exit of the roundabout center, where the robot stops to rotate around the roundabout, was calculated based on the encoder's measurements. Assume that the robot moves in a circular path as shown in Figure 15. The outer and inner circumferences of this roundabout that was likened to a disk were calculated based on the positions of the right and left side's encoders of the robot that constitutes the width (b) of the WMR.

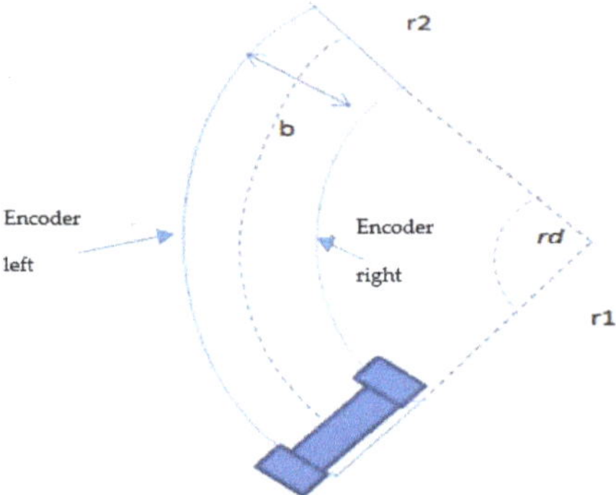

Figure 15. Robot rotation about the roundabout center to find the exit.

The following equations were used to calculate the angle of rotation *rd* in rad as presented in Equation (31):

$$r_2 = r_1 + b$$
$$encod\_right = rd\,r_1$$
$$encod\_left = rd r_2 \tag{31}$$
$$\frac{encod\_left}{rd} = \frac{encod\_right}{rd} + b$$
$$rd = \frac{encod\_left - encod\_right}{b}$$

Four *rd* conditions can be considered:

1. $rd > \pi/2$: robot will exit the roundabout at the first left turn.
2. $rd > \pi$: robot will exit the roundabout at the second left turn and in the straight direction.
3. $rd > 3\pi/2$: robot will exit the roundabout at the third left turn.
4. $rd > 2\pi$: robot will exit the roundabout at the fourth left turn.

5. Results and Discussion

Several robot's path scenarios have been performed in real road environments in both road following and roundabout cases as depicted in Figures 16–22, to show the capability of the suggested algorithms for road roundabout detection and navigation. Another set of experiments have been implemented in indoor and outdoor environments to test the performance of the suggested algorithm in terms of the lighting and weather conditions changings as shown in Figures 23–26.

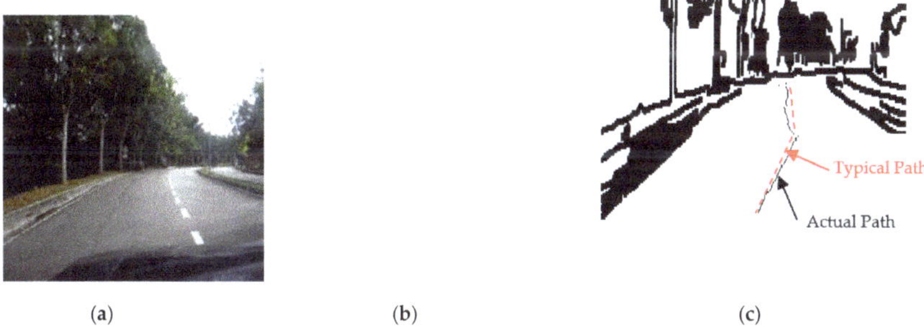

(a)   (b)   (c)

Figure 16. Autonomous detection and navigation of the proposed system in the road following with 5 m as width and 500 m as length: (**a**) original image, (**b**) image processing, and (**c**) generation of the path within the road following environment.

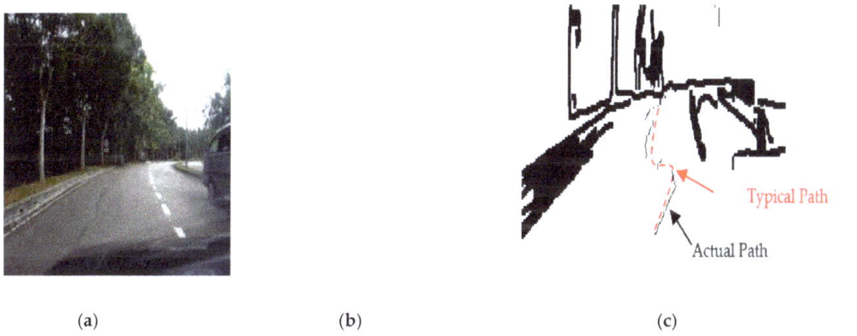

(a)   (b)   (c)

Figure 17. Autonomous detection and navigation of the proposed system in the road following (with 5 m as width and 500 m as length) with partial car on the side: (**a**) original image, (**b**) image processing, and (**c**) generation of the path within the road following environment.

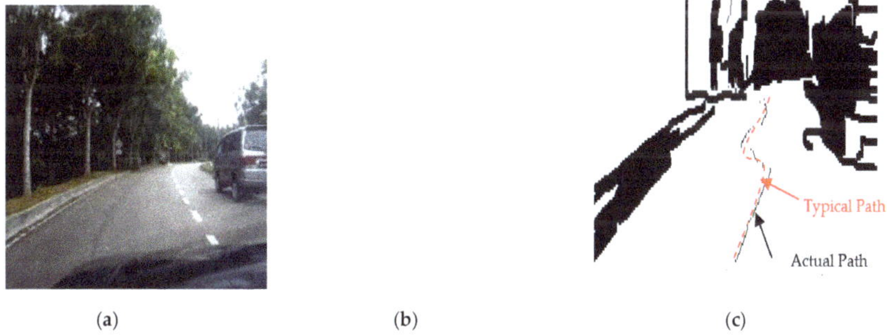

(a)   (b)   (c)

Figure 18. Autonomous detection and navigation of the proposed system in the road following (with 5 m as width and 500 m as length) a car partially presented on the side/in front: (**a**) original image, (**b**) image processing, and (**c**) generation of the path within the road following environment.

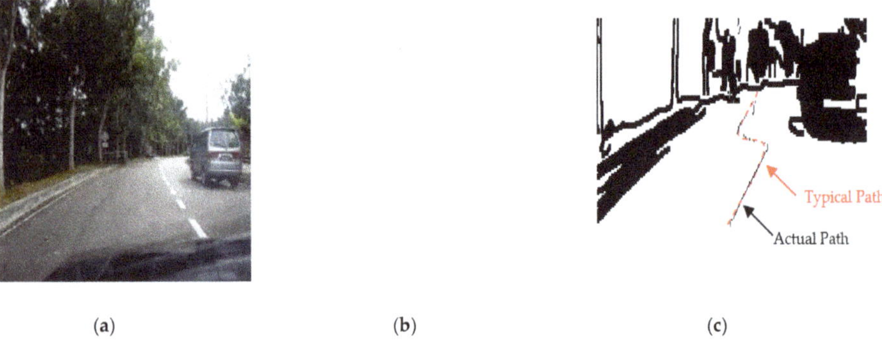

(a)   (b)   (c)

Figure 19. Autonomous detection and navigation of the proposed system in the road following (with 5 m as width and 500 m as length) with a car on the side/in front: (**a**) original image, (**b**) image processing, and (**c**) generation of the path within the road following environment.

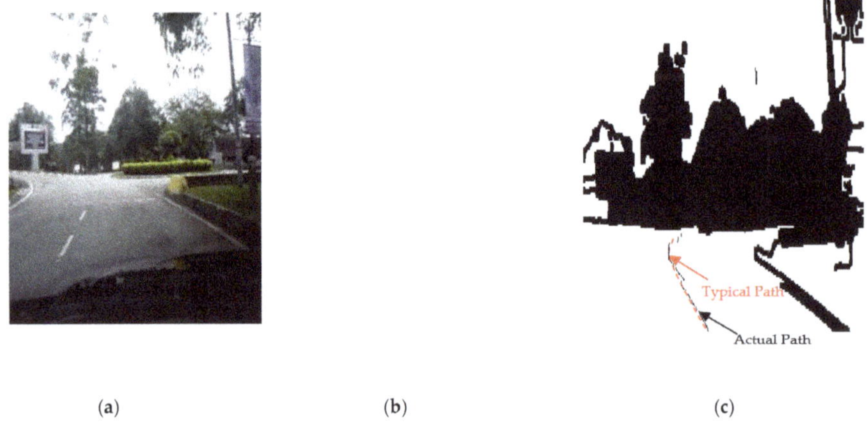

(a)   (b)   (c)

Figure 20. Autonomous detection and navigation of the proposed system in the road roundabout (with 5 m as diameter): (**a**) original image, (**b**) image processing, and (**c**) generation of the path within the road roundabout environment.

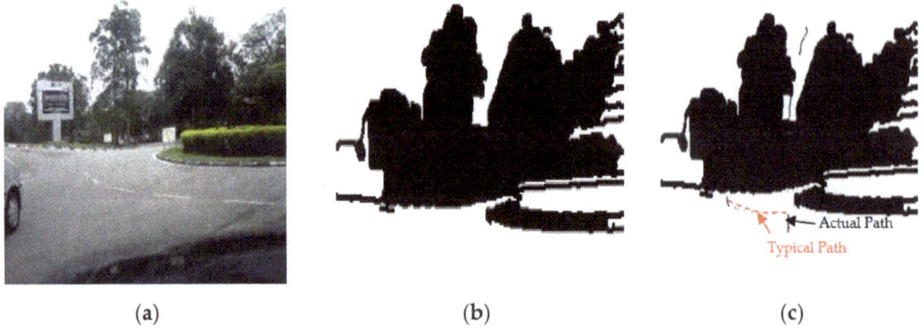

(a)   (b)   (c)

Figure 21. Autonomous detection and navigation of the proposed system in the road roundabout (with 5 m as diameter) with a car partially presented on the side/in front: (**a**) original image, (**b**) image processing, and (**c**) generation of the path within the road roundabout environment.

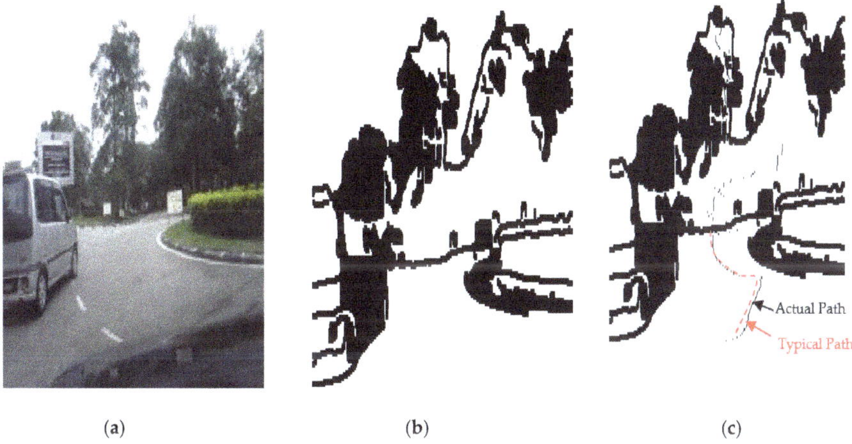

(**a**)   (**b**)   (**c**)

Figure 22. Autonomous detection and navigation of the proposed system in the road roundabout (with 5 m as diameter) with a car on the side/in front: (**a**) original image, (**b**) image processing, and (**c**) generation of the path within the road following environment.

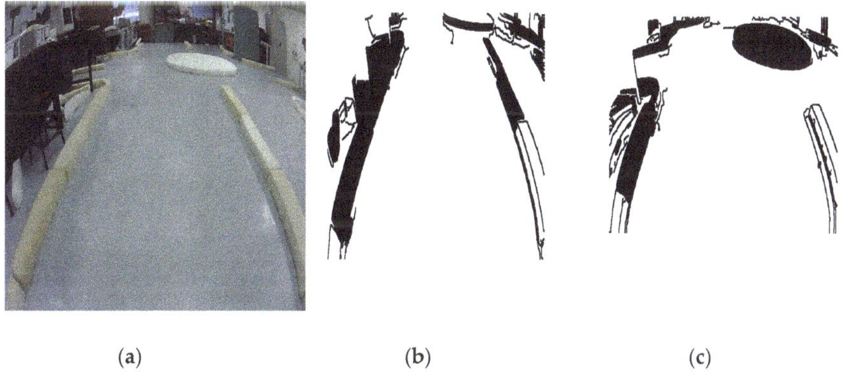

(**a**)   (**b**)   (**c**)

Figure 23. Camera sequence images: (**a**) original image when the WMR starts moving, (**b**) camera's local map when the WMR starts to move, and (**c**) camera's local map when the WMR detects the roundabout.

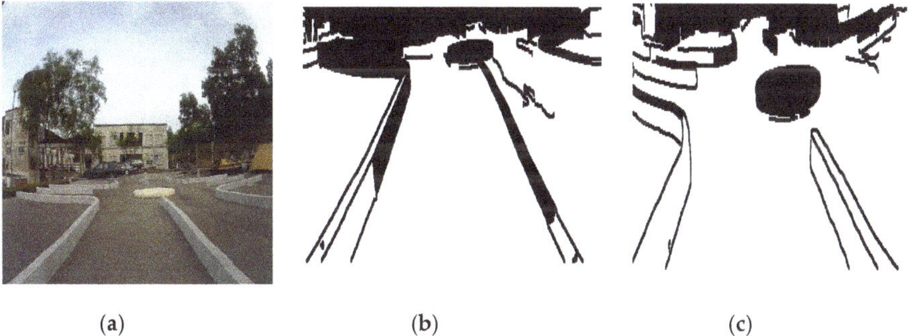

(**a**)   (**b**)   (**c**)

Figure 24. Outdoor camera sequences images: (**a**) original image when the WMR starts moving, (**b**) camera's local map when the WMR starts to move, and (**c**) camera's local map when the WMR detects the roundabout.

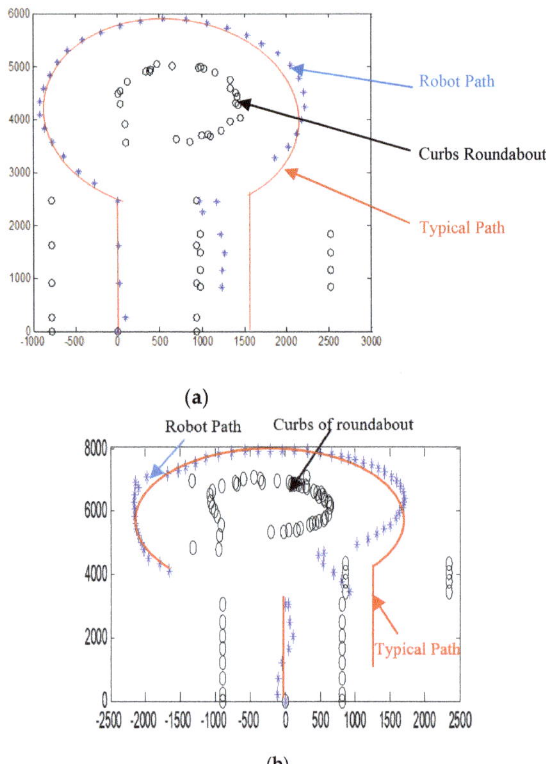

(**a**)

(**b**)

Figure 25. Robot path during navigation in a roundabout with 360° rotation. Note that blue '*' denotes the path, and black 'O' signifies the road environment. (**a**) Local mapping of the indoor environment acquired by sensors fusion. (**b**) Local mapping of the outdoor environment acquired by sensors fusion.

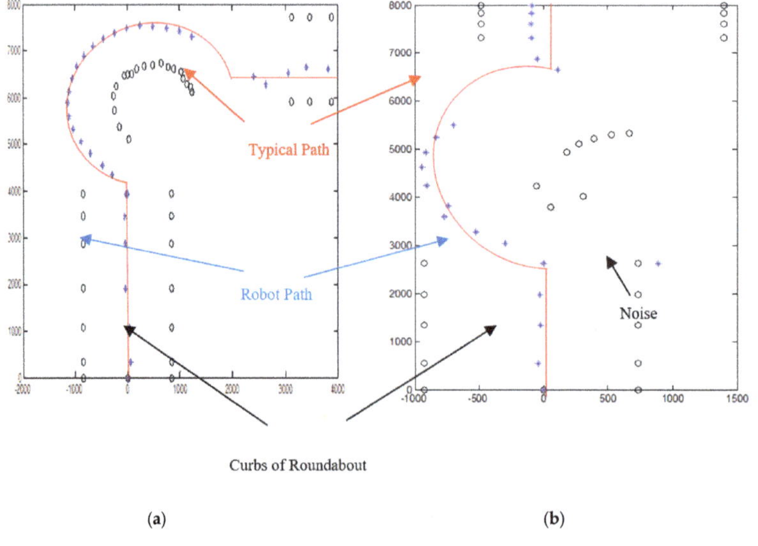

(**a**)

(**b**)

Figure 26. *Cont.*

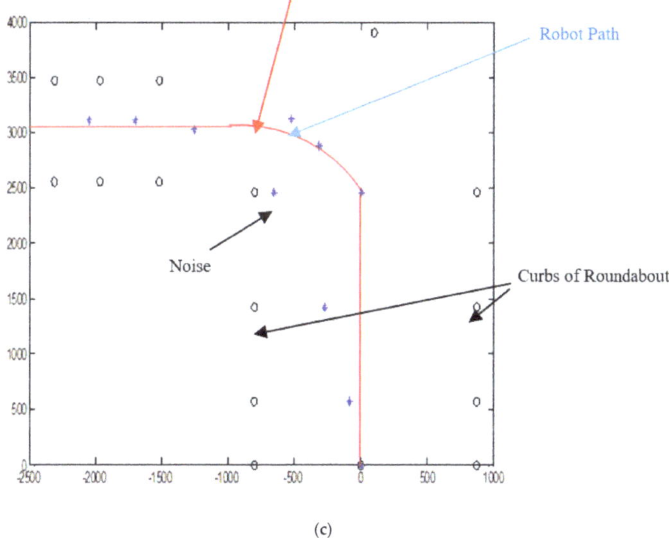

(c)

Figure 26. Robot path during navigation in a roundabout with 270° rotation. Note that blue '*' denotes the path, and black 'O' signifies the road environment for: (**a**) 270° rotation, (**b**) 180° rotation, and (**c**) 90° rotation.

In the discussion part, the results of the experiments that have been conducted in the real roads, indoor and outdoor setups are discussed based on the following aspects:

1. Accuracy of navigation system: It can be defined as the variation between the actual and typical paths during navigation in the road from start to goal position. For this purpose, the generated path (black dotted line as in Figures 16–22) is compared with the typical path (red dotted line as in Figures 16–22). The typical path in this work is considered as the path located in the middle of the road.
2. Efficiency (Reliability) of navigation system: It can be defined as the capability of the proposed algorithm to detect the road boundaries and borders among other surrounding environments of robot during autonomous navigation on the roads, in the presence of noise.
3. Cost of navigation system: One can differentiate between two kinds of costs, namely; fixed and operational costs. The fixed cost is the total cost of the hardware that has been used to perform the suggested algorithm, which is too low, in comparison with Tesla and Google autonomous vehicles. The total cost of robotic system in this project as shown in Figure 10 is around 5K USD; however, the cost of current autonomous vehicles such as Tesla or Google are in the range of 50–500K USD. The operational costs are varied during autonomous navigation on the roads based on the road conditions where the fuel consumption and electrical current profiles are changed during road navigation.

5.1. Road Following

The above-mentioned equations in Sections 3.2 and 4.2 were applied to detect the road roundabout and find the path of mobile robots in the real road following. The path was planned to be in the middle of the LRF measurements and the robot is able to track the middle of the roads as shown in Figures 16–19. Because of the accuracy of the LRF equal to 1 cm and the resolution (1 scan/100 ms), the robot path has a small deviation in the path as shown in Figures 16–19.

Several scenarios for autonomous navigation of the proposed algorithms in road following (with 5 m as width and 500 m as length) has been reported:

1. The autonomous vehicle is moving lonely on the road
2. The autonomous vehicle navigation system recognizes partially other vehicles on the side/in front of autonomous vehicle, as shown in Figure 17.
3. The autonomous vehicle navigation system recognizes complete vehicles on the side/in front of the autonomous vehicle, as shown in Figures 18 and 19.

The accuracy of the suggested autonomous robotic system is high, in the range of 1–3 cm, when it drives lonely on the road, as shown in Figure 16; however, it is in the range of 2–5 cm if there are obstacles on the sides or the edges cannot be detected, as shown in Figures 17–19, which can be increased to 3–10 cm if there are problems in the camera and LRF, such as deblurring, processing delay, or losses in LRF signals. The efficiency of the autonomous vehicle is good as it is able to detect well the path along the movement, no matter whether there are obstacles or not; with noting that the distance between the generated dotted-path is not constant due to losses of sensors measurements and long processing time. The maximum distance between two generated dotted-path is small, around 15 cm in real road, which does not present a bad impact to the robot's efficiency. Thus, the efficiency is in the range of 90–95% as shown in Figures 16–19. The operational cost increases when there are obstacles beside or in front of the autonomous vehicle as the path becomes a zigzag in this case; however, it is ok when the autonomous vehicle is moving alone.

Based on Figures 16–19, the average accuracy, efficiency, and operational cost of autonomous vehicles in road following are listed in Table 1.

Table 1. An average accuracy, efficiency, and operational cost for road following with 5 m as width and 500 m as length.

Condition/Property	Clear Road Curbs	Presence of Obstacle	Camera and LRF Problems	Missed Road Curbs
Accuracy	1–3 cm	2–5 cm	3–10 cm	2–5 cm
Efficiency	95%	90%	90%	90%
Operational Cost	decreased	increased	increased	Increased

5.2. Roundabout Intersection

The above-mentioned equations in Sections 3.2 and 4.2 were applied to detect the road roundabout and find the path of mobile robots in the real road following. The path was planned to be in the middle of the LRF measurements and the robot is able to track the middle of the roads as shown in Figure 10. Because the accuracy and resolution of the LRF is equal to 1 cm and 1 scan/100 ms, respectively, the robot path has a small deviation in the path as shown in Figures 20–22.

Similar to road following scenarios, the accuracy of the autonomous system when it is approaching the roundabout (5 m as diameter) is high, in the range of 2–3 cm, especially when it drives alone on the road as shown in Figure 20; however, it is in the range of 2–4 cm if there are obstacles on the sides, and the roundabout starts to be recognized or the edges cannot be detected as shown in Figures 21 and 22. The accuracy is located in the range of 3–8 cm if there are problems in the camera and LRF, such as deblurring, processing delay, or losses in LRF signals. The efficiency of the autonomous vehicle is also good, and it is in the range of 90–95%, as shown in Figures 20–22. The operational cost increases when the robot starts to recognize the roundabout or the obstacles are presented on side/in front of the autonomous vehicle.

According to Figures 20–22, the average of the accuracy, efficiency, and operational cost for the autonomous vehicle when it is passing through a roundabout are listed in Table 2.

Table 2. An average of the accuracy, efficiency, and operational cost of the autonomous vehicle when it is passing through a roundabout (with 5 m as diameter).

Condition/Property	Clear Road Curbs	Approaching to Roundabout	Presence of Obstacle	Missed Road Curbs	Camera and LRF Problem
Accuracy	2–3 cm	2–4 cm	2–4 cm	2–4 cm	3–8 cm
Efficiency	95%	90%	90%	90%	90%
Operational Cost	decreased	increased	increased	Increased	increased

To test the reliability of detection and navigation of the roundabout algorithm in the indoor and outdoor applications, another set of experiments have been conducted as shown in Figures 23–26.

Figures 23–25 show the sensors fusion based autonomous roundabout navigation from start to goal position with rotation angle equal to 360°. Figures 23 and 24 show the local map built from the camera sequence frames, where only the borders and intersections of the road remained in the images. The camera's local map was determined for each image in the sequences of the video frames as shown in Figure 23b,c and Figure 24b,c using the LS integrated with FL algorithm, which were applied to recognize the presence of roundabouts. Figures 23c and 24c show the last image in which the roundabout is detected.

By applying the algorithms described in Sections 3.2 and 4.2, the robot path for both indoor/outdoor roundabout environments can be determined as shown in Figure 25, where the series of blue points (*) is the path and a series of black points (O) denotes the road environment, where the entrance and exit curbs of the roundabout are located at the bottom-left and bottom-right of the figure, respectively. The roundabout center is located somewhere in the middle. The robot path looks smoother in the indoor environment as shown in Figure 25a; however, there are a cleared drifts in the outdoor one as shown in Figure 25b.

Similarly, the algorithms described in Sections 3.2 and 4.2 were applied to find the robot path in a roundabout with 270°, 180°, and 90° rotation as shown in Figure 26.

Because the Laser Range Finder is moving together with WMR platform and was simultaneously utilized to recognize (with measurements) the static road's curbs, it was observed that there is a slight shift on the right side at the exit as depicted in Figures 25 and 26. This is because the WMR platform exits from the roundabout outlet in an asymmetric manner just like when it first enters the inlet of roundabout. In addition, this drift could be related to the low accuracy of the LRF, which is about 3% of the measured distance; thus, if the measured distance is bigger than 1000 mm, the error should be in the region of 30 mm. The last reason is that the velocity of WMR is not completely controlled in these trials. In general, the main concern of this work is to present the robot trajectory rather than to show accurately the road intersections and curbs within the environment. It has been noticed that there is a noisy signal in sensor fusion measurements when they have been used to measure the distance between WMR and road curbs as depicted in Figures 25 and 26. In fact, those noises are not chosen as the robot trajectory since they are located far from the truth path as shown in Figure 20b. The typical path of the root is highlighted as red color in Figures 25 and 26.

5.3. Comparison with Other Related Works

The proposed Laser Simulator–Fuzzy Logic algorithm was compared with the work presented in Perez et al. [25] to navigate WMR on the roundabout environment. As has been discussed in Section 1, the roundabout navigation algorithm in Perez et al. [25] depends mainly on the maps and GPS signals to identify the road roundabout and find its dimension, which is almost similar to Tesla's and Google's cars navigation system on roundabouts [30]. Once the vehicle arrives at the entrance of the roundabout, the algorithm will generate a Bezier curve to navigate off-line the vehicle in the roundabout from the inlet until outlet of roundabout. The main disadvantage of the Bezier navigation approach is its dependencies on the off-line measurements that are coming from GPS and maps to find the dimensions of the roundabout setting, which may cause the vehicle to crash with the border of the roundabout, a scenario as shown in Figure 27. Other problems of such navigation system are coming from nonupdating of the maps' data, losing of GPS measurements in some areas, and nonregistered road roundabouts in the Google maps that could occur by making the offline path navigation in roundabout setting potentially dangerous.

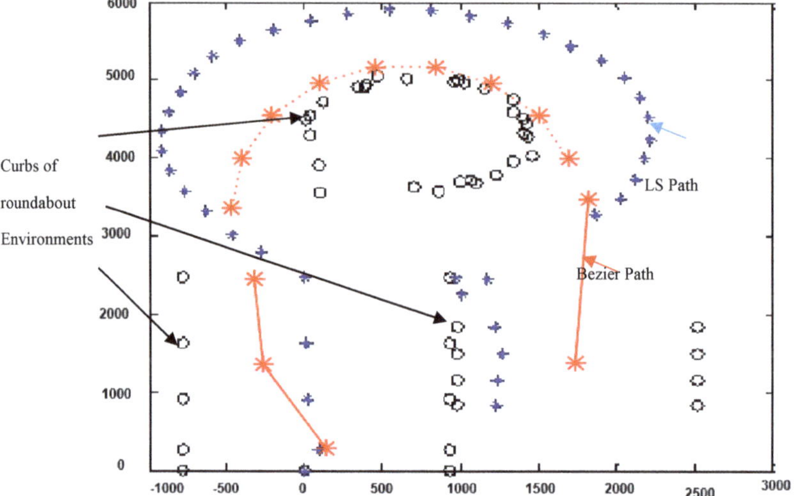

Figure 27. A comparison between Bezier roundabout navigation approach presented in Perez et al. [25] (red *) and the proposed roundabout navigation algorithm in this paper (blue *).

The proposed Laser Simulator–Fuzzy Logic approach in this paper could resolve such issues by utilizing an online roundabout navigation scheme. The comparison between Bezier roundabout navigation approach that has been presented in Perez et al. [25] and the proposed roundabout navigation algorithm in this paper, is depicted in Figure 27, where the Bezier algorithm crashes with roundabout border at the entrance of the roundabout.

6. Conclusions

The online path planning of the mobile robot has been fully derived in the road following and roundabout environments using LS integrated with an FL algorithm and sensor fusion technique. The proposed algorithm was used for roundabout detection through a camera's local map environments; while the sensor fusion was used for simultaneous planning of the robot path and building an accurate local map. The roundabout intersection was modeled, and the free-path collision was generated within its environment from the starting position located at a specific entrance to an appropriate exit of the roundabout. Results show the capability of the robot to effectively navigate in the road following and roundabout settings with multiple scenarios. Future work is to apply the signal stochastic and probabilistic methods like Kalman filter to eliminate the noise and improve the robot path. In addition, a low level control should also be applied to improve robot tracking.

Author Contributions: Conceptualization, M.A.H.A.; methodology, M.A.H.A.; software, M.A.H.A.; validation, M.A.H.A. and M.M.; formal analysis, M.A.H.A.; investigation, M.A.H.A.; resources, M.A.H.A.; data curation, M.A.H.A.; writing—Original draft preparation, M.A.H.A.; writing—Review and editing, M.M., W.A.J., K.M., W.A. and H.A.; visualization, M.A.H.A.; supervision, M.M.; project administration, M.A.H.A. and M.M.; funding acquisition, M.A.H.A. and M.M., W.A.J., K.M., W.A. and H.A. All authors regularly discussed the progress during the entire work. All authors have read and agreed to the published version of the manuscript.

Funding: This research was funded by Deanship of Scientific Research at King Saud University, grant number RG-1441-349. It was also supported by Universiti Malaysia Pahang, and Ministry of Higher Education (MOHE) under Research University Grants RDU 180323, RDU1803138 and RDU190804.

Acknowledgments: The authors extend their appreciation to the Deanship of Scientific Research at King Saud University for funding this work through Research group number, RG-1441-349. The authors would like also to thank Universiti Malaysia Pahang (UMP) and Ministry of High Education (MOHE) for providing the research grant and facilities.

Conflicts of Interest: The authors declare no conflict of interest.

References

1. Ali, M.A.H.; Mailah, M.; Tang, A.; Hing, H. Path Navigation of Mobile Robot in a Road Roundabout Setting. In Proceedings of the 1st International Conference on Systems, Control, Power and Robotics, Singapore, 11–13 March 2012.

2. Ali, M.A.H.; Mailah, M.; Tang, A.; Hing, H. Path Planning of Mobile Robot for Autonomous Navigation of Road Roundabout Intersection. *Int. J. Mech.* **2012**, *6*, 203–211.

3. Ali, M.A.H.; Mailah, M.; Tang, A.; Hing, H. A novel Approach for Visibility Search Graph Based Path Planning. In Proceedings of the 13th International conference on Robotics, Control and Manufacturing Systems, Kuala-Lumpur, Malaysia, 2–4 April 2013; pp. 44–49.

4. Lotufo, R.; Morgan, A.; Dagless, E.; Milford, D.; Morrissey, J.; Thomas, B. Real-Time Road Edge Following For Mobile Robot Navigation. *Electron. Commun. Eng. J.* **1990**, *2*, 35–40. [CrossRef]

5. Matsushita, Y.; Miura, J. On-line Road Boundary Modeling with Multiple Sensory Features, Flexible Road Modeland Particle Filter. *Robot. Auton. Syst.* **2011**, *59*, 274–284. [CrossRef]

6. Sotelo, M.A.; Rodriguez, F.J.; Magdalena, L.; Bergasa, L.M.; Boquete, L. A Color Vision-Based Lane Tracking System for Autonomous Driving on Unmarked Roads. *Auton. Robot.* **2004**, *16*, 95–116. [CrossRef]

7. Crisman, J.; Thorpe, C. SCARF: A Color Vision System that Tracks Roads and Intersections. *IEEE Trans. Robot. Autom.* **1993**, *9*, 49–58. [CrossRef]

8. Kluge, K.; Thorpe, C. The YARF System for Vision-Based Road Following. *Math. Comput. Model.* **1995**, *22*, 213–233. [CrossRef]

9. Okutomi, M.; Nakano, K.; Matsuyama, J.; Hara, T. Robust Estimation of Planar Regions for Visual Navigation Using Sequential Stereo Images. In Proceedings of the 2002 IEEE International Conference on Robotics and Automation, Washington, DC, USA, 11–15 May 2002; pp. 3321–3327.

10. Enkelmann, W.; Struck, G.; Geisler, J. ROMA—A System for Model-Based Analysis of Road Markings. In Proceedings of the Intelligent Vehicles '95. Symposium, Detroit, MI, USA, 25–26 September 1995; pp. 356–360.

11. Jochem, T.; Pomerleau, D.; Thorpe, C. *Vision-Based Neural Network Road and Intersection Detection and Traversal*; Robot. Inst., Carnegie Mellon Univ.: Pittsburgh, PA, USA, 1995.

12. Kim, S.; Roh, C.; Kang, S.; Park, M. Outdoor Navigation of a Mobile Robot Using Differential GPS and Curb Detection. In Proceedings of the 2007 IEEE International Conference on Robotics and Automation, Roma, Italy, 10–14 April 2007; pp. 3414–3419.

13. Georgiev, A.; Allen, P. Localization Methods for a Mobile Robot in Urban Environments. *IEEE Trans. Robot.* **2004**, *20*, 851–864. [CrossRef]

14. Narayan, A.; Tuci, E.; Labrosse, F.; Alkilabi, M.H. A Dynamic Colour Perception System for Autonomous Robot Navigation on Unmarked Roads. *Neurocomputing* **2018**, *275*, 2251–2263. [CrossRef]

15. Qian, C.; Shen, X.; Zhang, Y.; Yang, Q.; Shen, J.; Zhu, H. Building and Climbing based Visual Navigation Framework for Self-Driving Cars. *Mob. Netw. Appl.* **2017**, *23*, 624–638. [CrossRef]

16. Ali, M.A.H.; Mailah, M. Path Planning and Control of Mobile Root in Road Environments using Sensor Fusion and Active Force Control. *IEEE Trans. Veh. Techol.* **2019**, *68*, 2176–2195. [CrossRef]

17. Zhao, Z.; Chen, W.; Wu, X.; Liu, Z. Vehicle-Following Model Using Virtual Piecewise Spline Tow Bar. *J. Transp. Eng.* **2016**, *142*, 04016051. [CrossRef]

18. Zhao, Z.; Chen, W.; Peter, C.C.; Wu, X. A Novel Navigation System for Indoor Cleaning Robot. In Proceedings of the International Conference on Robotics and Biomimetics, Qingdao, China, 3–7 December 2016; pp. 2159–2164.

19. Battista, T.; Woolsey, C.; Perez, T.; Valentinis, F. A dynamic model for underwater vehicle maneuvering near a free surface. *IFAC OnLine* **2016**, *49*, 68–73.
20. Cossu, G.; Sturniolo, A.; Messa, A.; Grechi, S.; Costa, D.; Bartolini, A.; Scaradozzi, D.; Caiti, A.; Ciaramella, E. Sea-trial of optical ethernet modems for underwater wireless communications. *J. Light. Technol.* **2018**, *36*, 5371–5380. [CrossRef]
21. Ling, S.D.; Mahon, I.; Marzloff, M.; Pizarro, O.; Johnson, C.R.; Williams, S.B. Stereo-imaging AUV detects trends in sea urchin abundance on deep overgrazed reefs. *Limnol. Oceanogr. Methods* **2016**, *14*, 293–304. [CrossRef]
22. Jorge, F. Detecting Roundabout Manoeuvres Using Open-Street-Map and Vehicle State. Master's Thesis, Department Vehicle Safety, Chalmers University of Tech., Goteborg, Sweden, 2012.
23. Perez, J.; Milanes, V.; De Pedro, T.; Vlacic, L. Autonomous driving manoeuvres in urban road traffic environment: A study on roundabouts. *IFAC Proc. Vol.* **2011**, *44*, 13795–13800. [CrossRef]
24. Katrakazas, C.; Quddus, M.A.; Chen, W.-H.; Deka, L. Real-time Motion Planning Methods for Autonomous on Road Driving: State-of-Art and Future Research Directions. *Transp. Res. Part C Emerg. Technol.* **2015**, *60*, 416–442. [CrossRef]
25. Perez, J.; Godoy, J.; Villagra, J.; Onieva, E. Trajectory generator for autonomous vehicles in urban environments. In Proceedings of the 2013 IEEE International Conference on Robotics and Automation, Karlsruhe, Germany, 6–10 May 2013; pp. 1–7.
26. Okusa, Y. Navigation System. U.S. Patent 2007/0150182 A1, 28 June 2007.
27. Cuenca, L.G.; Sanchez-Soriano, J.; Puertas, E.; Fernández, J.; Aliane, N. Machine Learning Techniques for Undertaking Roundabouts in Autonomous Driving. *Sensors* **2019**, *19*, 2386. [CrossRef] [PubMed]
28. Chen, X.; Ma, H.; Wan, J.; Li, B.; Xia, T. Multi-view 3D Object Detection Network for Autonomous Driving. In Proceedings of the 2017 IEEE Conference on Computer Vision and Pattern Recognition (CVPR), Honolulu, HI, USA, 21–26 July 2017.
29. Yu, Y.; Li, J.; Wen, C.; Guan, H.; Luo, H.; Wang, C. Bag-of-visual-phrases and hierarchical deep models for traffic sign detection and recognition in mobile laser scanning data. *ISPRS J. Photogramm. Remote Sens.* **2016**, *113*, 106–123. [CrossRef]
30. Bimbraw, K. Autonomous cars: Past, present and future a review of the developments in the last century, the present scenario and the expected future of autonomous vehicle technology. In Proceedings of the 2015 12th International Conference on Informatics in Control, Automation and Robotics (ICINCO), Colmar, France, 21–23 July 2015; pp. 191–198.

 *sensors*

Article

An Elaborated Signal Model for Simultaneous Range and Vector Velocity Estimation in FMCW Radar

Sergei Ivanov *, Vladimir Kuptsov, Vladimir Badenko * and Alexander Fedotov

Institute of Physics, Nanotechnology and Telecommunications, Peter the Great St.Petersburg Polytechnic University, 29 Polytechnicheskaya str., Saint Petersburg 195251, Russia; vdkuptsov@yandex.ru (V.K.); afedotov@spbstu.ru (A.F.)
* Correspondence: ivanov_si@spbstu.ru (S.I.); badenko_vl@spbstu.ru (V.B.); Tel.: +7-9905-283-48-88 (S.I.); +7-921-309-41-00 (V.B.)

Received: 19 September 2020; Accepted: 13 October 2020; Published: 16 October 2020

Abstract: A rigorous mathematical description of the signal reflected from a moving object for radar monitoring tasks using linear frequency modulated continuous wave (LFMCW) microwave radars is proposed. The mathematical model is based on the quasi-relativistic vector transformation of coordinates and Lorentz time. The spatio-temporal structure of the echo signal was obtained taking into account the transverse component of the radar target speed, which made it possible to expand the boundaries of the range of measuring the range and speed of vehicles using LFMCW radars. An algorithm for the simultaneous estimation of the range, radial and transverse components of the velocity vector of an object from the observation data of the time series during one frame of the probing signal is proposed. For an automobile 77 GHz microwave LFMCW radar, a computer experiment was carried out to measure the range and velocity vector of a radar target using the developed mathematical model of the echo signal and an algorithm for estimating the motion parameters. The boundaries of the range for measuring the range and speed of the target are determined. The results of the performed computer experiment are in good agreement with the results of theoretical analysis.

Keywords: automotive LFMCW radar; radial velocity; lateral velocity; Doppler-frequency estimation; waveform; signal model

1. Introduction

In modern automobiles, microwave active locating systems are key elements of the widely adopted high-performance intelligent technologies [1,2]. The use of car microwave radars as part of the Advanced Driver Assistance System (ADAS) ensures the most comfortable and safe movement of the vehicle, and their use as part of the automatic control system provides a completely autonomous (unmanned) mode of vehicle movement.

The ever-increasing requirements for the advanced ADAS systems of the "intelligent" car pose challenges for the solution of which fundamentally different approaches are required. This includes the use of more sophisticated quasi-real-time signal processing algorithms, alternative modulation of sounding radar signals and the development of efficient microwave radar hardware [3–6].

Nowadays, it is generally accepted that for remote radio monitoring of the location and movement of surrounding objects in real time, i.e., simultaneously assessing the speed, range and bearing of targets in automotive applications, it is most effective to use microwave radars with linear frequency modulation (LFMCW) [1–6]. In active radar, the target speed is determined by the Doppler frequency shift, and the range is determined by the time delay of the radio signal propagation in the environment [7–9]. For automotive LFMCW radars, two main algorithms can be distinguished for simultaneous estimation of target speed and range. The first algorithm is based on a one-dimensional (1D) Fourier transform of the echo signal after it is converted to an intermediate frequency (IF) signal in the quadrature channels of the radar receiver [10,11]. The advantage of this algorithm is the relative simplicity of the technical implementation, the narrow operating frequency band of the radio frequency path and, as a consequence, the low sampling rate of the analog-to-digital converter. A significant disadvantage of the method is the appearance of false radar targets (false alarms) [12]. One of the ways to combat false targets is the use of a segmented structure of a sounding radio signal with variable modulation parameters: chirp duration, frequency deviation and initial generation frequency [13–17]. The creation of a segmented structure of the sounding radio signal can significantly reduce the likelihood of false targets. Phantom targets are eliminated in the process of monitoring the radar situation using tools of intelligent data analysis technologies (Data Mining), such as fuzzy logic, neural networks and Kalman filtering [18,19]. Solving the problem of false radar targets inevitably leads to an increase in the duration of observation and data analysis, which limits the application area of LFMCW radars that use the one-dimensional (1D) Fourier transform algorithm.

The second method of simultaneous estimation of target speed and range is based on the use of two-dimensional (2D) Fourier transform [20–22]. It is easy to show that the dynamic range of measured velocities and distances using this method is proportional to the sampling rate f_s of the analog IF signal:

$$f_s = 8f_0 \frac{R_{max}}{\delta R} \frac{V_{R\,max}}{c} = 4f_{D\,max} \frac{R_{max}}{\delta R}, \tag{1}$$

where R_{max} and δR—maximum range and range resolution respectively, c—speed of light, $V_{R\,max}$—maximum radial component of target velocity, $f_{D\,max}$—Doppler frequency shift, that corresponds to maximum velocity $V_{R\,max}$, and f_0—the initial frequency of the radiated chirp radio signal.

With typical parameter values for vehicle motion control tasks $R_{max}/\delta R = 300, f_0 = 80$ GHz and $V_{R\,max} = 300$ km/h, the frequency f_s reaches a value of more than 50 MHz, which presents certain difficulties for the technical implementation of LFMCW radars. To separate the "fast" and "slow" time scales in the received radio signal, which is necessary for an unambiguous estimate of the velocity, the condition for limiting the chirp duration T_c in a sequence of N chirps of one frame of the sounding signal must be met [23]:

$$T_c \leq \frac{1}{2f_{D\,max}}, \tag{2}$$

$$N \geq 2^{\left[\log_2\left(\frac{V_{R\,max}}{\delta V_R}\right)\right]}. \tag{3}$$

Relation Equation (3) defines the lower bound on the duration of a frame to achieve a given speed resolution δV_R, square brackets mean rounding up to the nearest integer.

In References [23], a model of the sounding signal and methods for estimating the radial velocity are proposed for the case when Equation (2) is not satisfied. The recommended displacement of the initial frequency of neighboring chirps with subsequent analysis of the phase or frequency relationships of two sequences of chirps with different parameters makes it possible to determine the Doppler frequency shift of the target in the Nyquist zones above the first. However, a rigorous mathematical analysis and self-noise accounting of the radar receiving system, carried out in Reference [24], showed that the resulting estimate of the radial velocity can be ineffective and unstable.

In recent years, leading manufacturers of microwave transceivers for LFMCW radars have been supplying equipment to enable future ADAS (advanced driver-assisted system) safety features, which are highly integrated, high-performance and cost-effective packaged SoC (system on chip) solutions [25]. The sampling rate f_s of analog-to-digital converters of automobile radars reaches 40 MHz and increasing it to 100–150 MHz is not a difficult technical problem. Thus, the functionality of modern microwave transceivers in a wide range of changes in the range and speed of the object is not limited to meeting the requirements of Equations (2) and (3). In this case, an adequate mathematical description of the reflected radio signal, free from the limitations and assumptions adopted in the "classical" mathematical model, plays a key role in the development of algorithms for estimating the range and speed of the target.

As a rule, the signal reflected from the target and arriving at the input of the radar receiving module is written in the form:

$$u(t,R) = s(t - \tau(t)), \tag{4}$$

where the time delay $\tau(t)$ of propagation of a radio signal in an environment with a speed c is determined from the solution of the nonlinear equation [26,27]:

$$\tau(t) = 2R(t - \tau(t))/c, \tag{5}$$

whose approximate solution for $c<<V$ has the form:

$$\tau(t) = 2\|R + Vt\|/c \approx 2(R + V_R t)/c. \tag{6}$$

Here, R is the radius vector that determines the position of the target at the initial time $t = 0$. When deriving Equation (5), the three-dimensional Galilean transformation of space-time for the inertial system is used. There are known special cases [28] of solving active location problems, when the results of calculating the Doppler frequency shift using the Lorentz space-time transformation [29,30] coincide with similar results following from the Galilean transformation. However, in this paper, in the generalized mathematical model of the echo signal, we use the three-dimensional vector form of the equations of the Lorentz transformations, as a more "physical" and adequate mathematical description of the propagation of a space-time sounding radio signal.

The use of linear approximation Equation (6) to calculate the norm of the vector $\|R + Vt\|$ at a high-speed V_R and a small distance R to the target leads to a significant error in estimating the values V_R and R for any algorithm for processing the input radio signal. The linear approximation Equation (6) does not take into account the transverse (lateral) movement of the target with a speed V_{TR}. At the same time, the V_{TR} velocity component is very important for determining the direction and overall speed of a radar target, and in the approximation Equation (6), can be determined only from the results of sequential processing of several frames [31] of the input radio signal. In References [32–34], to determine the transverse velocity component based on the analysis of data obtained over the time interval of one frame (frame) of a radio signal, parallel processing of echo radio signals reflected from several reflex zones of a radio monitoring object is proposed. This method makes rather stringent technical requirements for the characteristics of LFMCW radars, in particular, the angle resolution and the computational performance of the data processor.

In Reference [34], an algorithm for estimating the transverse velocity component based on the results of processing one frame (frame) of a radio signal is considered. However, estimates of the V_{TR} value were obtained under the assumption of "strictly" lateral movement of the radar target relative to the LFMCW radar, which limits the usage of the proposed algorithm in practical applications.

When solving problems of radar monitoring of vehicles, as a rule, an approximate calculation of the time delay is used $\tau(t)$:

$$\tau^2(t) \approx \tau_0^2 = 4R^2/c^2, \tag{7}$$

which is another important factor limiting the permissible range of speed and range measurements. In References [35–39], a refined mathematical model of the reflected radio signal $u(t,R)$ is proposed in which the delay is considered as a function of the slow time scale $\tau(kT_c)$. In the development of algorithms for estimating the speed and range of a target, in this case, methods of focusing, contrast optimization and discrete polynomial transformation of the signal phase are used [40]. In the mentioned works [35–39], the transverse component of the V_{TR} velocity is not taken into account.

The purpose of this study was to develop a software simulator that implements a signal that is identical to the real echo signal of the LFMCW radar for a single-point reflective radar target. The goal was also to develop an algorithm for estimating the range and velocity vector of a vehicle, including its transverse component, and to conduct a comparative computational experiment.

The structure of the paper is as follows. Section 2 of the paper is devoted to consideration of a generalized mathematical model of the reflected signal from a moving object based on the vector Lorentz transformation of coordinates and time. In Section 3, the spatio-temporal (ST) structure of the LFMCW radar echo signal, taking into account the transverse component of the radar target velocity vector, and considering the mathematical model of the signal at the I-Q output of the receiver's quadrature channels, is described. In Section 4, the derivation of the relationships for the simultaneous assessment of the distance and the velocity vector of an object, and also analyses of the permissible working range of measurement of motion parameters, are presented. In Section 5, the results of a computational experiment for estimating the parameters R, V_R and V_{TR} are analyzed. Finally, the concluding remarks and future research directions are presented in Section 6.

2. Universal Mathematical Model of the Space-Time Echo Signal at an Active Radiolocation of a Moving Target

As mentioned above, the Doppler effect plays a decisive role in assessing the range and speed of targets in automotive LFMCW radars. The mathematical description of the Doppler effect in a wide range of velocities and distances to the target has its own distinctive features that must be taken into account when assessing motion parameters using automotive LFMCW radars. In this section, a mathematical model of the space-time echo-signal with active radar of a moving target, taking into account the Doppler effect and in the absence of restrictions on the magnitude of the speed and range of a single target, is discussed. This makes it possible to use this model to implement a software simulator of the reflected signal when carrying out a computational experiment and testing the performance of algorithms for estimating the parameters of an object's motion.

It is assumed that the radar antenna is matched in polarization with the radio signal, which is vector electromagnetic radiation. In this case, the polarization effects can be disregarded and the electromagnetic radiation is described by a scalar complex-valued function $u(t,R)$ of two variables—time, t, and radius-vector, R, specifying the position of the observation point in space relative to an arbitrary point on the antenna aperture, Σ. We consider the case of one-point reflection of electromagnetic radiation, i.e., a radar target as a reflector has a dimension of 0.

We will assume that on the time interval T, equal to the signal observation time during one frame, the target M moves uniformly and rectilinearly at a speed, V, relative to the receiving-transmitting aperture, Σ, of the automobile LFMCW radar antenna. Consider two inertial reference systems [41] (Figure 1): the first fixed system (R,t) is associated with the antenna aperture with the origin of coordinates O at the reference fixed point Σ, and the second is the movable system (R',t'), "tightly" associated with the target M (the origin is at point O'). For the initial reference time, we will take the moment of the beginning of observation of the emission of the sounding signal. The origin O' of the moving system of coordinates is chosen so that when $t = 0$, it coincided with the beginning of the fixed coordinate system O.

Let the probing signal be emitted from the reference point O and its form is determined by the complex-valued function $s_0(t)$. Then, in the coordinate system associated with the aperture Σ, the spatio-temporal structure of the sounding signal at an arbitrary point R outside the near-field zone of the antenna has the form:

$$u_0(t,R) = \frac{C_0 U(t,\mathbf{R})}{\|\mathbf{R}\|} \dot{s}_0\left(t - \frac{\|\mathbf{R}\|}{c}\right), \tag{8}$$

where c—speed of light (group signal velocity in the medium), $U(t,R)$—signal amplitude, the value of which is determined by the radiation power, antenna gain and its radiation pattern over the field [42,43], and C_0—constant. In general, $U(t,R)$ is a function of time and radius vector R. In modern automotive LFMCW radars, a phased array antenna with a flat aperture and the ability to electronically control the directional pattern is used as an emitting structure [42,43].

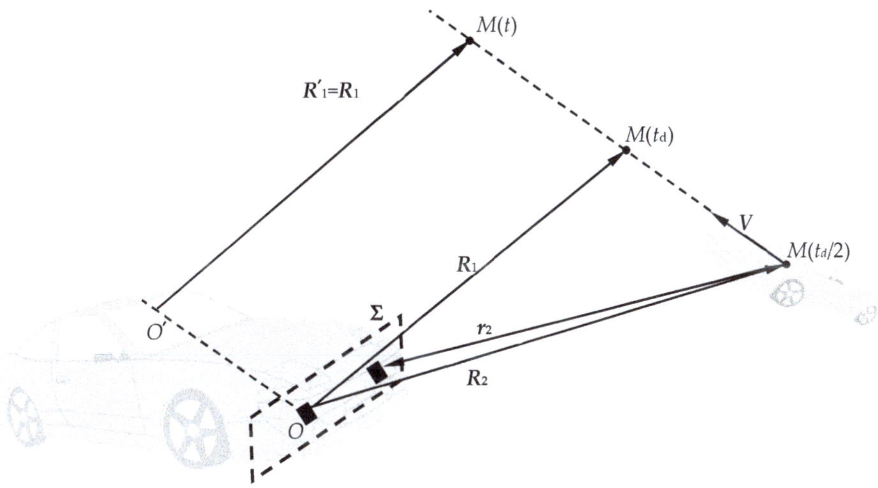

Figure 1. Coordinate system in the task of active location of a moving target.

The calculation of the echo signal at the antenna aperture includes two stages of the space-time transformation of the sounding radio signal Equation (8).

- With the help of Lorentz transformations [29,30], which put together coordinates and time in inertial systems (R,t) and (R',t'), we pass to a moving frame of reference associated with the target. Determine the field strength $u_0(R',t')$ at time t' at an arbitrary point R'. We take into account that in the frame of reference (R',t'), the location of the target M, regardless of time, is determined as $R' = R_1$, where R_1 is the position of the target in the system (R,t) at the time $t_d = 2\|R_2\|/c$—the time of arrival of the reflected signal at point O (Figure 1). Field-induced current at a target location in a moving coordinate system where the target is stationary:

$$i_t(t') = C_1 u_0(R',t') \exp\left[j\phi(t')\right],$$

where C_1—constant, and $\phi(t)$—phase shift between field and current. This current generates a secondary radiation field determined by the formula:

$$\dot{u}(t',R') = \frac{C_2(t',R')}{\| R' - R_1 \|} i_t\left(t' - \frac{\| R' - R_1 \|}{c}\right),\tag{9}$$

where $C_2(t, R')$—in the general case, the function of time and radius vector R' and the form of which is determined by the scattering indicatrix of the radar target [44,45].

- To determine the space-time structure of the field at the aperture Σ of the antenna, the reverse transition to the stationary frame of reference (R,t) is carried out using the Lorentz transformation. It is convenient to operate with the vector R_2, which determines the position of the target at the moment of the onset of re-emission, i.e., at time $t = t_d/2$. From Figure 1, it follows that $R_1 = R_2 + Vt_d = R_2 + VR_2/2$, and on the antenna aperture, the coordinates of an arbitrary point R are characterized by the radius vector ρ and, therefore, $R = \rho$ for the region Σ (Figure 2). It is advisable to set the vector $r_2 = \rho - R_2$ as a vector connecting the target with an arbitrary point ρ of the antenna aperture at time $t = t_d/2$ (see). It is efficient to decompose the target velocity vector V into two orthogonal components V_R and V_{TR}, one of which, V_R, is collinear to the radius vector R_2, the other, V_{TR}, is the transverse velocity component (see Figure 2). Also, the decomposition of the velocity vector V into two orthogonal components is possible with respect to the base radius vector r_2. In this case, we obtain the components V_r and V_{Tr}, respectively (see Figure 2). In Figure 2, ψ is the angle between the orthogonal components of the velocity V_R and V_{TR}, and the angles θ and φ characterize the angular position of the target M in a spherical coordinate system centered at point O and the reference plane Σ.

At a speed of the observed objects up to 350 km/h (upper segment of sports cars), the maximum value of the V_{max} speed module is $|V_{max}| < 2 \cdot 350 = 700$ km/h. The maximum duration of one frame of the sounding radio signal usually does not exceed 40 ms. In this case, for 77 GHz Automotive Radar Applications, as the calculations show, the relativistic factor in the Lorentz transformations can be ignored and "truncated" equations can be used:

$$R' = r + V[(\beta - 1)R \cdot r / V^2 - \beta t] \approx R - Vt; \qquad R \approx R' + Vt'; \tag{10}$$

$$t' \approx t - V \cdot R / c^2; \qquad t \approx t' + V \cdot R' / c^2, \tag{11}$$

$V = \|V\|$; $\beta = (1 - V^2/c^2)^{-1/2}$. The dot between the vectors hereinafter means the scalar product of the vectors.

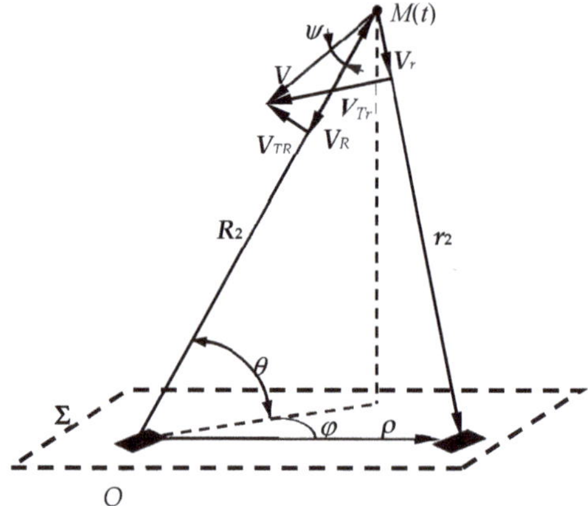

Figure 2. Radial and transverse components of the target velocity vector.

Calculation of the space-time structure of the echo radio signal $u(t,r_2)$ at the antenna aperture Σ in accordance with the above technique leads to the following expression:

$$u(t, r_2, R_2, V) = \dot{U}_{Rv} s_0(t - \tau(t, r_2, R_2, V)), \tag{12}$$

where the time delay $\tau(t,r_2,R_2,V)$ is equal to:

$$\tau(t, r_2, R_2, V) = \frac{V \cdot r_2}{c^2} - \frac{1}{c} \left\| r_2 - V\left(t - \frac{R_2}{c}\right) \right\| - \frac{1}{c} \left\| \left\{ R_2 + V\left[t - \frac{R_2}{c} - \frac{1}{c} \left\| r_2 - V\left(t - \frac{R_2}{c}\right) \right\| \right] \right\} \right\|, \tag{13}$$

where R_2 and r_2 are modules (norms) of the corresponding radius vectors. In Equation (13) for $\tau(t,r_2,R_2,V)$, the time is counted from the moment the sounding signal begins to be emitted.

Relations Equations (12) and (13) describe a universal mathematical model of the space-time structure of the signal with active location of moving targets using microwave radars. The versatility of the model implies an arbitrary functional time dependence of the sounding signal $s_0(t)$, the range of speeds $0 < V < 4$ km/s and distances R, covering the limiting parameters of the movement of existing and future land vehicles. The complex amplitude $\dot{U}_{Rv}$ is proportional to the product $U(t,R)C_2(R')\exp[j\phi(t)]$, i.e., is a complex-valued function of the variables t, R and R'. Typically, the form of this function is determined experimentally or by semi-empirical methods. However, for automotive LFMCW radars, further analysis of the mathematical model of the signal reflected from the target space-time structure of the signal is carried out in the approximation $U_{Rv} = \text{const}(t,R,R')$. The validity of such an approximation follows from the short duration of an individual chirp of the T_c signal.

The results obtained refine the results of Reference [41]. The obtained relations Equations (12) and (13) make it possible to implement a software simulator of the reflected signal during a computational experiment. On the basis of the considered universal mathematical model, we will develop a more accurate description of the spatio-temporal structure of the LFMCW radar echo signal, taking into account the transverse component of the radar target velocity vector, and which can be effectively used to implement signal processing algorithms.

3. Mathematical Model of Echo Radio-Signal Conversion in the LFMCW Radar Receiver

In this section, the developed general universal mathematical model of the spatio-temporal echo of a signal with active radar in the quasi-relativistic approximation for an arbitrary waveform is used to construct a mathematical model of the reflected radio signal for the case of LFMCW automotive radars. Consideration is carried out in the approximation of a distant location, when the value of the parameter equal to the ratio of the distance the target moves during processing within one frame to the distance to the target is less than one.

As noted above, modern automotive radar hardware makes it possible to generate and receive a continuous radio signal $s_0(t)$ in the form of a periodic sequence of segments with linear frequency modulation of the signal in Figure 3, with given operating parameters N, f_s and T_c, the values of which satisfy constraints Equations (1)–(3). The requirement for the radar range resolution δR imposes additional conditions on the frequency deviation Δf and the rate of change of the chirp frequency $\alpha = \Delta f / T_c$:

$$\Delta f \geq c/2\delta R; \quad \text{and} \quad \alpha T_c \geq c/2\delta R. \tag{14}$$

A mathematical description in the time domain of the sounding radio signal $S_{Tr}(t)$, shown in Figure 3, has the following form:

$$S_{Tr}(t) = U_{Tr} \cos\left\{2\pi f_0\left(t - \left[\frac{t}{T_c}\right]T_c\right) + \pi\alpha\left(t - \left[\frac{t}{T_c}\right]T_c\right)^2 + \phi_{Tr}\right\}\text{rect}\left(\frac{t - \left[\frac{t}{T_c}\right]T_c - \tau_m}{T_0} - \frac{1}{2}\right), \tag{15}$$

where the square brackets operator [∗] means rounding a number down to the nearest integer, $\text{rect}(x)$—rectangular function [46], ϕ_{Tr}—the initial phase of the chirp, τ_m—start of work of the ADC of the receiver within one chirp, and T_0—the working interval of data collection, during which the analog IF signal is converted into a digital code in the radar receiver. In Figure 3, the radio signal of the receiver is highlighted in blue, and the working area of the sounding radio signal is in red. For the received radio signal $S_{Rv}(t)$ in the time domain, we can write:

$$S_{Rv}(t) = S_{Tr}[t - \tau(t, r_2, R_2, V)], \tag{16}$$

where the time delay $\tau(t, r_2, R_2, V)$ in the general case is determined by the relation Equation (13). In Equation (16) and below, for brevity, the dependence of the signal is on variables r_2, R_2, V.

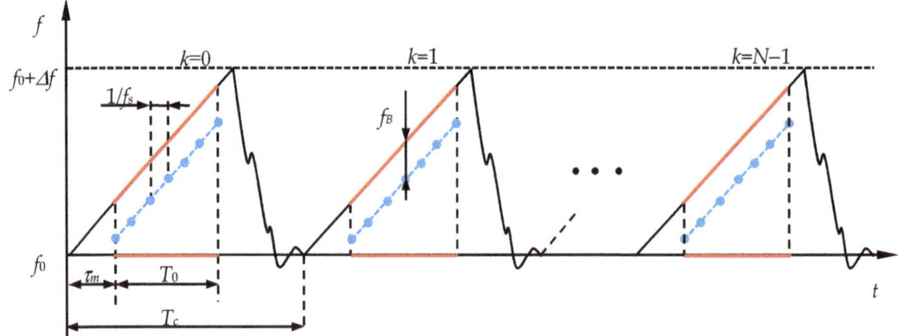

Figure 3. Radio signal of the transceiver. Blue—receiver radio signal, red—the working area of the sounding radio signal.

Further conversion of the frequency-modulated radio signal in the receiving path of the radar occurs in accordance with the coherent quadrature demodulation algorithm, in which the microwave signal is converted to the zero-frequency region using mixers in the I and Q quadrature channels. In this case, the signal at the output of the $S_{MIX}(t)$ mixers in the general case can be represented by a complex-valued function:

$$\dot{S}_{MIX}(t) = U_M \exp\ \{\arg\ [\dot{S}_{Tr}(t)] - \arg\ [\dot{S}_{Rv}(t)]\} = \dot{U}_M \exp\ \{\arg\ [\dot{S}_{Tr}(t)] - \arg\ [\dot{S}_{Tr}(t - \tau(t, r_2, R_2, V))]\}. \tag{17}$$

For further analysis, it is convenient to represent the current time t as the sum of the local time t_L (fast scale) and discrete time kT_c (slow scale):

$$t = t(t_L, k) = t_L + \tau_m + [t/T_c]T_c = t_L + \tau_m + kT_c,$$

$$\forall\ \{t_L \in \mathbb{R} : 0 \le t_L \le T_c\} \times \{k \in \mathbb{Z} : 0 \le k \le N - 1\}. \tag{18}$$

In the particular case of a sounding radio signal with linear frequency modulation Equation (15), the signal from the output of the mixer Equation (17), taking into account Equation (18), will have the following form:

$$\dot{S}_{MIX}(t_L, k) = \dot{U}_M \exp\left(j\left\{2\pi f_0(t_L, k) + 2\pi\alpha\tau(t_L, k)\left[t_L + \tau_m - \frac{\tau(t_L, k)}{2}\right] + \phi_{Tr} - \phi_{Rv}\right\}\right)\text{rect}\left(\frac{t_L}{T_0}\right). \tag{19}$$

Relations Equations (13) and (15)–(19) describe a universal mathematical model for transforming the spatio-temporal structure of the signal during active location of moving targets in microwave radars using a continuous sounding radio signal Equation (15) with linear frequency modulation. The range of permissible target speeds and ranges for this echo radio-signal model makes it applicable for analyzing the operation and prototyping of LFMCW radars that locate existing and future ground vehicles. However, the complex, nonlinear nature of the universal mathematical model of the echo radio signal makes it ineffective in the development of promising quasi-real-time systems for estimating the parameters of the movement of vehicles in the current state of development of software and hardware. Let us consider an approximate mathematical model of an echo radio signal, which makes it possible to significantly simplify the algorithms for processing and analyzing the input data of the radar.

Modes of operation of automotive LFMCW radars are subdivided into near and far location of the workspace. For near location (e.g., when the vehicle is parking), a short range and target speed are typical. For far radiolocation, the distance to the target ranges from several meters to several hundred meters, and the target speed ranges from fractions of a meter per second to 200 m/s. The further developed mathematical model of the spatio-temporal structure of the signal can be effectively used for the mode of long-range location of the working space with an automobile LFMCW radar. We take into account that the results of radar imaging should be promptly visualized on a user monitor, and the signal processing time should not exceed 20–40 m/s. In the FMCW long-range radar mode, the distance the target moves during processing usually does not exceed the distance to the target, and the Maclaurin power series expansion can be used for the vector norm $\|\boldsymbol{R} - \boldsymbol{V}t\|$ [46]:

$$
\begin{aligned}
\|\boldsymbol{R} - \boldsymbol{V}t\| &= \sqrt{R^2 - 2\boldsymbol{V}\cdot\boldsymbol{R}t + (Vt)^2} = R\sqrt{1 - \frac{2\boldsymbol{V}\cdot\boldsymbol{R}t}{R^2} + \frac{(Vt)^2}{R^2}} \\
&= R \sum_{n=0}^{\infty} \frac{(-1)^n (2n)!}{(1-2n)(n!)^2 (4^n)} \left(\frac{(Vt)^2}{R^2} - \frac{2\boldsymbol{V}\cdot\boldsymbol{R}t}{R^2} \right)^n \\
&\approx R - V_R t + \frac{V_{TR}^2 t^2}{2R}\left(1 + \frac{V_R t}{R} + \frac{V_R^2 t^2}{R^2} + O\left(\frac{V^3 t^3}{R^3}\right)\right),
\end{aligned}
\tag{20}
$$

where V_R—norm of velocity vector projection onto radius vector $\boldsymbol{R}$, $(V_R = \boldsymbol{V}\cdot\boldsymbol{R})$. Similarly, we can use V_r—the projection of the velocity on the radius vector $\boldsymbol{r}$, as well as the corresponding orthogonal components $\boldsymbol{V}_{TR}$ and $\boldsymbol{V}_{Tr}$ (see Figure 2).

The domain of convergence of a power series Equation (20):

$$
\left| \frac{(Vt)^2}{R^2} - \frac{2\boldsymbol{V}\cdot\boldsymbol{R}t}{R^2} \right| = \left| \frac{V^2 t^2}{R^2} - \frac{2V_R t}{R} \right| \le 1,
\tag{21}
$$

which must be taken into account when assessing the range and speed measurement limits of a radar target in this approximation. In microwave LFMCW radars of the millimeter wavelength range, the dimensions of the antenna aperture are significantly less than the minimum specified distance to the target. In this case, for the vector velocity, $\boldsymbol{V}$, projections norms with high accuracy $V_{TR} = V_{Tr}$ and $V_R = V_r$.

Power series expansion Equation (20) of the norm of the vector $\|\boldsymbol{R} - \boldsymbol{V}t\|$ accurate to the third order of smallness relative to Vt/R allows to find an approximate expression for calculating the time delay $\tau(t, r_2, R_2, V)$. From the general relation Equation (13), we obtain:

$$
\begin{aligned}
\tau(t, r_2, R_2, V) &\approx \tau_s(t_L, k) \\
&= \tau_R - \tfrac{1}{c} 2 V_R (t_L + \tau_m + kT_c) - \tfrac{\rho}{c}\cos\theta \\
&\quad + \frac{V_{TR}^2}{cR_2}[t_L + \tau_m + kT_c]^2 \left[1 - \frac{V_R(t_L + \tau_m + kT_c)}{R_2} + \frac{V_R^2(t_L + \tau_m + kT_c)^2}{R_2^2}\right] \\
&\quad + \frac{2V_R}{c}\tau_R = \tau_R - \tfrac{1}{c} 2 V_R (t_L + \tau_m + kT_c) - \tfrac{\rho}{c}\cos\theta + \Delta\tau_s(t_L, k)
\end{aligned}
\tag{22}
$$

where θ—the angle of inclination of the radio wave front to the antenna aperture (see Figure 2), and $\tau_R = 2R_2/c$—propagation delay of the sounding radio signal, determined by the doubled distance to the target at the time of its re-emission.

The first three terms in Equation (22) are the propagation delay of the sounding radio signal, considered in the traditional approach to the theoretical analysis of the characteristics of the automotive LFMCW radar. The dependence of the third term in Equation (22) on the ρ coordinate of the receiving element at the antenna aperture makes it possible to determine the direction to the radiation source—the target bearing when using a phased antenna array in the radar. The additional term $\Delta\tau_s(t_L, k)$ makes a significant contribution to the propagation delay of the sounding radio signal as the transverse velocity component increases and the distance to the target decreases.

A mathematical model of the complex-valued signal at the I-Q output of the quadrature channels of the LFMCW radar receiver, which takes into account the transverse component of the target velocity and, accordingly, retains its rigor in a wide range of changes in the speed and range to the observed target, is obtained by substituting the found value of the delay $\tau_s(t_L,k)$ Equation (22) into Equation (19), which determines the mixer signal in the general case:

$$
\begin{aligned}
\dot{S}_{MIX}(t_L,k) = U_M \exp\Big\{2\pi j\Big[&\alpha\tau_R + \tfrac{2V_R}{c}f_0 + \tfrac{2V_R}{c}\alpha kT_c \\
&+ \tfrac{V_{TR}^2}{cR_2}kT_c\Big(1 + \tfrac{V_R kT_c}{R_2}\Big)(2f_0 + \alpha kT_c)\Big]t_L \\
&+ 2\pi jkT_c\Big[\tfrac{2V_R}{c}f_0 - \tau_R\tfrac{2V_R}{c}\alpha + \tau_m\tfrac{2V_R}{c}\alpha + \tfrac{V_{TR}^2}{cR_2}kT_c\Big(1 + \tfrac{V_R kT_c}{R_2}\Big)f_0\Big] \\
&+ \Delta\phi\Big\}\mathrm{rect}\Big(\tfrac{t_L}{T_0}\Big),
\end{aligned}
\tag{23}
$$

here, $\Delta\phi$—component of the phase of the mixer signal, in the first approximation independent of the time t in the observation interval T:

$$
\Delta\phi = \phi_{Tr} - \phi_{Rv} + 2\pi\big[f_0\tau_R + \alpha(\tau_R - 2\tau_m)\tau_m\big].
$$

For the physical interpretation of the results obtained, it is advisable to introduce the phase $\Phi(t_L,kT_c)$ of the complex-valued signal of the $S_{MIX}(t_L,k)$ and rewrite Equation (23) in the following form:

$$
\begin{aligned}
\dot{S}_{MIX}(t_L,k) &= U_M \exp\big[j\Phi(t_L,kT_c)\big]\,\mathrm{rect}\Big(\tfrac{t_L}{T_0}\Big) \\
&= U_M \exp\{j[\Phi_B(t_L,kT_c) + \Phi_V(kT_c)] + \Delta\phi\}\mathrm{rect}\Big(\tfrac{t_L}{T_0}\Big) \\
&= U_M \exp\{2\pi j[f_B(kT_c)t_L + \textstyle\int_0^{kT_c} f_V(kT_c)d(kT_c)] + \Delta\phi\}\mathrm{rect}\Big(\tfrac{t_L}{T_0}\Big),
\end{aligned}
\tag{24}
$$

where $f_B(kT_c)$—instantaneous frequency of the complex-valued signal $S_{MIX}(t_L,k)$, in the technical literature called the beat frequency:

$$
\begin{aligned}
f_B(kT_c) &= \tfrac{\partial[\Phi_B(t_L,kT_c)/2\pi]}{\partial t_L} \\
&= \alpha\tau_R + \tfrac{2V_R}{c}f_0 + \tfrac{2V_R}{c}\alpha kT_c + \tfrac{V_{TR}^2}{cR_2}kT_c\Big(1 + \tfrac{V_R kT_c}{R_2}\Big)(2f_0 + \alpha kT_c) \\
&= \alpha\tau_R + \tfrac{2V_R}{c}f_0 + \Delta f_B(kT_c).
\end{aligned}
\tag{25}
$$

In Figure 3, f_B value is shown as the current difference between the frequencies of the sounding and echo radio signals within the duration of each k chirp of the radio signal. The one-dimensional Fourier transform makes it easy to calculate the beat frequency $f_B(kT_c)$ as a function of k. The term $\Delta f_B(kT_c)$ in Equation (23), which is additional in comparison with the classical method of analysis, is the result of using a more rigorous mathematical model of radio signal transformation in the radar receiver. A non-trivial conclusion is the nonlinear dependence of the beat frequency on the chirp number k, which must be taken into account when evaluating the parameters R and V at a large value of the transverse velocity and short distances to the target.

The Doppler frequency shift $f_V(kT_c)$ in Equation (23) is described by the formula:

$$
\begin{aligned}
f_V(kT_c) &= \tfrac{\partial[\Phi_V(kT_c)/2\pi]}{\partial(kT_c)} = \tfrac{2V_R}{c}f_0 - \tau_R\tfrac{2V_R}{c}\alpha + \tau_m\tfrac{2V_R}{c}\alpha + \tfrac{2V_{TR}^2}{cR_2}kT_c\Big(1 + \tfrac{3V_R kT_c}{2R_2}\Big)f_0 \\
&= -\tfrac{2V_R}{c}f_0 + \Delta f_V(kT_c),
\end{aligned}
\tag{26}
$$

where the term $\Delta f_V(kT_c)$ refines the proposed mathematical model of radio signal conversion. The dependence of the Doppler frequency shift $f_V(kT_c)$ on the slow time scale kT_c makes the traditional method of estimating the target velocity based on the 2D Fourier transform ineffective. The signal of the mixer $S_{MIX}(t_L,k)$ as a function of time kT_c with a strict mathematical description belongs to the class of frequency-modulated signals and, therefore, is not narrowband and the definition of the parameter V is not unambiguous. The spectrum width increases at a high transverse speed and small target ranges, which leads to an error in determining the parameters R and V.

The obtained relations Equations (23)–(25), which determine the temporal structure of the received signal by the LFMCW radar, make it possible to develop an algorithm for estimating the motion parameters V_R, V_{TR} and R_2.

4. Estimation of the Range and Velocity Vector of a Radar Target: Working Range of Parameters' Measurement

In this section, based on the analysis of the mathematical model of the complex-valued signal at the output of the I-Q quadrature channels of the LFMCW radar receiver, an algorithm is proposed for estimating such motion parameters as the radial and transverse components of the velocity, as well as the distance to the radar target. In addition, the boundaries of the range of variation of the variables V_R, V_{TR} and R_2 are calculated, where the proposed algorithm remains operational.

Let us write the phase $\Phi(t_L, kT_c)$ of the complex-valued signal of the mixer $S_{MIX}(t_L, k)$ Equation (23) as a third-degree polynomial in the variable kT_c:

$$\Phi(t_L, kT_c) = 2\pi\left[f_B(kT_c)t_L + a_0 + a_1 kT_c + a_2(kT_c)^2 + a_3(kT_c)^3\right], \tag{27}$$

the coefficients of the polynomial (a_0, a_1, a_2, a_3) are determined from the relation (23):

$$a_0 \equiv \Delta\phi; \quad a_1 = \frac{2V_R}{c}f_0\left(1 - \frac{\alpha\tau_R}{f_0} + \frac{\alpha\tau_m}{f_0}\right); \quad a_2 = \frac{V_{TR}^2}{cR_2}f_0; \quad a_3 = \frac{V_{TR}^2}{cR_2}\frac{V_R}{R_2}f_0. \tag{28}$$

In accordance with the developed mathematical model for converting the input signal in the radar receiver, the algorithm for estimating the range and speed of a radar target includes the following stages:

(1) Using the Fourier transform, the beat frequency $f_B(k)$ and the phase $\Phi(f_B, k)$ of the complex-valued spectral components of the signal of the receiver quadrature channels are calculated at each k-th local time interval $\forall \{t_L \in \mathbb{R}: 0 \leq t_L \leq T_c\}$ in a sequence of N chirps of the radio signal. The spectral component corresponds to the target in the radar monitoring area.

(2) The obtained time series of phase samples $\Phi(f_B, k)$ of the selected spectral component (target) is approximated by a polynomial of the third degree in the variable kT_c. As a result of the approximation procedure, we obtain estimates for the coefficients (a_0, a_1, a_2, a_3) of the polynomial.

(3) On the basis of the obtained estimates of the coefficients (a_0, a_1, a_2, a_3) together with the estimate of the beat frequency $f_B(k)$, the estimates of the motion parameters—the velocity vector and the distance to the target—are calculated.

The solution of the system of joint Equations (23), (25) and (27) leads to the following formulas for estimating the parameters of the movement of a radar target R and V:

Target distance estimation:

$$\hat{R}_2(k) = \frac{c}{2\alpha}\cdot\frac{\hat{f}_B(k) - \frac{\hat{a}_1}{2\pi}\left(1 + \frac{\alpha kT_c}{f_0}\right)\left(1 - \frac{\alpha\tau_m}{f_0}\right) - \frac{\hat{a}_2}{2\pi}\left(2 + \frac{\alpha kT_c}{f_0}\right)kT_c}{1 + \frac{\hat{a}_1}{2\pi}\left(1 + \frac{\alpha kT_c}{f_0}\right)\frac{1}{f_0}}, \tag{29}$$

hereinafter, the diacritical mark of the cap (circumflex) above the corresponding parameter means the estimate of this parameter obtained as a result of data processing according to the above algorithm.

Estimation of the radial component of the target's velocity:

$$\hat{V}_R(k) = \frac{c}{2f_0}\cdot\frac{\hat{a}_1}{2\pi\left(1 - \frac{2\alpha\hat{R}_2(k)}{cf_0} + \frac{\alpha\tau_m}{f_0}\right)}; \tag{30}$$

Estimation of the transverse component of the target speed:

$$\hat{V}_{TR}(k) = \sqrt{\frac{\hat{a}_2}{2\pi} \cdot \frac{c\hat{R}_2(k)}{f_0}}. \tag{31}$$

If the estimates of the coefficients $\{a_0, a_1, a_2, a_3\}$ of the polynomial and the beat frequency $f_B(k)$ coincide with the true values of the parameters, the estimates of the target range and velocity components will also be equal to their true values. In this case, the estimates do not depend on the chirp number k. If the error of estimates $\{a_0, a_1, a_2, a_3\}$ and $f_B(k)$ is not zero, then the algorithm for estimating the range and speed of the radar target should include the stage of calculating the sample mean of the corresponding estimates of the motion parameters.

Expansion of the range of measured motion parameters through the use of the developed algorithm for estimating the range and speed, in comparison with estimates based on traditional algorithms, determines the practical significance of the results obtained. Let us analyze the working range of motion parameters' measurement. The mathematical model of radio signal transformation, which takes into account, among other things, the transverse component of the velocity, is based on the necessary condition for the vector of measured parameters $\theta = \{V_R, V_{TR}, R_2\}$ belonging to the three-dimensional space $\Theta^{(3)}$:

$$\Theta^{(3)} = \Theta_1^{(3)} \cap \Theta_2^{(3)}; \tag{32}$$

$$\begin{aligned}\Theta_1^{(3)} &= \Theta_{11}^{(3)} \cup \Theta_{12}^{(3)} \\ &= \left\{\theta : \Delta f_B(NT_c)T_c \geq 1, \theta \in \mathbb{R}^3\right\} \cup \left\{\theta : \int_0^{NT_c} \Delta f_V(NT_c)d(NT_c) \geq 1, \theta \in \mathbb{R}^3\right\};\end{aligned} \tag{33}$$

$$\Theta_2^{(3)} = \left\{\theta : \Phi_V[(k+1)T_c] - \Phi_V(kT_c) \leq \pi, \forall k \in [0, N], \theta \in \mathbb{R}^3\right\}, \tag{34}$$

here, the signs $\cup$ and $\cap$ denote the union and intersection of two sets, respectively.

In relation Equation (33), the belonging of the point θ to the region $\Theta^{(3)}_{11}$ is determined by the noticeable influence of the correction $\Delta f_B(kT_c)$ on the beat frequency of the radio signal $f_B(kT_c)$, calculated on the local time interval $[0 \leq t_L \leq T_c]$. The belonging of point θ to the second region $\Theta^{(3)}_{12}$ is determined by the noticeable influence of the correction $\Delta f_V(kT_c)$ on the interval of the slow time scale $[0 \leq kT_c \leq NT_c]$.

For correct use of the method of two-dimensional Fourier transform [19–21] and an unambiguous estimate of the velocity vector of a radar target, it is necessary that the phase change $\Phi_V(kT_c)$, calculated on the local time interval $[0 \leq t_L \leq T_c]$, does not exceed the value π. This requirement determines that the point θ belongs to the three-dimensional subspace $\Theta^{(3)}_2$ of the measured parameters $\{V_R, V_{TR}, R_2\}$ in the relation Equation (34).

When constructing a mathematical model of radio signal transformation Equations (23)–(26), it is necessary to fulfill a number of conditions, such as the convergence of the Maclaurin power series expansion Equation (20) and the unambiguity of the parameter estimates Equations (1)–(3). The requirements for LFMCW radar design to the radar resolution in range δR and speed δV, maximum and minimum measured range R_{max}, R_{min} and speed V_{max}, V_{min} make the task of determining the boundaries of the region $\Theta^{(3)}$ a multiparameter problem.

Relations Equations (20) and (32)–(34) lead to the following system of inequalities as necessary conditions determining the boundaries of the domain $\Theta^{(3)}$:

$$\begin{cases} \dfrac{2V_R NT_c}{R_2} + \dfrac{V_{TR}^2 N^2 T_2^2}{R_2^2} \leq 1; \\[2ex] \dfrac{V_{TR}^2 N^2 T_2^2}{cR_2} \cdot \dfrac{V_R NT_c}{R_2} f_0 < 1 \ (\textit{more strictly} \ 0.3); \\[2ex] \dfrac{V_{TR}^2 N^2 T_2^2}{cR_2} f_0 \geq 1; \\[2ex] V_R + \dfrac{V_{TR}^2}{R_2} NT_c < V_{max}. \end{cases} \tag{35}$$

The first two inequalities in Equation (35) follow from the conditions for the convergence of the Maclaurin power series Equation (20) and the truncation of terms higher than the second order of smallness in the expansion of the norm of the vector $\|\mathbf{R} - \mathbf{V}t\|$. The third Equation (35) defines the lower bound for V_{TR_min}:

$$V_{TR\_min} = \frac{1}{NT_c}\sqrt{\frac{cR_2}{f_0}},\tag{36}$$

where the contribution of the transverse velocity component to the estimation of the parameters of the movement of the radar target must be taken into account in order to achieve the specified parameters of the resolution $\delta V \delta R$ in range and speed.

In Figure 4 in the V_{TR}-V_R plane, the curves of the contour of the section by the plane of constant fixed range R of the three-dimensional space $\Theta^{(3)}$ are presented, where the efficiency of the developed algorithm for estimating the parameters of the target movement is preserved. Points belonging to $\Theta^{(3)}$ lie between the boundary V_{TR_min} and the curve of the section contour, with a given value of the range R, which is determined by Equation (36).

For example, in Figure 4, 2D-sections of the region $\Theta^{(3)}$ are selected for the parameter values $R = 3$ m, $V1_{TR\_min} = 15.6$ km/h иR $= 50$ m, $V2_{TR\_min} = 63.8$ km/h. We must note that points in space $\Theta^{(3)}$ are the set of values of the measured parameters $\{V_R, V_{TR}, R_2\}$, the variance and bias of the estimate of which, based on the traditional 2D-FFT (Fast Fourier Transform) algorithm, can have a significant value.

The analysis of the graphs shown in Figure 4 shows that the operability area of the algorithm for estimating the motion parameters depends not only on the norm of the general velocity vector $\|\mathbf{V}\|$, but also on the ratio of the components of the velocity vector, that is, on the angle ψ of the angle between the velocity vector $\mathbf{V}$ and the radius-vector $\mathbf{R}_2$ (see Figure 2). For example, with $\psi = 45°$ and $V_{TR} = V_R = 200$ km/h (total speed $V = 283$ km/h), the lower limit of the range when estimating V_{TR} with a given speed resolution $\delta V = 1$ km/h is 45 m. In this case, all three conditions Equation (35) are satisfied (the second with a more stringent requirement, 0.3).

The estimates of the range and velocity vector of a radar target proposed in this section, as well as the calculation of the operating range for measuring the parameters V_R, V_{TR} and R_2, require experimental verification of the results obtained. The next section discusses the results of a computational experiment using the software simulator developed in the first section of the paper.

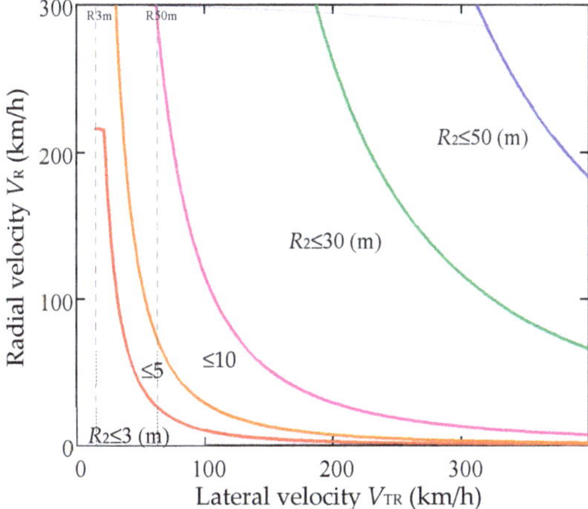

Figure 4. The limits of efficiency of the algorithm for estimating the parameters of the target movement.

5. Computational Experiment Results and Discussion

To check the obtained results of theoretical analysis, we carried out a computational experiment simulating the operation of the LFMCW radar in the mode of estimating the movement parameters V_R, V_{TR} иR$_2$ of a vehicle. Provided that the intrinsic noise and instability of the parameters of the radar receiving system, as well as the multiple reflections of the probing signal and in the approximation of a single-point reflective target, are neglected, the software simulator implements a signal that is identical to the real echo signal at the output of the quadrature channels of the LFMCW radar receiver. It can be used to test the effectiveness of the developed algorithm for estimating the range and velocity vector of the vehicle. In this sense, carrying out computer modeling using this signal simulator can be called a computational experiment. The parameters V_R, V_{TR} and R_2 were estimated in accordance with the algorithm developed above Equations (29)–(31).

The tasks of the computational experiment were as follows:

- For the developed algorithm, in the operating range of variation of V_R, V_{TR} and R_2, the determination of the parameter estimation error at different values of the angle ψ between the radial V_R and transverse V_{TR} components of the velocity.
- Comparison of the error in estimating the range and speed obtained using the developed algorithm and the traditional method based on two-dimensional FFT.
- Determination of the limits of the working range for measuring the parameters V_R, V_{TR} and R_2 as a result of a computational experiment and comparison of the obtained data with theoretical calculations of the fourth section.
- Carrying out a computational experiment for the physical interpretation of the results.

During the computational experiment, the following initial values of the parameters were used:

(1) The maximum sampling rate of an analog signal is about 55 MHz,
(2) Upper limit of range measurement—R_{max} = 400 m,
(3) Range resolution—δR = 1.6 m,
(4) Upper limit of speed measurement—V_{max} = ±300 km/h,
(5) Speed resolution—δV = 1 km/h,
(6) Frame rate—40 Hz,
(7) The number of counts at the working section of the chirp—512,
(8) The number of chirps in one frame—N = 2048.

We tested radar targets with a given dynamic range of changes in the values of motion parameters—radial and transverse speeds, as well as range. The ratio between the radial and transverse components of the velocity was regulated by changing the angle ψ between the vector of the full velocity V (see Figure 2) and the radius vector R_2. In this case, the norm of the full-velocity vector $\|V\|$ was fixed. In the computational experiment, the following motion modes were considered: $\psi = 0°$—strictly radial motion (moving away or approaching the target collinear to the radius vector R_2), $\psi = 90°$—strictly azimuthal (circular) movement, and $\psi = 45°$—motion in which the radial and transverse components of the velocity are equal. The results of the computational experiment for different angles, ψ, are shown in Figures 5 and 6.

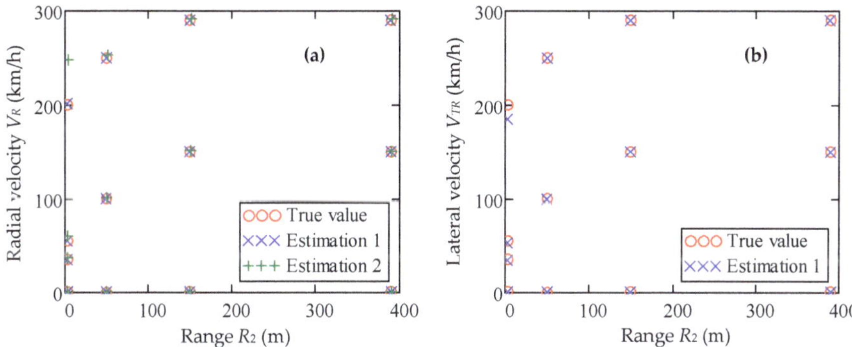

Figure 5. Estimation of motion parameters (**a**) V_R, (**b**) V_{TR} and R_2: estimation 1 is the result of applying the developed algorithm, and estimation 2 is the application of the traditional 2D-FFT algorithm. Velocity vector angle $\psi = 45°$.

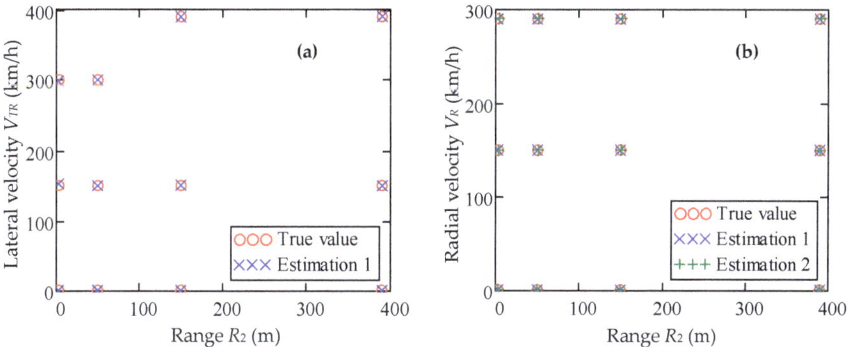

Figure 6. Estimation of motion parameters V_R, V_{TR} and R_2: (**a**) velocity vector angle $\psi = 90°$, (**b**) $\psi = 0°$.

Consider the case $\psi = 45°$, when the radial and transverse velocity components are equal. Figure 5a, on the coordinate plane V_R, R_2 (radial speed-range) shows the results of the estimation (Estimation 2) of the radial speed and range to the target calculated using the traditional two-dimensional FFT method for the case $\psi = 45°$. Also given are estimates (Estimation 1) of the same parameter received via the developed algorithm. In the traditional method for estimating the parameters V_R and R_2, the terms $\Delta f_B(kT_c)$ and $\Delta f_V(kT_c)$ in relations Equation (25) and Equation (26) are assumed to be equal to zero, which leads to an increase in the error in estimating the motion parameters with decreasing range.

It can be seen from Figure 5a that, in accordance with the results of theoretical analysis at short ranges ($R_2 = 3$ m) and high target speeds ($V_R = 200$ km/h), the estimate "Estimation 2" of the parameters V_R and R_2 has a large relative error in comparison with the estimate "Estimation 1". The error of the traditional measurement method reaches 25% and 40% to estimate the speed and range, respectively. The error of the measurement method Estimation 1 of the coordinates of a point (V_R, R_2) does not exceed the value of $\delta R \times \delta V$ in the range-radial velocity measurement area $[R_2 \times V_R] = [3$ m, 220 m] $\times$ [0 km/h, 220 km/h]. At the same time, at long range, both methods give the same estimate of the radial speed V_R, which coincides with the true speed of the radar target. This is clearly seen on the right side of Figure 5a, where all three points of the V_R value become indistinguishable.

In Figure 5b on the coordinate plane V_{TR}, R_2 (transverse speed-range) shows the results of Estimation 1 of the transverse component of the speed and range to the target calculated using the developed method for the case $\psi = 45°$. The error in determining the speed increases with decreasing range and increasing full speed, which is due to the approach of the coordinates of the point (V_{TR}, R_2)

to the boundary of the region $\Theta^{(3)}$ Equation (35). A computational experiment shows that already at $R_2 \geq 15$ m, the error of the method for measuring the coordinates of a point (V_{TR}, R_2) does not exceed the value of $\delta R \times \delta V$ in the entire working range of the V_{TR} and R_2 parameters.

Figure 6 show the results of estimating the parameters of the target movement, calculated using the developed method for the case of strictly azimuthal (circular) movement $\psi = 90°$ and strictly radial movement $\psi = 0°$, respectively. The analysis of the results of the computational experiment and the corresponding results of the theoretical analysis show that the relative error of the method for measuring the transverse velocity component V_{TR} is greater than the error in measuring the radial component V_R, but does not exceed 2% for the case $\psi = 0°$ and 8% for the case $\psi = 45°$ for $R_2 = 3$ m (see Figures 5b and 6a).

When assessing the motion parameters using the traditional method of two-dimensional FFT in the mode of strictly azimuthal (circular) target movement $\psi = 90°$, when the radial velocity component is absent, there is a large error in determining the value of V_R at a short range and high object speed. Consider the physical interpretation of this result. This phenomenon is due to the temporal structure of the signal, which is a frequency-modulated signal. The Fourier spectrum of such a signal according to Equation (26) for $R_2 = 3$ m and $V_{TR} = 290$ km/h is shown in Figure 7. The maximum spectral component is at a frequency of 21.6 kHz, which corresponds to a radial speed of 153 km/h with a true value of $V_R = 0$ km/h. At the same time, the estimate of the radial velocity, calculated using the developed algorithm, gives the value $V_R = 1.0$ km/h and $V_{TR} = 287$ km/h, which indicates the effectiveness of the algorithm.

An analysis of the results of estimating the range to a radar target with strictly radial movement $\psi = 0°$ shows that the greatest error of the traditional FFT Estimation 2 method for measuring R_2 is obtained at small ranges and high speeds of the observed object. So, with $R_2 = 3$ m and $V_R = 290$ km/h, Estimation 2 gives a value of 4.5 m, while Estimation 1 gives a value of 2.9 m, which is close to the true value.

Note that the computational experiment was carried out without the effect of noise on the radio signal. Therefore, the calculated results and the corresponding conclusions are correct for a large signal-to-noise ratio. In general, the results of numerical experiments agree well with the corresponding results of the theoretical analysis.

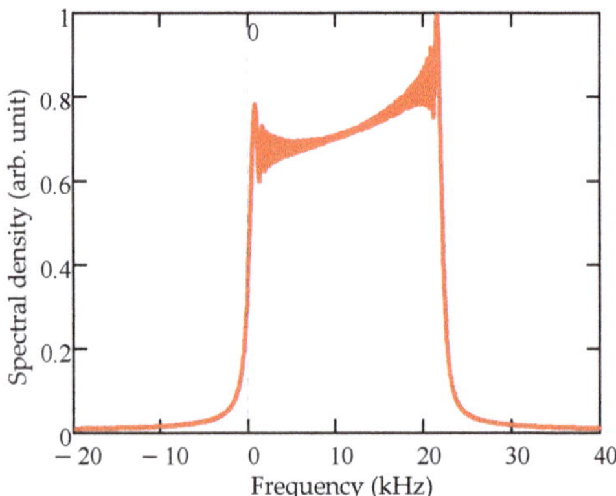

Figure 7. FFT conversion of radio echo signal in the slow time domain.

6. Conclusions

A rigorous mathematical, physically consistent model of the spatio-temporal structure of the echo signal of a radar operating in the active radar mode can be obtained using a quasi-relativistic vector transformation of Lorentz coordinates and time. This model is used to develop a software simulator of an echo-radio signal of a radar, which realizes the verification of the effectiveness of methods for assessing the parameters of vehicle movement by observation during one frame of a sample of an echo-radio signal of microwave LFMCW radars. The proposed model of the spatio-temporal structure of the echo signal of the LFMCW radar makes it possible to develop an algorithm for the simultaneous estimation of the velocity vector and the distance to the observed target, which remains operational in a wide range of changes in the parameters of the movement of vehicles. For the first time, an estimate of the radial and transverse components of the target velocity vector can be obtained by observing a number of data (sample) during one frame of the LFMCW radar sounding signal. For a 77 GHz microwave LFMCW radar with a sampling rate of about 55 MHz and a frame duration of 25 ms (frame rate 40 Hz), the range and speed measurement span with negligible noise level is from 3 to 400 m and from 1 to 290 km/h, respectively. The error of the proposed method for measuring the radial component of the speed and range of a radar target does not exceed the specified values of the resolution δV and δR in the entire specified working area of the range-radial velocity measurement. The error of the traditional method for measuring V_R and R_2 based on the two-dimensional Fourier transform is greater than the error of the proposed method. The relative error of the method for measuring the transverse velocity component V_{TR} increases with decreasing range and increasing target speed, but does not exceed 2% for the case $\psi = 0°$ and 8% for the case $\psi = 45°$ with $R_2 = 3$ m. The results of the computational experiment based on the developed rigorous mathematical model of the developed software simulator using the rigorous mathematical model of the LFMCW radar echo confirmed the corresponding theoretical analysis results. Thus, the previously set goal can be considered achieved.

For further work to validate the integrity of approaches to analyzing the performance of the developed algorithm by methods of computational or practical experiment, it is planned to develop an FMCW radar based on RF CMOS automotive transceiver TI AWR1243 and capture card DCA1000EVM (manufactured by Texas Instruments [47]) and make comparative test measurements.

Theoretically, for our further research, it is planned to tune the developed algorithm for estimating the parameters of movement in the presence of multiple radar targets and various kinds of radio interference. The mode of operation of the LFMCW radar in conditions of multiple radar targets and various kinds of radio interference is important in the practical use of the algorithm in modern systems of active driver assistance (ADAS—Advanced Driver Assistance System) to ensure the most comfortable and safe movement of the vehicle, including in the fully autonomous mode (unmanned) vehicle driving mode.

Author Contributions: S.I.: Conceptualization and Formal analysis. All authors discussed and designed the proposed algorithm. All authors have read and agreed to the published version of the manuscript.

Funding: The reported study was funded by RFBR, project number 19-29-06034.

Conflicts of Interest: The authors declare no conflict of interest.

References

1. Bengler, K.; Dietmayer, K.; Farber, B.; Maurer, M.; Stiller, C.; Winner, H. Three Decades of Driver Assistance Systems: Review and Future Perspectives. *IEEE Intell. Transp. Syst.* **2014**, *4*, 6–22. [CrossRef]
2. Meinel, H.H. Evolving automotive radar—From the very beginnings into the future. In Proceedings of the 8th European Conference. Antennas and Propagation, Hague, Netherlands, 6–11 April 2014; pp. 3107–3114. [CrossRef]
3. Bilik, I.; Longman, O.; Villeval, S.; Tabrikian, J. The Rise of Radar for Autonomous Vehicles: Signal Processing Solutions and Future Research Directions. *IEEE Signal Process.* **2019**, *5*, 20–31. [CrossRef]
4. Yang, B. High-Performance Automotive Radar: A Review of Signal Processing Algorithms and Modulation Schemes. *IEEE Signal Process. Mag.* **2019**, *5*, 32–44. [CrossRef]
5. Patole, M.; Torlak, M.; Wang, D.; Ali, M. Automotive radars: A review of signal processing techniques. *IEEE Signal Process. Mag.* **2017**, *2*, 22–35. [CrossRef]
6. Engels, F.; Heidenreich, P.; Zoubir, A.M.; Jondral, F.K.; Wintermantel, M. Advances in Automotive Radar: A framework on computationally efficient high-resolution frequency estimation. *IEEE Signal Process. Mag.* **2017**, *2*, 36–47. [CrossRef]

7.   Komarov, I.; Smolskiy, S. *Fundamentals of Short-Range FM Radar*; Artech House: Norwood, MA, USA, 2003.

8.   Stove, A.G. Linear FMCW radar techniques. *IEEE Proc. Radar Signal Process.* **1992**, *139*, 343–350. [CrossRef]

9.   Rohling, H.; Meinecke, M.-M. Waveform design principles for automotive radar systems. In Proceedings of the 2001 CIE International Conference on Radar (Cat No.01TH8559), Beijing, China, 15–18 October 2001; pp. 1–4. [CrossRef]

10.   Kim, G.; Mun, J.; Lee, J. A Peer-to-peer Interference Analysis for Automotive Chirp Sequence Radars. *IEEE Trans. Veh. Technol.* **2018**, *67*, 8110–8117. [CrossRef]

11.   Winkler, V. Range Doppler Detection for automotive FMCW Radars. In Proceedings of the 2007 European Radar Conference, Munich, Germany, 10–12 October 2007; pp. 166–169. [CrossRef]

12.   Kuptsov, V.D.; Ivanov, S.I.; Fedotov, A.A.; Badenko, V.L. Features of Multi-target Detection Algorithm for Automotive FMCW Radar. In *Internet of Things, Smart Spaces, and Next Generation Networks and Systems*; Springer: Cham, Switzerland, 2019; pp. 355–364. [CrossRef]

13.   Son, Y.S.; Sung, H.K.; Heo, S. Automotive Frequency Modulated Continuous Wave Radar Interference Reduction Using Per-Vehicle Chirp Sequences. *Sensors* **2018**, *18*, 2831. [CrossRef]

14.   Duan, Z.; Wu, Y.; Li, M.; Wang, W.; Liu, Y.; Yang, S. A novel FMCW waveform for multi-target detection and the corresponding algorithm. In Proceedings of the 2017 IEEE 5th International Symposium on Electromagnetic Compatibility (EMC-Beijing), Beijing, China, 28–31 October 2017; pp. 1–4. [CrossRef]

15.   Fan, Y.; Yang, Z.; Bu, X.; An, J. Radar Waveform Design and Multi-Target Detectionin Vehicular Applications. In Proceedings of the 2015 International Conference on Estimation, Detection and Information Fusion (ICEDIF), Harbin, China, 10–11 January 2015; pp. 286–289. [CrossRef]

16.   Kuptsov, V.D.; Ivanov, S.I.; Fedotov, A.A.; Badenko, V.L. Multi-target method for small unmanned vehicles parameters remote determination by microwave radars. *J. Phys. Conf. Ser.* **2020**, *1515*, 032045. [CrossRef]

17.   Kutsov, V.D.; Badenko, V.L.; Ivanov, S.I.; Fedotov, A.A. Millimeter Wave Radar for Intelligent Transportation Systems: A Case Study of Multi-Target Problem Solution. *E3S Web Conf.* **2020**, *157*, 05011. [CrossRef]

18.   Chen, B.; Pei, X.; Chen, Z. Research on Target Detection Based on Distributed Track Fusion for Intelligent Vehicles. *Sensors* **2020**, *20*, 56. [CrossRef] [PubMed]

19.   Ryu, I.; Won, I.; Kwon, J. Detecting Ghost Targets Using Multilayer Perceptron in Multiple-Target Tracking. *Symmetry* **2018**, *10*, 16. [CrossRef]

20.   Wang, W.; Du, J.; Gao, J. Multi-Target Detection Method Based on Variable Carrier Frequency Chirp Sequence. *Sensors* **2018**, *18*, 3386. [CrossRef]

21.   Kim, B.; Kim, S.; Jin, Y.; Lee, J. Low-Complexity Joint Range and Doppler FMCW Radar Algorithm Based on Number of Targets. *Sensors* **2020**, *20*, 51. [CrossRef]

22.   Saponara, S.; Neri, N. Radar Sensor Signal Acquisition and Multidimensional FFT Processing for Surveillance Applications in Transport Systems. *IEEE Trans. Instrum. Meas.* **2017**, *66*, 604–615. [CrossRef]

23.   Kronauge, M.; Rohling, H. New chirp sequence radar waveform. *IEEE Trans. Aerosp. Electron. Syst.* **2014**, *50*, 2870–2877. [CrossRef]

24.   Ivanov, S.I.; Kuptsov, V.D.; Fedotov, A.A. The signal processing algorithm of automotive FMCW radars with an extended range of speed estimation. *J. Phys. Conf. Ser.* **2019**, *1236*, 012081. [CrossRef]

25.   Mi, M.; Moallem, M.; Chen, J.; Li, M.; Murugan, R. Package Co-design of a Fully Integrated Multimode 76–81GHz 45nm RFCMOS FMCW Automotive Radar Transceiver. In Proceedings of the 2018 IEEE 68th Electronic Components and Technology Conference (ECTC), San Diego, CA, USA, 29 May–1 June 2018; pp. 1054–1061. [CrossRef]

26.   Gray, J.E.; Addison, S.R. Effect of nonuniform target motion on radar backscattered waveforms. *IEEE Proc. Radar Sonar Navig.* **2003**, *150*, 262–270. [CrossRef]

27.   Teal, P. Tracking Wide-Band Targets Having Significant Doppler Shift. *IEEE Trans. Audio Speech Lang. Process.* **2007**, *15*, 489–497. [CrossRef]

28.   Gupta, P.D. Exact derivation of the Doppler shift formula for a radar echo without using transformation equations. *Am. J. Phys.* **1977**, *45*, 674–675. [CrossRef]

29.   Fock, V. *The Theory of Space, Time and Gravitation*; Pergamon Press: London, UK, 1964.

30.   Madelung, E. *Die Mathematischen Hilfsmittel des Physikers*; Springer: Berlin, Germany, 1964.

31.   Fölster, F.; Rohling, H.; Lübbert, U. An automotive radar network based on 77 GHz FMCW sensors. In Proceedings of the IEEE International Radar Conference, Arlington, VA, USA, 9–12 May 2005; pp. 871–876. [CrossRef]

32. Fölster, F.; Rohling, H. Observation of lateral moving traffic with an automotive radar. In Proceedings of the 2006 International Radar Symposium, Krakow, Poland, 24–26 May 2006; pp. 1–4. [CrossRef]
33. Rohling, H.; Fölster, F.; Ritter, H. Lateral velocity estimation for automotive radar applications. In Proceedings of the 2007 IET International Conference on Radar Systems, Edinburgh, UK, 15–18 October 2007; p. 181. [CrossRef]
34. Haderer, A.; Wagner, C.; Feger, R.; Stelzer, A. Lateral velocity estimation using an FMCW radar. In Proceedings of the 2009 European Radar Conference (EuRAD), Rome, Italy, 30 September–2 October 2009; pp. 129–132.
35. Berizzi, F.; Corsini, G. Autofocusing of inverse synthetic aperture radar images using contrast optimization. *IEEE Trans. Aerosp. Electron. Syst.* **1996**, *32*, 1185–1191. [CrossRef]
36. Haderer, A.; Scherz, P.; Stelzer, A. Precise radial velocity estimation using an FMCW radar. In Proceedings of the 7th European Radar Conference, Paris, France, 30 September–1 October 2010; pp. 164–167.
37. Asuzu, P.; Thompson, C. A more exact linear FMCW radar signal model for simultaneous range-velocity estimation. In Proceedings of the 2018 IEEE Radar Conference (RadarConf18), Oklahoma, OK, USA, 23–27 April 2018; pp. 7–11. [CrossRef]
38. Asuzu, P.; Thompson, C. A Signal Model for Simultaneous Range-Bearing Estimation in LFMCW Radar. In Proceedings of the 2019 IEEE Radar Conference (RadarConf), Boston, MA, USA, 22–26 April 2019; pp. 1–6. [CrossRef]
39. Asuzu, P. On the Measurement of Range and its Time-Derivatives in LFMCW Radar. In Proceedings of the 2020 IEEE International Radar Conference (RADAR), Washington, DC, USA, 28–30 April 2020; pp. 494–499. [CrossRef]
40. Peleg, S.; Friedlander, B. The discrete polynomial-phase transform. *IEEE Trans. Signal Process.* **1995**, *43*, 1901–1914. [CrossRef]
41. Kramer, G.Y.; German, A.M. About correctly accounting for the Doppler effect in coherent processing of space-time radio signals. *Radio Eng. Electron. Phys.* **1982**, *27*, 97–100.
42. Xu, J.; Hong, W.; Zhang, H.; Wang, G.; Yu, Y.; Jiang, Z.H. An Array Antenna for Both Long- and Medium-Range 77 GHz Automotive Radar Applications. *IEEE Trans. Antennas Propag.* **2017**, *65*, 7207–7216. [CrossRef]
43. Ku, B.-H.; Inac, O.; Chang, M.; Yang, H.-H.; Rebeiz, G.M. A High-Linearity 76–85-GHz 16-Element 8-Transmit/8-Receive Phased-Array Chip with High Isolation and Flip-Chip Packaging. *IEEE Trans. Microw. Theory Tech.* **2014**, *62*, 2337–2356. [CrossRef]
44. Motomura, T.; Uchiyama, K.; Kajiwara, A. Measurement results of vehicular RCS characteristics for 79GHz millimeter band. In Proceedings of the 2018 IEEE Topical Conference on Wireless Sensors and Sensor Networks (WiSNet), Anaheim, CA, USA, 14–17 January 2018; pp. 103–106. [CrossRef]
45. Matsunami, I.; Ryohei, N.; Kajiwara, A. Target state estimation using RCS characteristics for 26GHz short-range vehicular radar. In Proceedings of the 2013 International Conference on Radar, Adelaide, SA, Australia, 9–12 September 2013; pp. 304–308. [CrossRef]
46. Korn, G.A.; Korn, T.M. *Mathematical Handbook for Scientists and Engineers*; McGraw-Hill Book Company: New York, NY, USA, 1968.
47. AWR1243 Single-Chip77- and 79-GHz FMCW Transceiver. Available online: https://www.ti.com/lit/ds/symlink/awr1243.pdf (accessed on 5 October 2020).

Publisher's Note: MDPI stays neutral with regard to jurisdictional claims in published maps and institutional affiliations.

Article

Hybrid Solution Combining Kalman Filtering with Takagi–Sugeno Fuzzy Inference System for Online Car-Following Model Calibration

Mădălin-Dorin Pop [1], Octavian Proștean [1,*], Tudor-Mihai David [2] and Gabriela Proștean [3]

[1] Automation and Applied Informatics Department, Politehnica University of Timisoara, Bvd. V. Parvan, No. 2, 300223 Timisoara, Romania; madalin.pop@student.upt.ro

[2] Faculty of Automation and Computers, Politehnica University of Timisoara, Bvd. V. Parvan, No. 2, 300223 Timisoara, Romania; tudor.david@student.upt.ro

[3] Management Department, Politehnica University of Timisoara, Bvd. M. Viteazu, No. 1, 300223 Timisoara, Romania; gabriela.prostean@upt.ro

* Correspondence: octavian.prostean@upt.ro

Received: 24 August 2020; Accepted: 24 September 2020; Published: 27 September 2020

Abstract: Nowadays, the intelligent transportation concept has become one of the most important research fields. All of us depend on mobility, even when we talk about people, provide services, or move goods. Researchers have tried to create and test different transportation models that can optimize traffic flow through road networks and, implicitly, reduce travel times. To validate these new models, the necessity of having a calibration process defined has emerged. Calibration is mandatory in the modeling process because it ensures the achievement of a model closer to the real system. The purpose of this paper is to propose a new multidisciplinary approach combining microscopic traffic modeling theory with intelligent control systems concepts like fuzzy inference in the traffic model calibration. The chosen Takagi–Sugeno fuzzy inference system proves its adaptive capacity for real-time systems. This concept will be applied to the specific microscopic car-following model parameters in combination with a Kalman filter. The results will demonstrate how the microscopic traffic model parameters can adapt based on real data to prove the model validity.

Keywords: fuzzy inference; calibration; car-following; Takagi–Sugeno; Kalman filter; microscopic traffic model; continuous-time model

---

1. Introduction

Traffic modeling is one of the research domains receiving a higher interest from scientists. Traffic congestion [1] affects everyone and introduces a dependence on this issue even if we consider the movement of people or the delivery of products or services. In this context, a concept has appeared regarding smart mobility that tries to use ITS (intelligent transportation systems)-specific algorithms to optimize road traffic flow.

In the traffic modeling process, there are several subsystems that contribute to the determination of a closer-to-reality road traffic model. For each new model proposal, a validation system ensures the comparison of those model simulation results and real traffic data. The calibration process will minimize the differences between simulation and real data. This calibration is responsible for the improvement in simulation parameters by ensuring a better fitting of them compared to real data.

The current paper aims to deliver a new calibration method for road traffic parameters at the microscopic level. This method consists of a Kalman filter integrated with a Takagi–Sugeno FIS (fuzzy inference system) for the car-following model in continuous-time. The Takagi–Sugeno FIS using the Kalman-filtered values as input for the car-following model consists of our calibration system proposal.

This paper starts with an overview of actual relevant research about road transportation improvements and fuzzy logic applications.

The third section of this paper presents a description of the car-following model, considered the most known microscopic road traffic model. After highlighting some general features of the calibration process, the paper introduces the application of a continuous-time Kalman filter for the online calibration of the car-following model.

Section 4 describes the FIS using the Kalman-filtered values as inputs for the relative velocity between FV (follower vehicle) and LV (leader vehicle) and simulates the inter-vehicle spacing estimation error relative to the measurement error. This FIS aims to provide the specific offset for inter-vehicle spacing until the system is calibrated.

The last three sections consist of a case study presentation, results analysis, discussions and conclusions related to our research.

2. Literature Overview

Recent trends in improvements in urban road traffic provide a direction of research to common platforms creation for the intelligent management of sensor networks, having as the main objective the assurance of appropriate real-time decisions in different environmental conditions [2]. To improve the sensor data collection and transmission, new technologies can be used such as 5G and VLC (visible-light communication) [3]. The energy consumption is one of the major current problems in road monitoring systems, and smart solutions for energy savings are welcomed [4].

Currently, ITS development involves the usage of Bluetooth sensors in the TMS (traffic monitoring systems). This approach applied in several cities can enhance their chance to become smart cities and, implicitly, to improve the citizens' quality of life [5]. This sensing technology proved its utility in travel times estimation and traffic signal control. An interesting approach in this regard uses a modifiedmax-pressure algorithm based on travel times collection from Bluetooth sensors. The results from an experiment consisting of a real implementation of this approach in a signalized intersection from Jerusalem demonstrated its validity and brought many advantages. The benefits of this technology compared to video-based systems arose from the easier installation and maintenance processes and the availability of measurements of turning ratios [6]. On the same direction of Bluetooth technology usage, researchers propose in [7] a real-time road traffic monitoring system based on the purchased data from private companies. To improve the quality of these data, an augmentation framework that uses Bayesian inference is proposed and uses the ability of Bluetooth sensing technology to present the bimodal traffic flow pattern that ensures compliance with the speed limits. In this regard, an experiment on a principal arterial corridor in Orlando, USA achieved good results.

During the past few years, fuzzy logic has been finding a fast-growing number of applications in different fields [8]. The road traffic forecasting represents an interesting direction in ITS research. The usage of atmospheric and air pollution data for this reason in approaches based on recurrent neural networks [9] shows the high level of intelligent systems implications in solving road traffic problems. Another possible solution of traffic flow performance forecasting is to use fuzzy neural networks [10]. This approach of forecasting methods expects a wide mixture of factors like information from drivers that have an important impact on traffic concentration and capacity of the urban road network.

The installation of the sensors appointed for traffic data collection is crucial for traffic lights management and traffic jam detection, and can bring improved travel times for the vehicles crossing the road network [11]. To achieve this goal of short travel times, a solution is to set the green and red intervals by using a fuzzy controller designed to dynamically update the maximum values associated with the green intervals [12]. Another approach is to use an adaptive traffic light system based on the fuzzy Q-learning control method using the intelligent agents concept [13].

A major problem related to the traffic flow is the determination and the control of the dynamic distance between two vehicles. A good approach in this regard is to use the fuzzy interpolation for distance-gap computation, considering that planning for collision has a pre-set constant time [14].

The use of automate driving can minimize this risk of collision, through a method that can determine an adaptive velocity for the vehicles participating in the traffic process.

The driver behavior consists of one of the most challenging things in transportation modeling and represents the main source of uncertainty. Video surveillance systems prove their efficiency in road traffic patterns identification and behaviors classification. In this regard, the usage of a combination between the PAM (Pachinko allocation model) with the SVM (support vector machine) achieved better performances than the LDA (latent Dirichlet allocation) approach [15]. An innovative proposal shows that traffic patterns can be retrieved from social media applications that enable user geo-location data sharing. From this perspective, individual activities durations can be retrieved and correlated with the geo-location data to reconstruct the user mobility trajectories [16].

Besides the microscopic traffic models, there are two more types of models like macroscopic and mesoscopic. Macroscopic traffic models focus on the complete road flow, involving vehicle distributions and traffic density, and mesoscopic models consider the vehicles as vehicle groups in the modeling process. Regarding the calibration of the microscopic parameters, some researchers consider that this can be avoided by adopting the macroscopic modeling approach. In this case, the existence of an MFD (macroscopic fundamental diagram) can easily estimate the space-mean flows and trip completion rates. The experiment used fixed detectors and floating vehicles as sensors for data collection [17]. Nevertheless, a disadvantage of this macroscopic evaluation is a lack of inter-vehicle behavior evaluation that can provide individual driving patterns.

3. Microscopic Traffic Models Calibration

Microscopic traffic models [18] give more attention to the details of traffic flow and are vital for traffic analysis, especially in the presence of ITS. Initial model calibration is necessary to identify the parameter values. It requires the activities of all participants in traffic in order to have feedback of the traffic with parameters like vehicle position, accelerations/decelerations, and vehicle speed. Moreover, the interaction between the vehicles involved in the movement process needs special attention [19].

3.1. Car-Following Model—General Description

Car-following is a microscopic traffic model that analyzes how two vehicles interact during movement. One of these vehicles is the LV and the other is the FV. The second one shall adapt its movement parameters based on the vehicle ahead behavior [20]. Further, starting from MIMO (multiple-input multiple-output) systems theory, the paper presents the state-space representation of this road traffic model in continuous-time.

The inputs of the state-space system for the linear continuous model of the car-following are the vehicles' accelerations/decelerations u_1 and u_2, and the standard safety distance between the vehicles S. The state of this system is represented by the vehicles' velocities noted as x_1 and x_3, and the running distances of the vehicles, x_2 and x_4. Based on these, the system output y consisting of the dynamic safety distance between FV and LV will be computed and will be further used to adapt the FV parameters to ensure compliance with the established standard safety distance.

The linear continuous car-following model previously presented can be characterized using the following system equations, as it was stated in [21–23]:

$$\begin{cases} \begin{bmatrix} \dot{x}_1 \\ \dot{x}_2 \\ \dot{x}_3 \\ \dot{x}_4 \end{bmatrix} = \begin{bmatrix} 0 & 0 & 0 & 0 \\ 1 & 0 & 0 & 0 \\ 0 & 0 & 0 & 0 \\ 0 & 0 & 1 & 0 \end{bmatrix} \cdot \begin{bmatrix} x_1 \\ x_2 \\ x_3 \\ x_4 \end{bmatrix} + \begin{bmatrix} 1 & 0 \\ 0 & 0 \\ 0 & 1 \\ 0 & 0 \end{bmatrix} \cdot \begin{bmatrix} u_1 \\ u_2 \end{bmatrix} \\ y = \begin{bmatrix} 0 & -1 & 0 & 1 \end{bmatrix} \cdot \begin{bmatrix} x_1 \\ x_2 \\ x_3 \\ x_4 \end{bmatrix} + S \end{cases} \tag{1}$$

The standard safety distance S computation considers the vehicle average length L:

$$S = L \cdot \left(1 + \frac{x_3}{16.10}\right) \tag{2}$$

To obtain the classic structure of MIMO state-space representation, some notations according to (3) shall be defined:

$$\begin{aligned}
\bar{x}_1 &= x_3 - x_1, \\
\bar{x}_2 &= x_4 - x_2 = s, \\
\bar{y} &= \bar{x}_2
\end{aligned} \tag{3}$$

where s is the dynamic distance between LV and FV.

Rewriting (1) by using the notations from (3), the MIMO state-space representation of the linear continuous car-following model looks like:

$$\begin{cases} \dot{\bar{x}}(t) = A \cdot \bar{x}(t) + B \cdot u(t) \\ \bar{y}(t) = C \cdot \bar{x}(t) + S \end{cases} \tag{4}$$

The vectors and matrices extended forms are: $\bar{x} = \begin{bmatrix} \bar{x}_1 \\ \bar{x}_2 \end{bmatrix}$, $u = \begin{bmatrix} u_1 \\ u_2 \end{bmatrix}$, $A = \begin{bmatrix} 0 & 0 \\ 1 & 0 \end{bmatrix}$, $B = \begin{bmatrix} -1 & 1 \\ 0 & 0 \end{bmatrix}$, and $C = \begin{bmatrix} 0 & 0 \\ 0 & 1 \end{bmatrix}$. Moreover, the system eigenvalues equal zero, the concatenated matrix $\begin{bmatrix} A & B \end{bmatrix}$ is controllable, and the matrix $\begin{bmatrix} A & C \end{bmatrix}$ is observable.

3.2. Calibration Process

The calibration process finds many issues in the case of microscopic traffic modeling. One of the common problems is the measurement of traffic characteristics like velocity, travel times, and the distance between vehicles because of the influence introduced by each vehicle behavior or travel conditions [19]. The calibration data shall consider this uncertainty in driver decisions and need to adapt to it.

Figure 1 illustrates a description of road traffic systems at the microscopic level. The following three main components ensure the existence of a model closer to reality: microscopic traffic data, modeled microscopic traffic system, and model validation.

The first component is responsible for collecting traffic data from the real world. The second component simulates the real system behavior and uses the estimated values for road traffic parameters as inputs to calculate the simulation output data. A comparison between the simulated data and the real microscopic traffic data consists of a validation step. The output of this component is a decision based on the similarity between the real and simulated model and consists of the calibration component input. The calibration step shall establish the offset values to be applied to the model inputs to reduce the difference compared to the real data. Calibration is performed until these offset values become equal to zero.

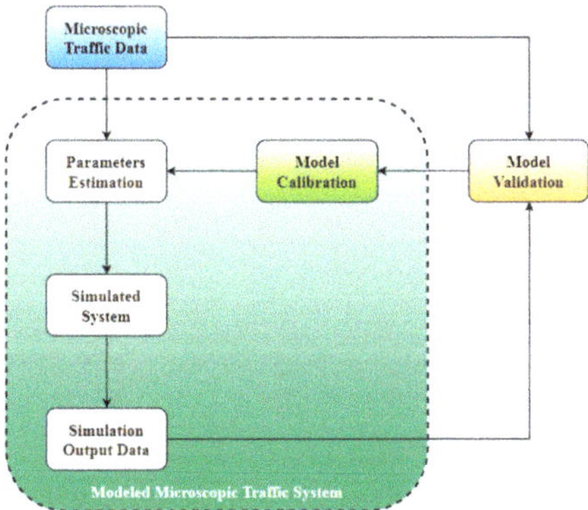

Figure 1. Microscopic traffic system.

3.3. Kalman Filter for Online Calibration

Online calibration is specific for systems that are using real-time data in order to adapt the model, by changing the simulated parameters to be close to the real system. This category of calibration includes the Kalman filter method [24–26].

Kalman filters are popular in the calibration of state-space models characterizing a system in discrete-time, but also applies to continuous-time systems [27], as in our case. For the microscopic road traffic, this method proved its efficiency even if the traffic conditions were nonstationary or stationary. This approach permits a real-time continuous update of studied parameters by updating them through the specific offsets. Further, the application of this method will be presented for car-following model calibration [26].

The state equations for the calibration system of the car-following model according to the Kalman filtering concept are shown in (5), where $\gamma_i(t)$, $i = \{1, 3, s\}$ are the parameters that need calibration.

$$\begin{cases} \dot{x}_1(t) = x_1(t) + \gamma_1(t) \\ \dot{x}_3(t) = x_3(t) + \gamma_3(t) \\ \dot{s}(t) = s(t) + (x_1(t) - x_3(t)) \cdot T + \gamma_s(t) \end{cases} \tag{5}$$

Considering that $\zeta_i(t)$, $i = \{1, 3, s\}$ are the measurement errors for the real traffic parameters, the output equations of the calibration system are computed by using relation (6).

$$\begin{cases} x_1^{obs}(t) = x_1(t) + \zeta_1(t) \\ x_3^{obs}(t) = x_3(t) + \zeta_3(t) \\ s^{obs}(t) = s(t) + \zeta_s(t) \end{cases} \tag{6}$$

Based on previous assumptions, the Kalman filter state-space representation can be obtained as follows:

$$\begin{cases} \dot{x}(t) = A_k \cdot x(t) + B_k \cdot u(t) + D_k \cdot \gamma(t) \\ y(t) = C_k \cdot x(t) + \zeta(t) \end{cases} \tag{7}$$

where the matrices specific for the continuous Kalman filter have the following values: $A_k = \begin{bmatrix} 1 & 0 & 0 \\ 0 & 1 & 0 \\ T & -T & 1 \end{bmatrix}$, $B_k = [0]$, $C_k = \begin{bmatrix} 1 & 0 & 0 \\ 0 & 1 & 0 \\ 0 & 0 & 1 \end{bmatrix}$, and $D_k = \begin{bmatrix} 1 & 0 & 0 \\ 0 & 1 & 0 \\ 0 & 0 & 1 \end{bmatrix}$.

The continuous-time Kalman filter estimates the values according to (8):

$$\hat{x}(t) = \hat{x}_{pr}(t) + K_k \cdot \left(y(t) - C_k \cdot \hat{x}_{pr}(t) \right) \tag{8}$$

where $\hat{x}_{pr}(t)$ is a prediction obtained based on the previous knowledge using the following relation:

$$\dot{\hat{x}}_{pr}(t) = A_{k_{t-1}} \cdot \hat{x}_{pr}(t) + B_{k_{t-1}} \cdot \hat{x}_{pr}(t) \tag{9}$$

where $A_{k_{t-1}}$ and $A_{k_{t-1}}$ represent the values of A_k and B_k at time $t-1$, respectively.

The gain matrix K_k incorporates the compromise to adjust the estimated parameters by using real measured data but, at the same time, to avoid the propagation of measurement errors:

$$K_k = P_{pr_k} \cdot C_k^T \cdot \left[C_k \cdot P_{pr_k} \cdot C_k^T + R_k \right]^{-1} \tag{10}$$

The matrix P_{pr_k} is the covariance matrix that contains the estimation error and will be updated at each step by applying the following relation:

$$P_{pr_{k+1}} = A_k \cdot P_{F_k} \cdot A_k^T + D_k \cdot Q_k \cdot D_k^T \tag{11}$$

where $P_{pr_{k+1}}$ represents the value of P_{pr_k} at time $t+1$, and P_{F_k} is defined as:

$$P_{F_k} = [I - K_k \cdot C_k] \cdot P_{pr_k} \tag{12}$$

The Kalman filter covariance matrix of errors R_k has the definition from (13).

$$R_k = \begin{bmatrix} \sigma_{x_1(t),\, x_1(t)} & 0 & \sigma_{x_1(t),\, s(t)} \\ 0 & \sigma_{x_3(t),\, x_3(t)} & \sigma_{x_3(t),\, s(t)} \\ \sigma_{s(t),\, x_1(t)} & \sigma_{s(t),\, x_3(t)} & \sigma_{s(t),\, s(t)} \end{bmatrix} \tag{13}$$

with the propagation of variances and covariances expressed as:

$$\sigma_{x_n s} = \sum_{i,\, j} \frac{\partial x_n}{\partial x_i} \cdot \frac{\partial s}{\partial x_j} \cdot \sigma_{x_{ij}} \tag{14}$$

for the functions of interests defined as: $x_n = (x_1, x_2, \cdots, x_i, \cdots)$ and $s - (x_1, x_2, \cdots, x_j, \cdots)$, with the explanation that $\sigma_{x_n s}$ is the covariance of x_n and s, and the term $\sigma_{x_{ij}}$ represents the covariance of x_i and x_j.

4. Fuzzy Calibration of Microscopic Traffic Models

The FIS usually solves problems related to the nonlinearity of systems or in the cases of systems that have time delays, but is also suitable for continuous-time systems [28].

Microscopic traffic modeling is a complex problem because of the uncertainty in drivers' decisions for lane changes or for the acceleration/deceleration behavior. The driver behavior influences all corresponding dynamic traffic parameters, introducing, in some cases, a swap in role between LV and FV. These reasons make the microscopic traffic modeling problem suitable for implementation with FIS, especially for the calibration process where all mentioned uncertainties need filtering in order to establish the best offset values. The application of these values ensures the dynamical adaptation of the modeled system to the received real traffic conditions.

In the following, we will show the particularities of the FIS, especially on the use of Takagi–Sugeno. A new calibration method will be issued based on this theoretical background in combination with the Kalman filtering concept previously presented. Compared to the simple use of Kalman filters, this hybrid approach tries to cover the learning of patterns in an offset setting.

4.1. Takagi–Sugeno FIS

Probably the most known model to implement an FIS is to use the Takagi–Sugeno approach [29–33]. Similar to other fuzzy methods, this model description uses the fuzzy specific IF-THEN rules. These input–output associations of the nonlinear modeled system characterize the dynamics of each fuzzy rule by creating a linear system model.

A general Takagi–Sugeno FIS for continuous-time models can be described by the following fuzzy rules [29,30]:

$$\text{IF } z_1(t) \text{ is } F_{i1} \text{ AND } \cdots \text{ AND } z_p(t) \text{ is } F_{ip} \text{ THEN } \begin{cases} \dot{x}(t) = (A_i + \Delta A_i) \cdot x(t) + (B_i + \Delta B_i) \cdot u(t) \\ y(t) = (C_i + \Delta C_i) \cdot x(t) \end{cases} \tag{15}$$

where the notations have the following meaning:

- $z_j(t)$, $j = \{1, 2, \cdots, p\}$ are the premise variables;
- $F_{ij}(t)$, $i = \{1, 2, \cdots, r\}$, and $j = \{1, 2, \cdots, p\}$ represent the fuzzy sets;
- r is the number of defined fuzzy rules;
- $x(t) \in \mathbb{R}^n$ is the state vector;
- $u(t) \in \mathbb{R}^m$ is the input vector;
- $A_i \in \mathbb{R}^{n \times n}$ is the state matrix;
- $B_i \in \mathbb{R}^{n \times m}$ is the input matrix;
- $C_i \in \mathbb{R}^{q \times n}$ is the output matrix where q is the number of output parameters;
- ΔA_i, ΔB_i, and ΔC_i are the matrices that incorporate the uncertainties.

The previous assumptions lead to the inferred model expression:

$$\begin{cases} \dot{x}(t) = \sum_{i=1}^{r} h_i(z(t)) \cdot ((A_i + \Delta A_i) \cdot x(t) + (B_i + \Delta B_i) \cdot u(t)) \\ y(t) = \sum_{i=1}^{r} h_i(z(t)) \cdot (C_i + \Delta C_i) \cdot x(t) \end{cases} \tag{16}$$

where $h_i(z(t))$ is the normalized grade of membership for each rule and is compliant with (17).

$$\begin{cases} \sum_{i=1}^{r} h_i(z(t)) = 1 \\ 0 < h_i(z(t)) < 1 \end{cases} \tag{17}$$

These equations represent the fundamentals for the implementation of an optimized calibration method for the special case of car-following models.

4.2. Car-Following Online Calibration based on Hybrid Takagi–Sugeno FIS Combined with Kalman Filtering

Figure 2 shows our proposal for the calibration system internal structure. Moreover, its relations with the other external subsystems are also represented. As part of the modeled microscopic traffic system, the calibration model has the feature of providing the necessary data to adapt the internal model parameters based on the evaluation received as input from the system responsible with the model validation. Together with the validation result, two sets of parameters consisting of simulation values and microscopic traffic real data will be sent to the calibration subsystem.

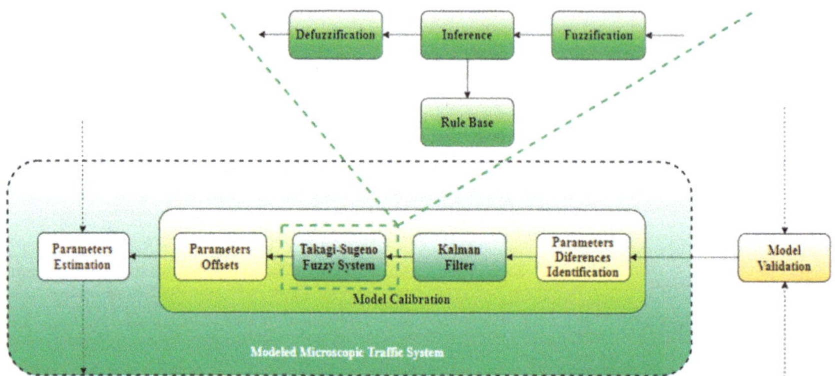

Figure 2. Proposed approach for microscopic traffic model calibration.

The first internal subsystem of a calibration subsystem is intended for identifying the differences between real and simulation data. These will be forwarded to the Kalman filter. The filtered values will be forwarded to the final decision step regarding the offset values consisting of a Takagi–Sugeno FIS. This last subsystem embeds the fuzzy specific components. The fuzzification ensures that the identified differences between the real and simulation parameters are converted to a fuzzy variable. Further, after the fuzzy rules have been defined, the output fuzzy variables will consider them in establishing the connections with the input fuzzy variables through an inference step. Defuzzification will convert the fuzzy variables to the corresponding types of analyzed parameters. These values consisting of offsets that shall be applied to the simulation parameters will also be saved by a subsystem that will build a knowledge base for the parameters offset.

After the model parameters have been updated with the offsets provided by the calibration subsystem, the process will be resumed until the offsets tend to zero. As the values of the simulation are closer to the real ones, the chances of the model to be validated increase. When the offsets are equal to zero, the model is considered as validated.

The inter-vehicle spacing offset for the next run of the simulation can be considered the same as the new estimated simulation error. From this point of view, the calibration module will be responsible for estimating the values that can reduce the difference between simulation values and the values retrieved from real traffic measurements. The new estimation error shall be computed as it is presented below and will be further applied directly by the subsystem that estimates the model parameters based on retrieved real traffic data, depending on corresponding conditioned parameters:

$$\begin{cases} \Delta s(t) = \Delta s_0(t) + \zeta_s(t) - \gamma_s(t), \ \gamma_s(t) \neq \zeta_s(t) \\ \Delta s(t) = \Delta s_0(t) + \gamma_s(t) = \Delta s_0(t) + \zeta_s(t), \ \gamma_s(t) = \zeta_s(t) \end{cases} \tag{18}$$

where $\Delta s_0(t)$ is considered an initial offset that can be defined based on FV and LV behavior for velocity evolution, as can be seen in (19). This correction step shall be applied to ensure that it maintains the standard safety distance s between LV and FV:

$$\Delta s_0(t) = \begin{cases} 0, \ x_1(t) \geq x_3(t) \\ S, \ x_1(t) < x_3(t) \end{cases} \tag{19}$$

The model calibration will be done for the linguistic variables defined in Table 1. The inputs and outputs are defined in a manner that simplifies the fuzzy rules writing process. More than that, the output is defined for each possible situation that can describe the car-following model behavior from the LV and velocity evolution perspectives. Additional information that is taken into account is related to the measurement and simulation errors of dynamic safety distance between LV and FV (the inter-vehicle spacing). The output consisting of the necessary offsets that shall be applied to the simulation values for LV and FV vehicles will be defined by equations that use the Kalman filter-specific data. These offsets Δs will be computed according to (18).

Table 1. Linguistic variables for Takagi–Sugeno fuzzy inference system (FIS) based on Kalman-filtered values.

Parameter Role	Variable Name	Variable	Definition
Input 1	LV velocity relative to FV velocity (REL_SPEED)	LOW EQUAL HIGH	$x_1(t) < x_3(t)$ $x_1(t) = x_3(t)$ $x_1(t) > x_3(t)$
Input 2	Simulated inter-vehicle spacing estimation error relative to measurement error (REL_ERR)	LOW EQUAL HIGH	$\gamma_s(t) < \zeta_s(t)$ $\gamma_s(t) = \zeta_s(t)$ $\gamma_s(t) > \zeta_s(t)$
Output	Inter-vehicle spacing offset Δs (OFFSET)	REDUCE MAINTAIN INCREASE	Equation (18)

Considering the introduced linguistic variables and the possible offset values defined by (18), the fuzzy rules are according to (20).

$$
\begin{aligned}
&\text{IF REL\_SPEED} = \text{LOW} && \text{AND REL\_ERR} = \text{LOW} && \text{THEN OFFSET} && = \text{INCREASE}\\
&\text{IF REL\_SPEED} = \text{LOW} && \text{AND REL\_ERR} = \text{EQUAL} && \text{THEN OFFSET} && = \text{MAINTAIN}\\
&\text{IF REL\_SPEED} = \text{LOW} && \text{AND REL\_ERR} = \text{HIGH} && \text{THEN OFFSET} && = \text{REDUCE}\\
&\text{IF REL\_SPEED} = \text{EQUAL} && \text{AND REL\_ERR} = \text{LOW} && \text{THEN OFFSET} && = \text{INCREASE}\\
&\text{IF REL\_SPEED} = \text{EQUAL} && \text{AND REL\_ERR} = \text{EQUAL} && \text{THEN OFFSET} && = \text{MAINTAIN}\\
&\text{IF REL\_SPEED} = \text{EQUAL} && \text{AND REL\_ERR} = \text{HIGH} && \text{THEN OFFSET} && = \text{REDUCE}\\
&\text{IF REL\_SPEED} = \text{HIGH} && \text{AND REL\_ERR} = \text{LOW} && \text{THEN OFFSET} && = \text{INCREASE}\\
&\text{IF REL\_SPEED} = \text{HIGH} && \text{AND REL\_ERR} = \text{EQUAL} && \text{THEN OFFSET} && = \text{MAINTAIN}\\
&\text{IF REL\_SPEED} = \text{HIGH} && \text{AND REL\_ERR} = \text{HIGH} && \text{THEN OFFSET} && = \text{REDUCE}
\end{aligned}
\tag{20}
$$

The fuzzy matrix F_k that considers the fuzzy rules and (19) is shown in (21).

$$
F_k =
\begin{bmatrix}
S + \zeta_s(t) - \gamma_s(t) & S + \gamma_s(t) & S + \zeta_s(t) - \gamma_s(t)\\
\zeta_s(t) - \gamma_s(t) & \gamma_s(t) & \zeta_s(t) - \gamma_s(t)\\
\zeta_s(t) - \gamma_s(t) & \gamma_s(t) & \zeta_s(t) - \gamma_s(t)
\end{bmatrix}
\tag{21}
$$

5. Simulation and Results

This section describes the model implementation using Simulink from MATLAB R2020a. A case study for a real crossroad from Timișoara (Romania) will validate the proposed approach for car-following model calibration.

5.1. Simulation Model

Figure 3 shows the Simulink (MATLAB R2020a) implementation model. The implementation is a simplified car-following model that studies the impact of velocities in the FV strategy to adapt its running distance based on the FV moving behavior. In this case, the acceleration was not taken into account.

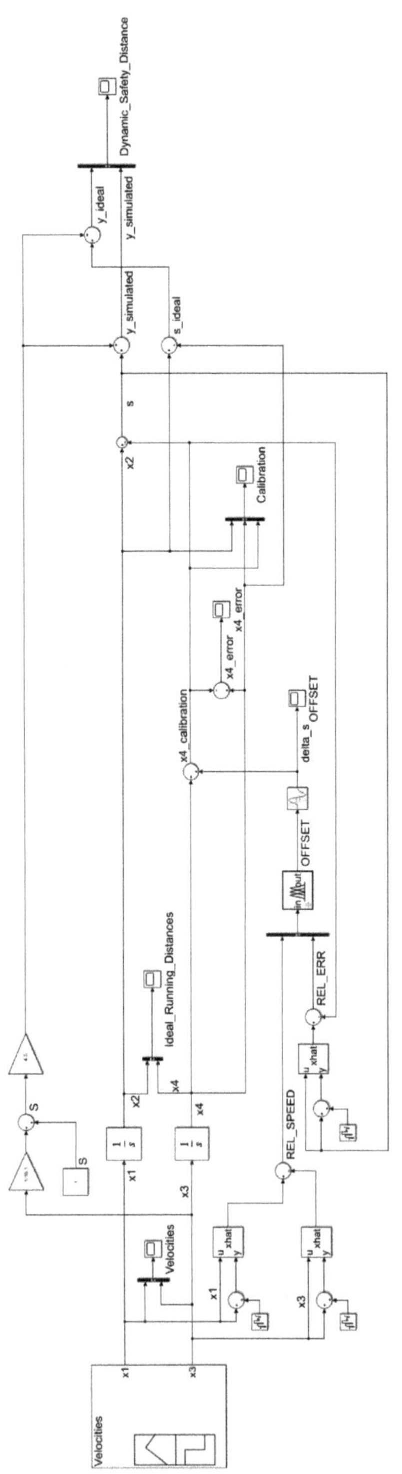

Figure 3. Simulation model—implementation using Simulink (MATLAB R2020a).

The computation of the standard safety distance S is done continuously considering Equation (2). This value is needed to ensure collision avoidance in the case of the increase in FV velocity. For this computation, an average vehicle length L $=$ 4.50 m was used. The output of the modeled system also considers S in the dynamic safety distance profile.

The inputs of the model and also the internal states can be affected by noise in the case of model implementation. To simulate this behavior, a Band-Limited White Noise Simulink block was added to the simulation for the velocities and inter-vehicle spacing to see their impact on the running distances for the LV and FV, and the dynamic safety distance y. The applied noise consists of normally distributed random numbers with the following characteristics that are the default values for the mentioned Simulink block:

- noise power: 0.10;
- sample time: 0.10;
- seed: 23341.

An adaptive system based on Kalman filtering and Takagi–Sugeno FIS will ensure the removal of the introduced noises by the simulated model. The Fuzzy Controller Simulink block uses the Kalman-filtered values as input for the relative speed between the FV and LV and also the simulated inter-vehicle spacing estimation error relative to the measurement error. The implementation details of the Takagi–Sugeno FIS are available in the second part of the current section.

Scope Simulink blocks were added in some points of interest for the following reasons:

- to monitor the input values;
- to obtain the running distances profile for the real behavior of the LV and FV;
- to capture the influences in the simulated running distances introduced by the simulated system;
- to correlate the calculated offset values for inter-vehicle spacing with the calibration process evolution;
- to create an overview of the calibration process impact on the system output.

5.1.1. Input Data

To prove the proposed model validity, the simulation uses, as input data, the data provided by Timișoara City Hall-General Directorate of Roads, Bridges, Parking and Utility Networks-Traffic Monitoring Office, Timișoara, Romania. Figure 4 shows, marked in red, the chosen piece of road between two crossroads: Liviu Rebreanu-Calea Șagului and Liviu Rebreanu-Gheorghe Ranetti. These crossroads have a high daily traffic flow and consists of a good input for our study. The data were collected by using inductive loop sensors that were placed on the studied road network to monitor the vehicle numbers and the velocities.

Figure 4. Real mapping of the studied piece of road (Source: *Open Street Map* view).

5.1.2. Takagi–Sugeno Model Implementation Details

The implementation Takagi–Sugeno FIS using the Kalman-filtered values as inputs, according to the assumptions from the previous section, uses *Fuzzy Logic Toolbox* from MATLAB R2020a. Figure 5 shows the implemented calibration step as it looks in the chosen simulation tool.

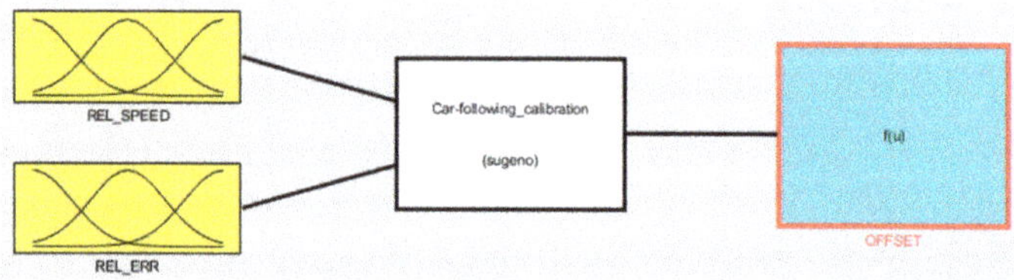

Figure 5. Takagi–Sugeno FIS calibration model implementation using Kalman-filtered values as input.

The established fuzzy rules presented in (20) were implemented for the car-following model calibration using the rule editor for Takagi–Sugeno FIS provided by the simulation tool, as it can be seen in Figure 6.

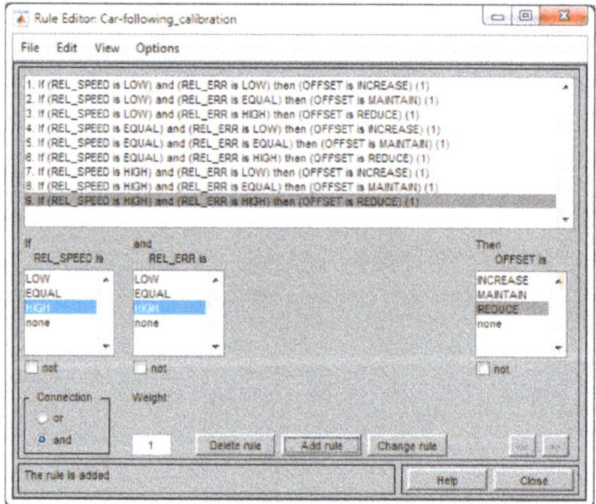

Figure 6. Fuzzy rules implementation using rule editor.

A manner to analyze the simulation results specific to MATLAB R2020a is the usage of rule viewer. Each input of the calibration system and the output are shown in Figure 7. The output values of the car-following calibration system will be modified interactively based on chosen inputs. Membership functions were defined based on historical traffic data evolution.

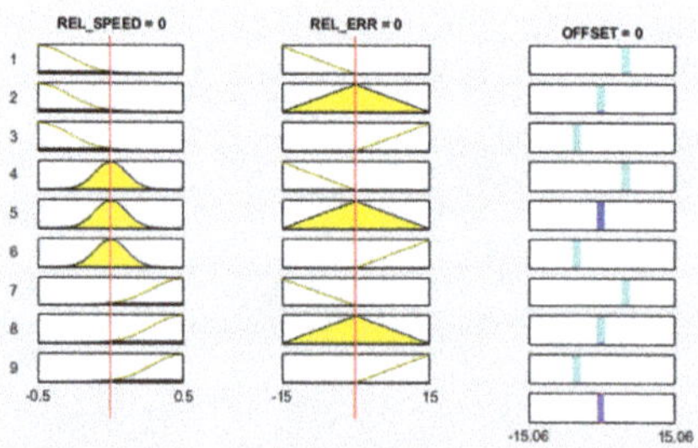

Figure 7. Rule viewer for Takagi–Sugeno FIS car-following calibration system.

Another option to analyze the results of fuzzy rules evaluation is through the surface viewer. Figure 8 illustrates the system performance for the relationship between the linguistic variables used as inputs or as output for the modeled microscopic traffic calibration system.

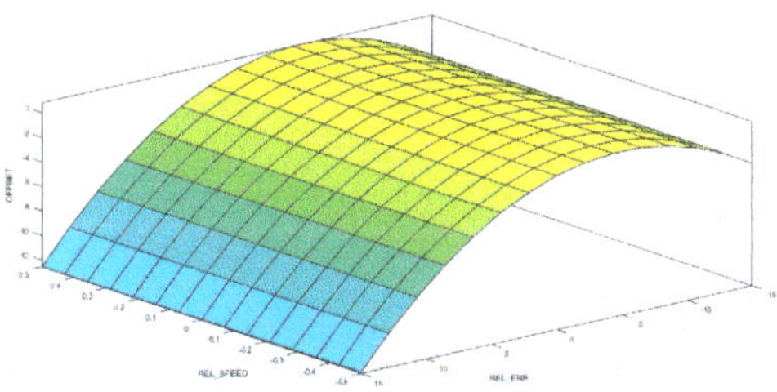

Figure 8. Surface viewer for Takagi–Sugeno FIS car-following calibration system.

5.2. Simulation Results

Figure 9 shows the input data for the proposed traffic calibration method. The evolution in time of the LV and FV velocities for the chosen piece of road between two intersections can be observed.

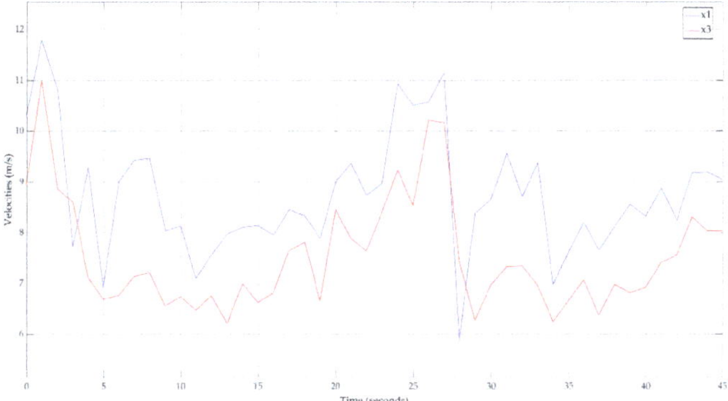

Figure 9. Input data—velocities for leader vehicle (LV) (x_1) and follower vehicle (FV) (x_3).

Based on the input values, the ideal evolution was computed for the LV and FV running distances. Figure 10 shows this result after the application of the car-following approach for the real traffic data.

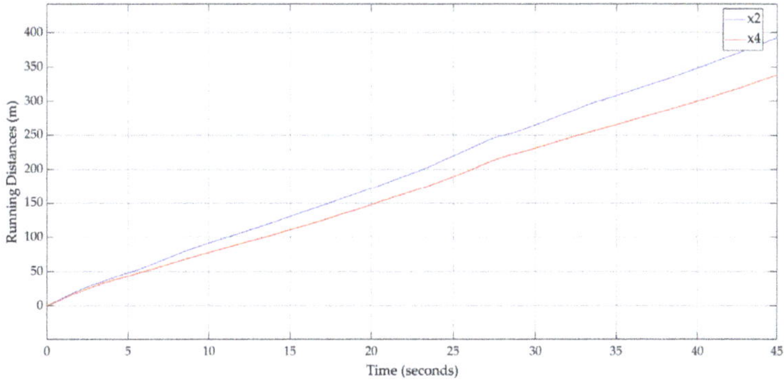

Figure 10. Running distances for LV (x_2) and FV (x_4)—ideal evolution.

The result of the proposed hybrid approach for calibration of the car-following model in continuous-time is available in Figure 11. The modeled running distance of the FV that is affected by noise succeeds at reproducing the real behavior of the FV.

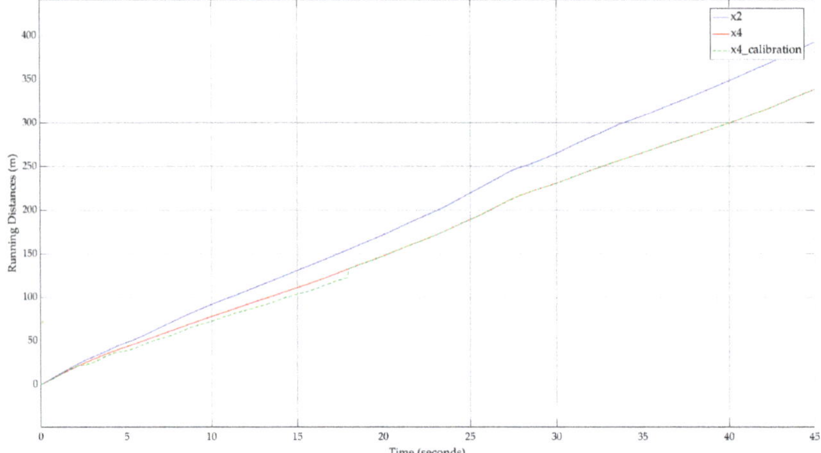

Figure 11. Running distances for LV (x_2) and FV (x_4)—calibration result for FV (x_4).

A better way to see the utility of the proposed approach is to analyze the calculated offset values. Figure 12 illustrates the evolution of the calculated offset values for the inter-vehicle spacing applied to the FV. Before the system joins the calibrated state, both positive and negative offset values are applied. After approximately t = 18 s, the system succeeded at learning different patterns of velocities and simulated inter-vehicle spacing error patterns and could reproduce the real behavior. After that time, the system was calibrated and could reproduce the real behavior.

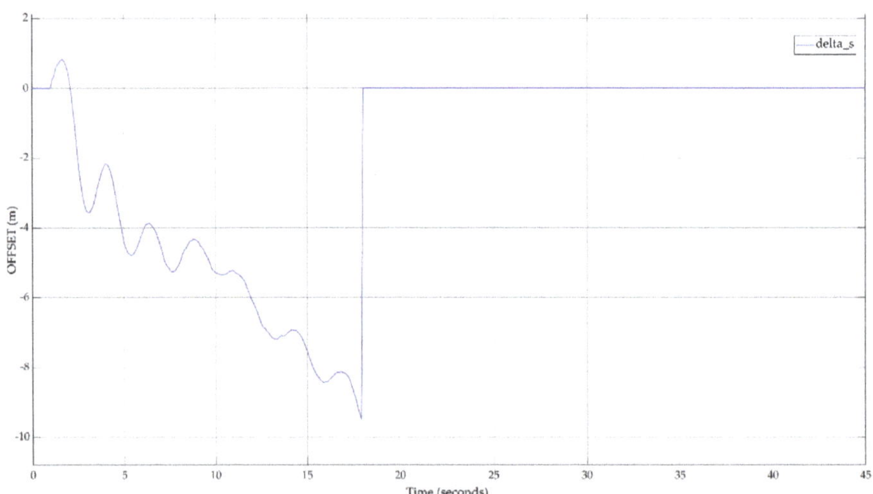

Figure 12. Inter-vehicle spacing offset applied to FV (x_4).

The calibration result can also be observed by analyzing the dynamic safety distance that includes the standard safety distance. Figure 13 depicts the overview of the calibration result for the system output. The safety level is ensured by the direct application of safety length on the output, and safety is guaranteed by the car-following model even in cases of vehicles with different car lengths.

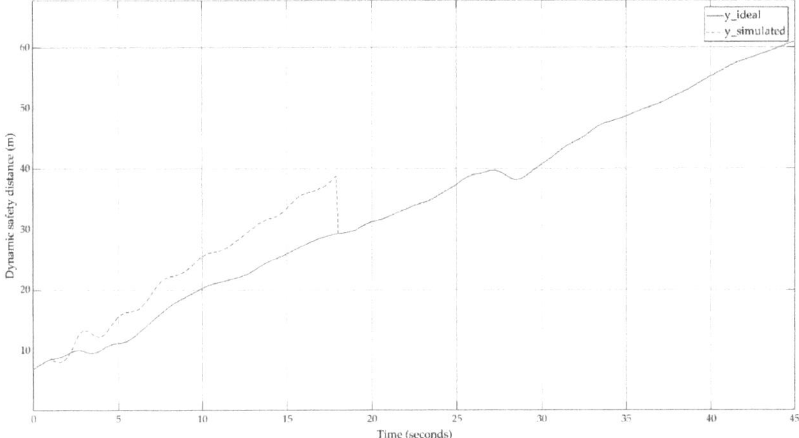

Figure 13. Dynamic safety distance—calibration overview.

6. Discussions

To show the motivation of using the hybrid proposal of Kalman filtering with Takagi–Sugeno FIS instead of a solution based on Kalman filtering only, a simulation is needed for this comparison. Figure 14 depicts the implementation using Simulink (MATLAB R2020a) for the mentioned reason. The simulation aims to analyze the behavior of FV for both cases of calibration methods. In addition to the simulation blocks from Figure 3, an extension introduces the computation of the FV running distance based on the Kalman filtering-only approach. In this case, the model uses only the filtered values to control the FV movement strategy. A big disadvantage, in this case, is the neglect of the FV relative velocity that can result in wrong offset values that will negatively influence the movement strategy. The calibration system shall ensure the capability for time-varying offset values application.

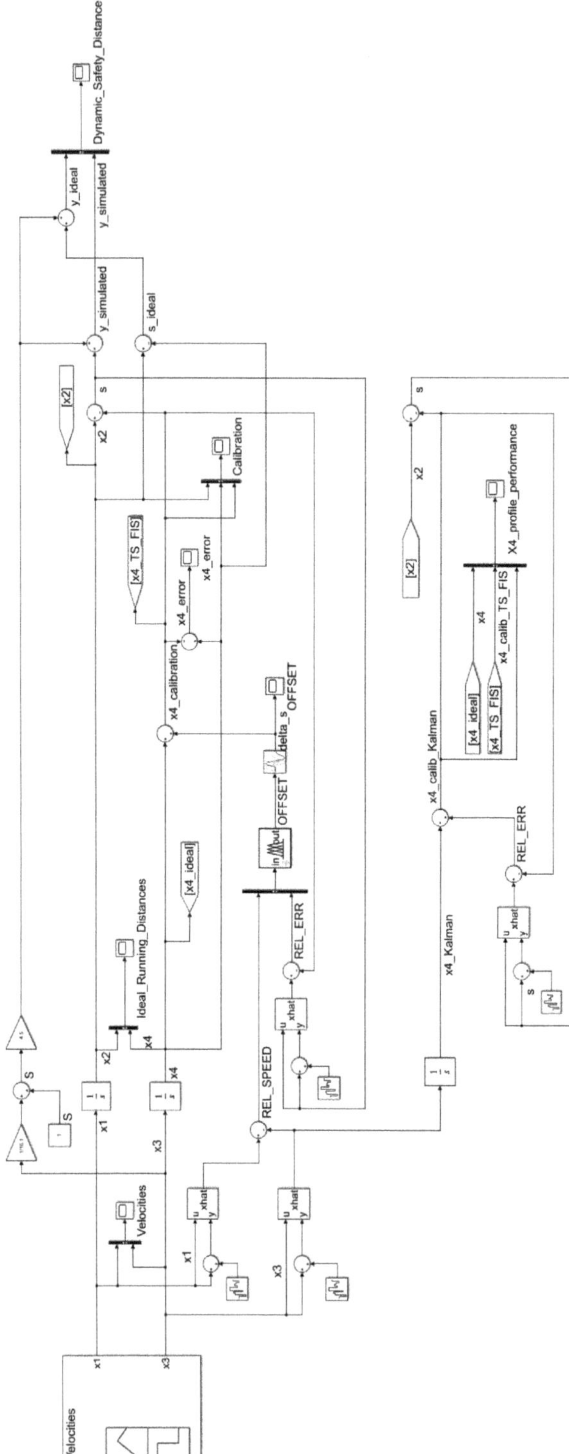

Figure 14. Simulation model to compare the Kalman filtering-only approach with the hybrid Kalman filtering and Takagi–Sugeno FIS approach-implementation using Simulink (MATLAB R2020a).

The simulation results for this comparison (Figure 15) illustrate that both approaches succeeded in filtering the noises and provided a similar-to-real-case trajectory evolution. The Kalman filtering-only approach cannot reproduce the real behavior through its neglect of inter-vehicle interaction from a relative velocity perspective. The hybrid approach takes advantage of this interaction between FV and LV and succeeds at identifying the time-varying appropriate offset value that reproduces the real behavior. This advantage can be visually observed from trajectory evolution where, after t = 18 s, the system is calibrated using the hybrid approach. In the same time, the Kalman filter-only approach introduces a uniform increase in computation error that leads to a scaled running distance compared to real traffic conditions. From a computational complexity perspective, both approaches fit to real-time processing. The Takagi–Sugeno FIS does not introduce computational delays that can lead to a major increase in the real-time data processing timings.

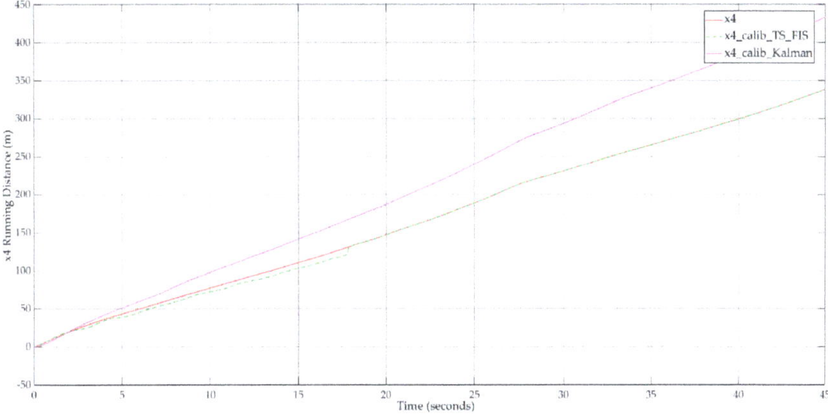

Figure 15. Running distance for FV (x_4—comparison of calibration result for the Kalman filtering-only approach and the hybrid Kalman filtering and Takagi–Sugeno FIS approach.

Another problem to address is the variety of vehicles with different lengths on the road network. The various vehicle lengths do not have a direct influence on the proposed method. The method intends to provide a calibration strategy based on the relative velocity and internal computed inter-vehicle spacings (s according to Equation (3)). The safety road assurance is done by a final step of adding a standard safety distance that depends on the LV length. The extension of this model for a chain of moving vehicles is already included even if we talk about pairs of vehicles where LV and FV are continuously changing. The data are retrieved from inductive loops sensors placed on the road network. The first vehicle that passes over the inductive loop is the LV and the second one is the FV. If we talk about a chain of vehicles that are moving on the road network, the FV from the first pair becomes the LV for the third vehicle and this behavior continues until the end of the chain of moving vehicles.

7. Conclusions

The purpose of this research was to propose a new approach of traffic models calibration at the microscopic modeling level. Our proposal consists of a combined solution that uses Kalman filtering and Takagi–Sugeno FIS for the online calibration of the car-following model.

A review of relevant works on ITS and fuzzy systems was needed to show the current trends in these research fields. Moreover, this created the necessary path to the introduction of our proposal.

A general presentation of the microscopic traffic modeling and calibration process, with special attention to the particular case of the car-following model, precedes our proposal. From an FIS perspective, our approach started with an introduction of the Takagi–Sugeno FIS followed with a mapping of the issues related to car-following models.

The proposed method was validated by the simulation results obtained from the Simulink (MATLAB R2020a) implementation of the continuous-time car-following model together with the system responsible for the calibration process. The input data of this study consisted of real road traffic data from Timișoara, Romania. Takagi–Sugeno FIS proved again its utility in adaptive systems through the optimization of the parameter offset-establishing process.

A comparison between a Kalman filtering-only approach and hybrid Kalman filtering with the Takagi–Sugeno FIS was conducted in Simulink (MATLAB R2020a). The comparison results show that the hybrid approach could provide a closer model to the real model. The big advantage of the hybrid approach was that it can provide the time-varying offset based on real-time road traffic parameters. Moreover, this proposal implies the specific interaction between FV and LV in the offset computation according to the microscopic traffic modeling theory.

Further works can extend this approach at the mesoscopic traffic modeling level where the conditions for velocities evaluation can be assigned to vehicle groups instead of individual vehicles. A big challenge in that direction will be the safety distance assurance inside a group of vehicles because mesoscopic modeling does not offer enough granularity compared to microscopic traffic models.

Author Contributions: Conceptualization, M.-D.P.; software, M.-D.P.; validation, O.P. and G.P.; investigation, M.-D.P., O.P. and T.-M.D.; writing—original draft preparation, M.-D.P. and T.-M.D.; writing—review and editing, O.P. and G.P.; supervision, O.P. and G.P. All authors have read and agreed to the published version of the manuscript.

Funding: This research received no external funding.

Acknowledgments: The data used for simulations were received from Timișoara City Hall-General Directorate of Roads, Bridges, Parking and Utility Networks—Traffic Monitoring Office, Timișoara, Romania (Romanian official institution name: Primăria Municipiului Timișoara—Direcția Generală Drumuri, Poduri, Parcaje și Rețele Utilitare–Birou Monitorizare Trafic, Timișoara, Romania) based on the approved request RE2019-002611/18.12.2019. The support is gratefully acknowledged.

Conflicts of Interest: The authors declare no conflict of interest.

References

1. Luo, Z.; Zhang, Y.; Li, L.; He, B.; Li, C.; Zhu, H.; Wang, W.; Ying, S.; Xi, Y. A Hybrid Method for Predicting Traffic Congestion during Peak Hours in the Subway System of Shenzhen. *Sensors* **2019**, *20*, 150. [CrossRef] [PubMed]

2. Merenda, M.; Praticò, F.G.; Fedele, R.; Carotenuto, R.; Della Corte, F.G. A Real-Time Decision Platform for the Management of Structures and Infrastructures. *Electronics* **2019**, *8*, 1180. [CrossRef]

3. Marabissi, D.; Mucchi, L.; Caputo, S.; Nizzi, F.; Pecorella, T.; Fantacci, R.; Nawaz, T.; Seminara, M.; Catani, J. Experimental Measurements of a Joint 5G-VLC Communication for Future Vehicular Networks. *J. Sens. Actuator Netw.* **2020**, *9*, 32. [CrossRef]

4. Fedele, R.; Praticò, F.G.; Carotenuto, R.; Della Corte, F.G. Energy savings in transportation: Setting up an innovative SHM method. *Math. Model Eng. Probl.* **2018**, *5*, 323–330. [CrossRef]

5. Chaves-Diéguez, D.; Pellitero-Rivero, A.; García-Coego, D.; González-Castaño, F.; Rodríguez-Hernández, P.; Piñeiro-Gómez, Ó.; Gil-Castiñeira, F.; Costa-Montenegro, E. Providing IoT Services in Smart Cities through Dynamic Augmented Reality Markers. *Sensors* **2015**, *15*, 16083–16104. [CrossRef]

6. Mercader, P.; Uwayid, W.; Haddad, J. Max-pressure traffic controller based on travel times: An experimental analysis. *Transp. Res. Part C Emerg. Technol.* **2020**, *110*, 275–290. [CrossRef]

7. Gong, Y.; Abdel-Aty, M.; Park, J. Evaluation and augmentation of traffic data including Bluetooth detection system on arterials. *J. Intell. Transp. Syst.* **2019**, 1–13. [CrossRef]

8. Zadeh, L.A. Foreword: Fuzzy logic. *J. Intell. Inf. Syst.* **1993**, *2*, 309–310. [CrossRef]

9. Awan, F.M.; Minerva, R.; Crespi, N. Improving Road Traffic Forecasting Using Air Pollution and Atmospheric Data: Experiments Based on LSTM Recurrent Neural Networks. *Sensors* **2020**, *20*, 3749. [CrossRef]

10. Dmitry, S.; Sergey, V. Use of Fuzzy Neural Networks for a Short Term Forecasting of Traffic Flow Performance. In *Creativity in Intelligent Technologies and Data Science*; Kravets, A.G., Groumpos, P.P., Shcherbakov, M., Kultsova, M., Eds.; Springer International Publishing: Cham, Switzerland, 2019; Volume 1083, pp. 392–405, ISBN 9783030297428. [CrossRef]

11. Cruz-Piris, L.; Rivera, D.; Fernandez, S.; Marsa-Maestre, I. Optimized Sensor Network and Multi-Agent Decision Support for Smart Traffic Light Management. *Sensors* **2018**, *18*, 435. [CrossRef]

12. Shirvani Shiri, M.J.; Maleki, H.R. Maximum Green Time Settings for Traffic-Actuated Signal Control at Isolated Intersections Using Fuzzy Logic. *Int. J. Fuzzy Syst.* **2017**, *19*, 247–256. [CrossRef]

13. Daeichian, A.; Haghani, A. Fuzzy Q-Learning-Based Multi-agent System for Intelligent Traffic Control by a Game Theory Approach. *Arab. J. Sci. Eng.* **2018**, *43*, 3241–3247. [CrossRef]

14. Balas, M.M.; Balas, V.E.; Duplaix, J. Optimizing the distance-gap between cars by constant time to collision planning. In Proceedings of the 2007 IEEE International Symposium on Industrial Electronics, Vigo, Spain, 4–7 June 2007; pp. 304–309. [CrossRef]

15. Huynh-The, T.; Banos, O.; Le, B.-V.; Bui, D.-M.; Yoon, Y.; Lee, S. Traffic Behavior Recognition Using the Pachinko Allocation Model. *Sensors* **2015**, *15*, 16040–16059. [CrossRef] [PubMed]

16. Hasan, S.; Ukkusuri, S.V. Urban activity pattern classification using topic models from online geo-location data. *Transp. Res. Part C Emerg. Technol.* **2014**, *44*, 363–381. [CrossRef]

17. Geroliminis, N.; Daganzo, C.F. Existence of urban-scale macroscopic fundamental diagrams: Some experimental findings. *Transp. Res. Part B Methodol.* **2008**, *42*, 759–770. [CrossRef]

18. Ferrara, A.; Sacone, S.; Siri, S. Microscopic and Mesoscopic Traffic Models. In *Freeway Traffic Modelling and Control*; Springer International Publishing: Cham, Switzerland, 2018; pp. 113–143, ISBN 9783319759593. [CrossRef]

19. Barceló, J. (Ed.) Models, traffic models, simulation, and traffic simulation. In *Fundamentals of Traffic Simulation*; International Series in Operations Research & Management Science; Springer: New York, NY, USA, 2010; pp. 1–62, ISBN 9781441961426. [CrossRef]

20. Rothery, R.W. Car following models. In *Traffic Flow Theory: A State-of-the-Art Report*; Gartner, N., Messer, C.J., Rathi, A.K., Eds.; Turner-Fairbank Highway Research Center: McLean, VA, USA, 2001.

21. Khodayari, A.; Kazemi, R.; Ghaffari, A.; Manavizadeh, N. Modeling and intelligent control design of car following behavior in real traffic flow. In Proceedings of the 2010 IEEE Conference on Cybernetics and Intelligent Systems, Singapore, 28–30 June 2010; pp. 261–266. [CrossRef]

22. Pan, D.; Zheng, Y. Optimal control and discrete time-delay model of car following. In Proceedings of the 2008 7th World Congress on Intelligent Control and Automation, Chongqing, China, 25–27 June 2008; pp. 5657–5661. [CrossRef]

23. Pop, M.-D.; Proştean, O.; Proştean, G. Multiple Lane Road Car-Following model using bayesian reasoning for lane change behavior estimation: A smart approach for smart mobility. In Proceedings of the 3rd International Conference on Future Networks and Distributed Systems—ICFNDS '19, Paris, France, 1–2 July 2019; pp. 1–8. [CrossRef]

24. Emami, A.; Sarvi, M.; Asadi Bagloee, S. Using Kalman filter algorithm for short-term traffic flow prediction in a connected vehicle environment. *J. Mod. Transport.* **2019**, *27*, 222–232. [CrossRef]

25. Kalman, R.E.; Bucy, R.S. New Results in Linear Filtering and Prediction Theory. *J. Basic Eng.* **1961**, *83*, 95–108. [CrossRef]

26. Punzo, V.; Formisano, D.J.; Torrieri, V. Nonstationary Kalman Filter for Estimation of Accurate and Consistent Car-Following Data. *Transp. Res. Rec.* **2005**, *1934*, 2–12. [CrossRef]

27. Shi, P.; Boukas, E.-K.; Agarwal, R.K. Kalman filtering for continuous-time uncertain systems with Markovian jumping parameters. *IEEE Trans. Autom. Control* **1999**, *44*, 1592–1597. [CrossRef]

28. Lam, H.K. A review on stability analysis of continuous-time fuzzy-model-based control systems: From membership-function-independent to membership-function-dependent analysis. *Eng. Appl. Artif. Intell.* **2018**, *67*, 390–408. [CrossRef]

29. Abdelkrim, R.; Gassara, H.; Chaabane, M.; El Hajjaji, A. Stability approaches for Takagi-Sugeno systems. In Proceedings of the 2015 IEEE International Conference on Fuzzy Systems (FUZZ-IEEE), Istanbul, Turkey, 2–5 August 2015; pp. 1–6. [CrossRef]

30. Bouyahya, A.; Manai, Y.; Haggege, J. New Lyapunov function for Takagi-Sugeno discrete time uncertain systems. In Proceedings of the 2015 7th International Conference on Modelling, Identification and Control (ICMIC), Sousse, Tunisia, 18–20 December 2015; pp. 1–5. [CrossRef]

31. Hadjili, M.L.; Kara, K. Modelling and control using Takagi-Sugeno fuzzy models. In Proceedings of the 2011 Saudi International Electronics, Communications and Photonics Conference (SIECPC), Riyadh, Saudi Arabia, 24–26 April 2011; pp. 1–6. [CrossRef]

32. Petritoli, E.; Leccese, F.; Cagnetti, M. Takagi-Sugeno Discrete Fuzzy Modeling: An IoT Controlled ABS for UAV. In Proceedings of the 2019 II Workshop on Metrology for Industry 4.0 and IoT (MetroInd4.0 IoT), Naples, Italy, 4–6 June 2019; pp. 191–195. [CrossRef]
33. Takagi, T.; Sugeno, M. Fuzzy identification of systems and its applications to modeling and control. *IEEE Trans. Syst. Man. Cybern.* **1985**, *SMC-15*, 116–132. [CrossRef]

Letter

Computationally Efficient Cooperative Dynamic Range-Only SLAM Based on Sum of Gaussian Filter [†]

Jung-Hee Kim [1] and Doik Kim [2,*]

[1] Department of Electronic Engineering, Hanyang University, Seoul 04763, Korea; jhkim.sp.0901@gmail.com
[2] Center for Intelligent and Interactive Robotics, Korea Institute of Science and Technology, Seoul 02792, Korea
* Correspondence: doikkim@kist.re.kr
† This paper is an extended version of our paper published in Kim, J.-H.; Kim, D. Cooperative Range-only SLAM based on Sum of Gaussian Filter in Dynamic Environments. In Proceedings of the 2019 IEEE/RSJ International Conference on Intelligent Robots and Systems (IROS), Macau, China, 3–8 November 2019.

Received: 14 May 2020; Accepted: 8 June 2020; Published: 10 June 2020

Abstract: A cooperative dynamic range-only simultaneous localization and mapping (CDRO-SLAM) algorithm based on the sum of Gaussian (SoG) filter was recently introduced. The main characteristics of the CDRO-SLAM are (i) the integration of inter-node ranges as well as usual direct robot-node ranges to improve the convergence rate and localization accuracy and (ii) the tracking of any moving nodes under dynamic environments by resetting and updating the SoG variables. In this paper, an efficient implementation of the CDRO-SLAM (eCDRO-SLAM) is proposed to mitigate the high computational burden of the CDRO-SLAM due to the inter-node measurements. Furthermore, a thorough computational analysis is presented, which reveals that the computational efficiency of the eCDRO-SLAM is significantly improved over the CDRO-SLAM. The performance of the proposed eCDRO-SLAM is compared with those of several conventional RO-SLAM algorithms and the results show that the proposed efficient algorithm has a faster convergence rate and a similar map estimation error regardless of the map size. Accordingly, the proposed eCDRO-SLAM can be utilized in various RO-SLAM applications.

Keywords: simultaneous localization and mapping (SLAM); range-only SLAM; sum of Gaussian (SoG) filter; cooperative approach

---

1. Introduction

Simultaneous localization and mapping (SLAM), that is, localizing a robot while building a map of unknown environments at the same time, has been a popular topic. It has been developed in several forms, e.g., visual SLAM [1,2], bearing-only SLAM [3], range-only SLAM (RO-SLAM) [4–9], and their combinations [10,11], etc.

Compared with other types of SLAMs, the RO-SLAM, which uses range sensors only, has some distinguishing features. First, the range measurements are highly ambiguous. The potential location of a node can be anywhere on a ring shape, as shown in Figure 1a. Such ambiguity progressively disappears with the additional range measurements at different locations of a robot, as illustrated in Figure 1b,c. Second, the map itself simply shows the locations of nodes. Therefore, the data association issue, which is one of the main difficulties in other types of SLAMs, does not need to be considered. Due to this simplicity, the RO-SLAM has been applied in a wide range of applications such as submarine autonomous vehicles, search and rescue, etc., [6,12].

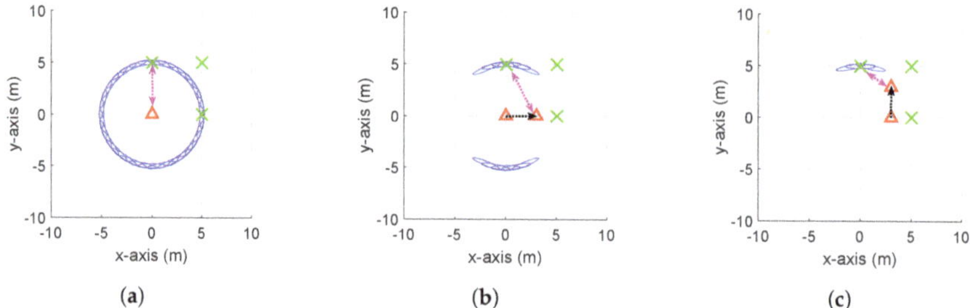

Figure 1. Explanation of the conventional RO-SLAM algorithms. Red triangles and green crosses indicate robots and neighbor nodes, respectively. Black and magenta dotted lines are the movement of the robot and range measurements, respectively. Blue ellipse is a possible location of nodes. The conventional RO-SLAM algorithms involve (**a**) the initial localization when receiving the first measurement, (**b**) movement of a robot to disambiguate the x-axis, and (**c**) movement to disambiguate the y-axis.

The Extended Kalman filter (EKF) has been widely employed as a standard solution to the RO-SLAM problem, along with some beacon initialization methods, e.g., trilateration [4], the probability grid [5], the particle filter [7], and the sum of Gaussian (SoG) filter [5,6]. The trilateration method (or multilateration) is simple, but its performance can be seriously damaged by measurement noise. The probability grid can perform better than the trilateration, but its performance depends on the size and resolution of the grid. The particle filter can represent an arbitrary probability density function (pdf) with a number of particles, such as a non-Gaussian ring shape for the beacon location in the RO-SLAM; however, it has a high computational burden due to the large number of particles. Compared with the particle filter approach, the SoG filter [5,6] can give a similar estimation accuracy in a more computationally efficient way, because it can efficiently cover almost the same area of beacon locations with a small number of Gaussian distributions. In this paper, the multilateration and SoG approaches are utilized to determine the initial position to give a fast and accurate localization result.

To improve the estimation accuracy, several RO-SLAM algorithms employ a cooperative approach, in which there are not only direct measurements between a robot and its neighboring nodes, but also, other inter-node measurements are integrated for localization [8,9]. In Ref. [8], the inter-node measurements were simply incorporated into the standard EKF based framework, and the efficiency and scalability of the large scale sensor network were further improved by using the sparse extended information filter (SEIF) based cooperative algorithm [9,13]. Recently, the SoG filter based cooperative approach, called the cooperative dynamic RO-SLAM (CDRO-SLAM) [14], was introduced by using the advantages of the SoG filter, such as the high estimation accuracy, as mentioned above. In addition, the CDRO-SLAM can cope with moving nodes by appropriately manipulating the SoG filter, unlike other RO-SLAM algorithms which assume that nodes are fixed.

In this paper, the efficient implementation of the CDRO-SLAM (eCDRO-SLAM) is proposed by introducing several efficient implementation techniques to mitigate the high computational burden of the CDRO-SLAM due to the inter-node measurements. Furthermore, its detailed computational complexity analysis is also given, and the results show that the computational efficiency of the eCDRO-SLAM is significantly improved over the CDRO-SLAM, which is shown in the experiment results.

The main properties of the proposed eCDRO-SLAM can be summarized, where the first two properties are obtained by inheriting from the advantages of the CDRO-SLAM:

- Accurate and fast localization results are achieved by incorporating the cooperative approach and the SoG filter.
- Sensor nodes can be applied to dynamic environments where installed nodes are not fixed or moving such as unstable and unstructured disaster sites, moving landmarks, interactive information with human and non-fixed objects to analyze human activity, etc.

- To mitigate the high computational cost induced by the inter-node measurements, several efficient techniques are introduced, which greatly reduces the required computational complexity of the proposed eCDRO-SLAM.

This paper is organized as follows: Section 2 derives the CDRO-SLAM, including a cooperative scheme using inter-node measurements and a tracking scheme of non-fixed nodes. In Section 3, an efficient implementation of the CDRO-SLAM (eCDRO-SLAM) is presented to mitigate the computational burden due to the cooperative scheme. Section 4 verifies the performance of the eCDRO-SLAM with several experimental results, and Section 5 gives the concluding remarks.

2. Cooperative Dynamic RO-SLAM

2.1. Overview

This section summarizes our previous work on the CDRO-SLAM [14], aimed at developing a cooperative RO-SLAM under dynamic environments. The CDRO-SLAM algorithm is composed of two stages: (i) the initialization stage estimates the initial locations of a robot and map by using inter-node measurements, which improves and accelerates the map estimation and (ii) the movement stage further refines the map generated in the initialization stage or tracks nodes in motion.

In more detail, the initialization stage is described in Algorithm 1. (i) Once the initial positions of nodes are guessed with the help of iterative multilateration, it is refined by employing the sum of Gaussian (SoG) filter approach. (ii) The SoG variables are first generated to find the candidate locations of neighbors, and (iii) the SoG variables are updated by exploiting the inter-node measurements to quickly reduce the number of candidates. (iv) The weighted SoG variables are merged into a single Gaussian distribution to estimate the new positions of the nodes. (v) Furthermore, a method to transfer the weight of the SoG variables is applied to resolve the inherent non-convexity in the anchor-free localization. (vi) The above procedures are repeated until the positions of all nodes converge.

In the movement stage, a robot and/or some nodes can be moved. Firstly, if a robot moves, its motion is updated using raw odometry data. Then, after receiving a new distance from the neighbor node, the locations of the corresponding neighbor node and the robot are further refined by using the EKF procedure, like in the usual RO-SLAM algorithms (see [15] for details). Secondly, for the case where some nodes move, their movement can be tracked by properly manipulating the SoG filter, which is illustrated in Algorithm 2 and Figure 2. (i) If j denotes an index of a moving node, the movement of the jth node is first detected by one of motion detection techniques (e.g., the norm of the accelerometer). (ii) When getting a newly modified distance due to the motion of the jth node, the corresponding SoG variables are reset (see Figure 2a). After that, the new position of the jth node is estimated by (iii) updating and (iv) merging the SoG variables (see Figure 2b). Finally, the SoG variables from the jth node to its neighbor nodes are also recalculated by (v) resetting and (vi) updating them to reflect the newly estimated position of the jth node (see Figure 2c).

Differences between the conventional SoG based RO-SLAM algorithms and the proposed CDRO-SLAM are discussed in terms of the weight update scheme and computational complexity. Figures 1 and 3 show the weight update processes of both approaches, respectively. For both algorithms, the weight update starts when the first measurement between a robot and one of its neighbor nodes is received, and then the potential location is assumed to be anywhere within a ring shape whose radius is equal to the measurement (see Figures 1a and 3a). For conventional algorithms, the ambiguity progressively disappears by moving a robot along with x and y axes, i.e., a non-collinear motion (see Figure 1b,c). On the other hand, the proposed algorithm updates the weight by using inter-node measurements without moving a robot. The ambiguity is gradually removed as more measurements from the non-collinear adjacent nodes arrive (see Figure 3b,c).

The computational cost of the CDRO-SLAM is increased due to the inclusion of the SoG variables for inter-node measurements. To reduce this computational burden, several efficient implementation techniques such as the weight symmetry of the SoG distributions are introduced later in Section 3.

Algorithm 1: Initialization stage of the CDRO-SLAM.
(M for the number of nodes, and i, j, m for the indices of nodes)

---

1 (i) Iterative multilateration
2 **for** $i \leftarrow 0$ **to** $M - 1$ **do**
3 | Initialize the location of ith node by using (1)–(4).
4 **end**

5 (ii) Generating SoG
6 **for** $i, j \leftarrow 0$ **to** $M - 1, (j \neq i)$ **do**
7 | Generate the mean $\mathbf{m}_{ij}^k$, covariance $\mathbf{C}_{ij}^k$, and weight w_{ij}^k by using (5)–(7).
8 **end**

9 (iii) Updating SoG
10 Update the mean $\mathbf{m}_{ij}^k$ and covariance $\mathbf{C}_{ij}^k$ using EKF.
11 /* Algorithms 3 and 4 */
12 **for** $i, j, m \leftarrow 0$ **to** $M - 1, (j \neq i, m \neq i, m \neq j)$ **do**
13 | Compute and normalize the likelihood $l_{ij,m}^k$.
14 | Update and normalize the weight w_{ij}^k.
15 **end**

16 (iv) Merging SoG
17 **for** $i \leftarrow 0$ **to** $M - 1$ **do**
18 | Compute the new location of the ith node with (8)–(9).
19 **end**

20 (v) Transferring weight
21 **for** $i, j \leftarrow 0$ **to** $M - 1, (j \neq i)$ **do**
22 | Transfer the weight w_{ij}^k by using (11).
23 **end**

24 (vi) Determining convergence
25 **for** $i \leftarrow 0$ **to** $M - 1$ **do**
26 | Determine the convergence of the ith node.
27 **end**

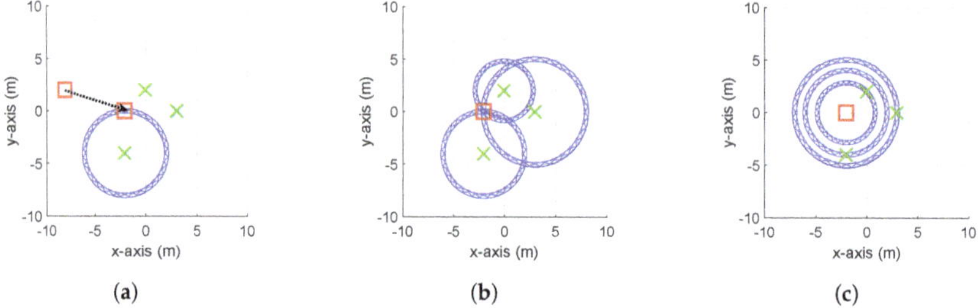

(a)   (b)   (c)

Figure 2. Description of the updated procedures for tracking moving nodes. (**a**) Reset the SoG variables from the neighbor node to the moving node, (**b**) update and merge the SoG variables, and (**c**) reset the SoG variables from the moving node to its neighbor nodes and update the corresponding weights. The red squares and green crosses represent the moving nodes and neighbor nodes, respectively. The black dotted line represents the movement of a node.

Algorithm 2: Moving stage of the cooperative dynamic range-only simultaneous localization and mapping (CDRO-SLAM).
(M represents the number of nodes, and i, j, m represent the indices of nodes)

---

1 (i) Detecting movement
2 Detect whether the jth node is moving or not.

3 (ii) Resetting SoG (Figure 2a)
4 Reset the mean $\mathbf{m}_{ij}^k$, covariance $\mathbf{C}_{ij}^k$, and weight w_{ij}^k using Equations (5)–(7).

5 (iii) Updating the SoG (Figure 2b)
6 Update the mean $\mathbf{m}_{ij}^k$ and covariance $\mathbf{C}_{ij}^k$ using EKF.
7 /* Algorithms 3 and 4 */
8 **for** $i, m \leftarrow 0$ **to** $M - 1, (i \neq j, m \neq i, m \neq j)$ **do**
9   Compute and normalize the likelihood $l_{ij,m}^k$.
10   Update and normalize the weight w_{ij}^k.
11 **end**

12 (iv) Merging the SoG (Figure 2b)
13 Estimate the new location of the jth node.

14 (v) Resetting the SoG (Figure 2c)
15 **for** $i \leftarrow 0$ **to** $M - 1, (i \neq j)$ **do**
16   Reset the mean $\mathbf{m}_{ji}^k$, covariance $\mathbf{C}_{ji}^k$, and weight w_{ji}^k using Equations (5)–(7).
17 **end**

18 (vi) Updating the weight (Figure 2c)
19 **for** $i, m \leftarrow 0$ **to** $M - 1, (i \neq j, m \neq i, m \neq j)$ **do**
20   Compute and normalize the likelihood $l_{ji,m}^k$.
21   Update and normalize the weight w_{ji}^k.
22 **end**

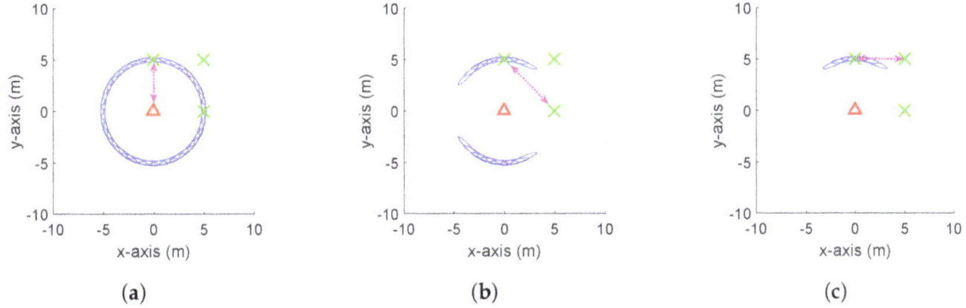

(a)   (b)   (c)

Figure 3. Explanation of the weight update scheme in the CDRO-SLAM. Red triangles and green crosses indicate robots and neighbor nodes, respectively. Black and magenta dotted lines are the movement of the robot and range measurements, respectively. Blue ellipse is a possible location of nodes. The CDRO-SLAM involves (**a**) the initialization of the SoG variables when receiving the first measurement, (**b**) determining the SoG distribution after updating the weight with the first inter-node measurement, and (**c**) the second inter-node measurement.

2.2. Initialization Stage

2.2.1. Iterative Multilateration

The initial estimation is done by employing the modified iterative multilateration method so that errors are not propagated in each iteration. As stated in [16], the first three nodes are supposed to be placed at the origin, on the positive x-axis, and in the upper half plane, respectively, as follows:

$$\mathbf{p}_0(n) = (x_0(n), y_0(n)) = (0, 0) \tag{1}$$

$$\mathbf{p}_1(n) = (x_1(n), y_1(n)) = (d_{01}, 0) \tag{2}$$

$$\mathbf{p}_2(n) = (x_2(n), y_2(n))$$
$$= \left(\frac{d_{01}^2 + d_{02}^2 - d_{12}^2}{2x_1(n)}, \sqrt{d_{02}^2 - x_2^2(n)} \right). \tag{3}$$

In Equations (1)–(3), $\mathbf{p}_i(n)$ is defined as the position of the ith node at iteration n, and d_{ij} is the distance measurement between the ith and jth nodes. The next node is localized by exploiting the positions of the first two nodes as in $\mathbf{p}_2(n)$, resulting in the following two candidates due to its ambiguity:

$$\mathbf{p}_m(n) = (x_m(n), y_m(n))$$
$$= \left(\frac{d_{01}^2 + d_{0m}^2 - d_{1m}^2}{2x_1(n)}, \pm\sqrt{d_{0m}^2 - x_m^2(n)} \right). \tag{4}$$

The sign of $y_m(n)$ can be chosen as the one whose estimated distance from $\mathbf{p}_m(n)$ to $\mathbf{p}_2(n)$ is closer to the measurement d_{2m}. Note that $\mathbf{p}_m(n)$ is affected by the position error of $\mathbf{p}_1(n)$, as shown in Equation (4). The above procedures (i.e., Equations (1)–(4)) are used in Algorithm 1.

2.2.2. Generating SoG

After guessing the initial positions of a robot and map, SoG variables are generated as shown in Figure 4. Note that in the conventional SoG based RO-SLAM methods, only the SoG variables between robot and its neighbor nodes are generated, while in the proposed CDRO-SLAM, those between any pair of two nodes are generated. For two different nodes (let us say the ith and jth nodes, $i \neq j$), the mean $\mathbf{m}_{ij}^k$, covariance $\mathbf{C}_{ij}^k$, and weight w_{ij}^k are generated as described in [5,6]:

$$\mathbf{m}_{ij}^k = \begin{bmatrix} x_i(n) + d_{ij}\cos\left(\frac{2\pi k}{N}\right) \\ y_i(n) + d_{ij}\sin\left(\frac{2\pi k}{N}\right) \end{bmatrix}, \tag{5}$$

$$\mathbf{C}_{ij}^k = \begin{bmatrix} \mathbf{v}_r & \mathbf{v}_t \end{bmatrix} \begin{bmatrix} \sigma_r^2 & 0 \\ 0 & \sigma_t^2 \end{bmatrix} \begin{bmatrix} \mathbf{v}_r^T \\ \mathbf{v}_t^T \end{bmatrix}, \tag{6}$$

$$w_{ij}^k = \frac{1}{N}, \tag{7}$$

where $k = 0, 1, \ldots, N-1$, is an index of SoG, and N is number of SoG, respectively. Note that the procedures of generating the SoG variables (i.e., Equations (5)–(7)) is utilized in Algorithms 1 and 2.

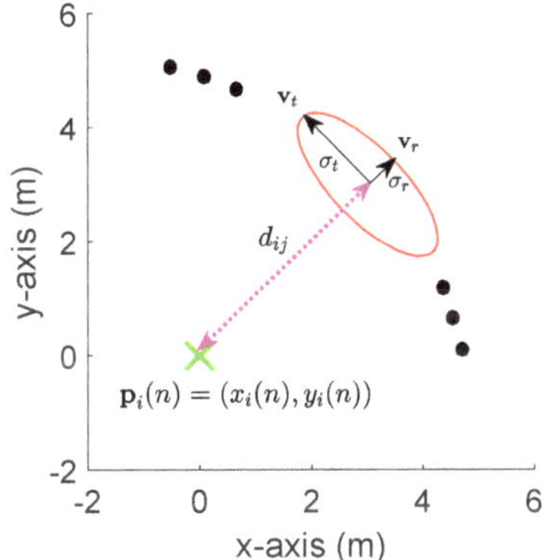

Figure 4. Generation of SoG variables. The dotted magentaline represents the measurement. $\mathbf{v}_r$ and $\mathbf{v}_t$ are, respectively, the radial and tangential unit vectors, and σ_r and σ_t denote, respectively, the radial and tangential standard deviations.

2.2.3. Updating SoG

The update scheme of SoG variables consists of two parts. When collecting the current measurement d_{ij}, the corresponding mean $\mathbf{m}_{ij}^k$ and covariance $\mathbf{C}_{ij}^k$ are first updated using the standard EKF procedure (see [15]). After that, the weights w_{ij}^k between all inter-nodes are updated by using the inter-node measurements to remove the ambiguity.

A detailed update procedure for the weight w_{ij}^k is summarized in Algorithm 3. In order to update w_{ij}^k between the ith and jth nodes, the likelihood $l_{ij,m}^k$ is first computed with respect to the mth nodes connected with the jth node (in lines 2–5 of Algorithm 3), where m is an index of the neighbor nodes. After the normalization of $l_{ij,m}^k$ in lines 6–8, the weight w_{ij}^k is updated in lines 9–11 and normalized in lines 12–14. The above procedures are performed for all possible mth neighbor nodes, i.e., $m \neq i$, $m \neq j$. As more measurements d_{mj} are used, the updated weight w_{ij}^k becomes a larger value if $d_{ij,m}^k$ is close to d_{mj}; otherwise, it will converge rapidly to zero.

Algorithm 3: Weight update procedure.

---

1 **for** $m \leftarrow 0$ **to** $M - 1$, $m \neq i$, $m \neq j$ **do**

2   **for** $k \leftarrow 0$ **to** $N - 1$ **do**

3          $d_{ij,m}^k = \left\| \mathbf{m}_{ij}^k - \mathbf{p}_m(n) \right\|$

4          $l_{ij,m}^k = \frac{1}{\sqrt{2\pi\sigma^2}} \exp\left(-\frac{(d_{ij,m}^k - d_{mj})^2}{2\sigma^2} \right)$

5   **end**

6   **for** $k \leftarrow 0$ **to** $N - 1$ **do**

7          $l_{ij,m}^k = \frac{l_{ij}^k}{\sum_k l_{ij}^k}$

8   **end**

9   **for** $k \leftarrow 0$ **to** $N - 1$ **do**

10          $w_{ij}^k = w_{ij}^k \times l_{ij,m}^k$

11   **end**

12   **for** $k \leftarrow 0$ **to** $N - 1$ **do**

13          $w_{ij}^k = \frac{w_{ij}^k}{\sum_k w_{ij}^k}$

14   **end**

15 **end**

2.2.4. Merging SoG

After updating the SoG variables, the weighted SoG is merged into a single Gaussian distribution to estimate the new locations of nodes. A closed-form solution [17,18] for the merged SoG is employed for the new location $\mathbf{p}_i(n+1)$ and its covariance $\mathbf{C}_i(n+1)$, as follows:

$$\mathbf{p}_i(n+1) = \frac{1}{\sum_{j,k}\left(w_{ji}^k\right)} \sum_{j,k}\left(w_{ji}^k \mathbf{m}_{ji}^k\right), \tag{8}$$

$$\mathbf{C}_i(n+1) = \frac{1}{\sum_{j,k}\left(w_{ji}^k\right)} \sum_{j,k} w_{ji}^k(\mathbf{C}_{ji}^k + \bar{\mathbf{C}}_{ji}^k), \tag{9}$$

$$\bar{\mathbf{C}}_{ji}^k = \left(\mathbf{m}_{ji}^k - \mathbf{p}_i(n+1)\right)\left(\mathbf{m}_{ji}^k - \mathbf{p}_i(n+1)\right)^T. \tag{10}$$

2.2.5. Transferring Weight

Due to the inherent non-convex nature of anchor-free localization, a local solution can be obtained after merging SoG variables. To prevent falling into a local solution due to a small weight being ignored, a weight transfer method is introduced here by distributing weights around the maximum weight. If $k_{\max}$ indicates an index of the maximum weight from the SoG distribution, the weight can be heuristically distributed as a Gaussian distribution at the center of $k_{\max}$. More specifically, with the mean distance between two indices $k_{\max}$ and k, defined by $d_{k_{\max}-k} = \|\mathbf{m}_{ij}^{k_{\max}} - \mathbf{m}_{ij}^k\|$, the weight is distributed as follows:

$$w_{ij}^k = \frac{1}{\sqrt{2\pi\sigma^2}}\exp\left(-\frac{d_{k_{\max}-k}^2}{2\sigma^2}\right), \tag{11}$$

where σ is experimentally determined (e.g., 0.5). Finally, the weight w_{ij}^k is normalized to ensure that its sum is 1.

The effectiveness of the proposed weight transfer technique was verified with a simulation, as shown in Figure 5. In the simulation, five nodes were deployed, where four nodes were placed in a rectangular shape and a robot was moved inside the rectangular shape. The noise variance was set to 0.3–0.8. The RMSE performance was obtained by ensemble averaging over 20 independent trials. As shown in Figure 5, the RMSE performance improved by about 50% with the proposed weight transfer method because it gives a chance to avoid a local minimum.

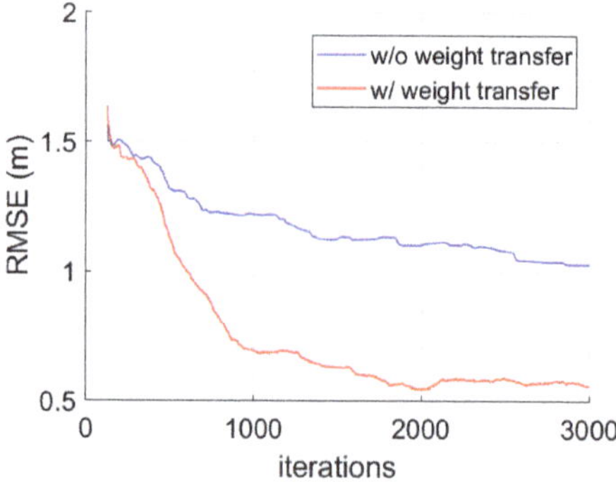

Figure 5. The root-mean-square error (RMSE) performance of map estimation with and without the weight transfer technique.

2.2.6. Convergence

The convergence of each node is determined by using $\tilde{\mathbf{p}}_i(n)$, the time-average of $\mathbf{p}_i(n)$, as follows:

$$\tilde{\mathbf{p}}_i(n+1) = \alpha \tilde{\mathbf{p}}_i(n) + (1-\alpha)\mathbf{p}_i(n). \tag{12}$$

If $\|\tilde{\mathbf{p}}_i(n+1) - \tilde{\mathbf{p}}_i(n)\| < \varepsilon$, then the corresponding node is considered to be converged. Here, ε is a small positive constant threshold. If not converged, the procedure from Section 2.2.3 to Section 2.2.4 is repeated when a new measurement is available.

2.3. Moving Stage

As mentioned in Section 2.1, the movement of any node can be tracked in the moving stage. Unlike the conventional RO-SLAM algorithms, the proposed CDRO-SLAM can deal with the movement of the neighbor nodes as well as the robot's movement. In particular, the tracking procedure during the movement of neighbor nodes was derived in Algorithm 2 by adopting a detection and tracking moving object (DATMO) strategy [19] into a SoG approach. As shown in Algorithm 2, the same procedures as in the initialization stage are recombined; thus, the detailed procedure for the moving stage is not given here. Before concluding this section, it should be noted from Algorithm 2 that the indices of SoG variables in (ii) and (iii) of Algorithm 2 are symmetric counterparts of those in (v) and (vi) (see Figure 2b,c), which will be utilized when developing an efficient implementation, as described in the following section.

3. Efficient Implementation

Due to the extra SoG variables between the inter-nodes, the CDRO-SLAM [14] has a greater computational cost compared with the conventional SoG based RO-SLAM algorithms which use only the direct measurements between a robot and its neighbor nodes. To mitigate this incurred computational burden, several efficient implementation techniques are proposed here.

Efficient implementation can be applied to the corresponding computation parts shown in Algorithms 1–3. Especially, the weight update procedure that causes the main computational burden is summarized in Algorithm 3 for the original one and Algorithm 4 for the efficient implementation to show the computational efficiency.

Algorithm 4: Efficient weight update procedure.

---

1   **for** $m \leftarrow 0$ **to** $M-1$, $m \neq i$, $m \neq j$ **do**

2   **for** $k \leftarrow 0$ **to** $N-1$ **do**

3       $d_{ij,m}^k = \left\| \mathbf{m}_{ij}^k - \mathbf{p}_m(n) \right\|$

4       $l_{ij,m}^k = \dfrac{1}{\sqrt{2\pi\sigma^2}}\exp\left(-\dfrac{(d_{ij,m}^k - d_{mj})^2}{2\sigma^2}\right)$

5       $w_{ij}^k = w_{ij}^k \times l_{ij,m}^k$

6   **end**

7   **end**

8   **for** $k \leftarrow 0$ **to** $N-1$ **do**

9     $w_{ij}^k = \dfrac{w_{ij}^k}{\sum_k w_{ij}^k}$

10   **end**

11   **for** $k \leftarrow 0$ **to** $N-1$ **do**

12     $r = \mathrm{mod}(k + N/2, N)$

13     $w_{ji}^r = w_{ij}^k$

14   **end**

3.1. Reduced Normalization

When computing the likelihood $l^k_{ij,m}$, its normalization can be omitted without damaging the performance. Similarly, the normalization of the weight w^k_{ij} can be carried out only once after the end of for-loop m.

3.2. Symmetric SoG Distribution

Obviously, a higher weight represents a higher probability of the existence of a node. Let us consider two weights w^k_{ij} and w^k_{ji} for $i \neq j$. As shown in Figure 6a, the high weight area for each node is symmetrical, i.e., a high w^k_{ij} occurs in the direction of a negative y-axis for the w^k_{ij} distribution, and a high w^k_{ji} occurs in the direction of the positive y-axis for the w^k_{ji} distribution.

In Figure 6b, the calculated distribution of two weights w^k_{ij} and w^k_{ji} is described when $N = 32$ without considering symmetry. If the symmetry property is used, once w^k_{ij} is computed (the highest probability occurred in $k = 25$), its counterpart w^k_{ji} can be simply copied from w^k_{ij} by using the modulo operation (shown in lines 11–14 of Algorithm 4), which gives $k = 9$ for the highest probability and is equal to the index with the calculated distribution of w^k_{ji} in Figure 6b.

Furthermore, the symmetry property can be applied in the movement stage. As discussed in the previous section, the SoG distributions of (iii) and (vi) in Algorithm 2 are symmetrical counterparts of each other. Therefore, the weight w^k_{ji} of (vi) of Algorithm 2 is simply symmetrically copied from the weight w^k_{ij} that was previously computed in (iii) of Algorithm 2.

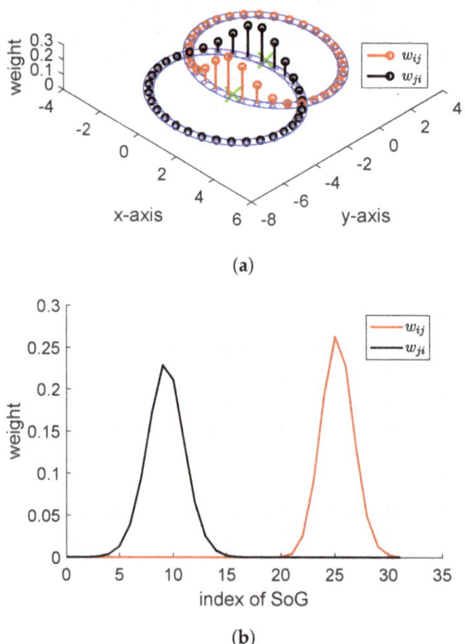

Figure 6. Weight symmetry between w^k_{ij} and w^k_{ji}. The green crosses represent the estimated positions of the ith and jth nodes. (**a**) The SoG variables are distributed in the xy plane, and the weight is shown on the z-axis. (**b**) The weights w^k_{ij} and w^k_{ji} are depicted against the SoG index k.

3.3. Removal of Low Probability

As investigated in several SoG based RO-SLAM algorithms, the merging procedure can be efficiently realized by dropping small Gaussian weights in Equations (8)–(9). The threshold value can be set as small enough, e.g., $0.00001/N$ [20], where N is the number of SoG.

3.4. Efficiency Analysis

In order to show the efficiency, two versions of the proposed algorithm are considered: one is the original CDRO-SLAM and the other is its efficient implementation version (eCDRO-SLAM) with the aforementioned efficient techniques. The computational efficiency of CDRO-SLAM and eCDRO-SLAM is also compared with that of several conventional RO-SLAM algorithms, i.e., the efficient probabilistic RO-SLAM (EPRO-SLAM) [6] and the RBPF-based RO-SLAM (RBPF-RO-SLAM) algorithm [7].

The required computational complexity for updating the weight is considered in terms of the computing likelihood number $l^k_{ij,m}$, which is cost-dominant and is summarized in Table 1. Here, M, Q, and N are the numbers of nodes, particles, and SoG, respectively. Note that RBPF-RO-SLAM requires a higher computational cost than EPRO-SLAM, since $Q = \beta \cdot d_{ij}$ is usually determined to be much larger than N, e.g., $Q \geq 1000$ and $N = 32$, where $\beta = [400, 2000]$ [7].

Table 1. Computational complexity of updating the weight.

Algorithms	Likelihood
EPRO-SLAM [6]	N
RBPF-RO-SLAM [7]	Q
CDRO-SLAM	$N(M-2)$
eCDRO-SLAM	$N(M-2)/2$

For CDRO-SLAM, Algorithm 3 requires $N(M-2)$ computations, and for eCDRO-SLAM, Algorithm 4 needs $N(M-2)/2$ computations, on average. The computation of Algorithm 4 is reduced by half compared with Algorithm 3.

It is not easy to compare non-cooperative and cooperative algorithms at the same level because their approaches to estimating the location are totally different, i.e., the cooperative algorithms use many inter-node measurements for the map estimation, but the non-cooperative algorithms use robot movement to measure independent and direct ranges between the robot and its neighbors. As illustrated in Figure 1a–c, the EPRO-SLAM requires several direct measurements at the non-collinear locations to remove the ambiguity in the location of nodes, which results in αN computations in total, where α is the number of required direct measurements at the non-collinear locations. On the other hand, the proposed CDRO-SLAM can yield an accurate location even with one iteration, as described in Figure 3a–c. In this case, the computational burden ratio of the proposed eCDRO-SLAM over EPRO-SLAM is $\frac{(M-2)/2}{\alpha}$ at each iteration. When α is minimal (i.e., $\alpha = 3$ as shown in Figure 1), the computational complexity ratio can be approximately $\frac{M}{6}$. In the case of (at least) $\alpha \approx M$, to achieve a similar ambiguity level to the proposed eCDRO-SLAM, the ratio can be approximately $\frac{1}{2}$.

Furthermore, when updating all nodes, the total computational burden of the proposed eCDRO-SLAM is $NM(M-1)(M-2)/2$, since there are $M(M-1)$ SoG variables between inter-nodes; thus, the total computational burden ratio is about $\frac{NM(M-1)(M-2)/2}{NM\alpha} \approx M/2$ for $\alpha \approx M$, i.e., the total computational burden increases linearly as the number of nodes, M, increases.

4. Experiments

As shown in Figure 7, two sets of experiments were conducted with the same number of nodes ($M = 9$) but different sizes (3.6 m × 4.8 m for Experiment 1 and 6.6 m × 8.4 m for Experiment 2). In the experiments, the motion of the robot was arbitrary controlled remotely by a person, and it was represented by using the measured odometry in Figure 7. Also, TurtleBot3 Burger and Pozyx [21] were used in the experiments for the ultra-wide-band (UWB) measurements . For the acquisition of measurements, all the inter-node measurements were initially collected while a robot remained still, and then the direct measurements were collected while the robot moved, in order to clearly show the difference of weight update scheme between the conventional RO-SLAM and the proposed eCDRO-SLAM. The performance of the eCDRO-SLAM with the inter-node measurements was verified with real experimental data, and it was compared with EPRO-SLAM [6] and RBPF RO-SLAM [7], which use direct measurements only. Note that the proposed eCDRO-SLAM was considered only because it gave almost same performance when compared with the CDRO-SLAM, except the computational efficiency. The computational complexity comparison of the eCDRO-SLAM and CDRO-SLAM will be given in Section 4.3. The map estimation and the tracking performance of moving nodes were also investigated.

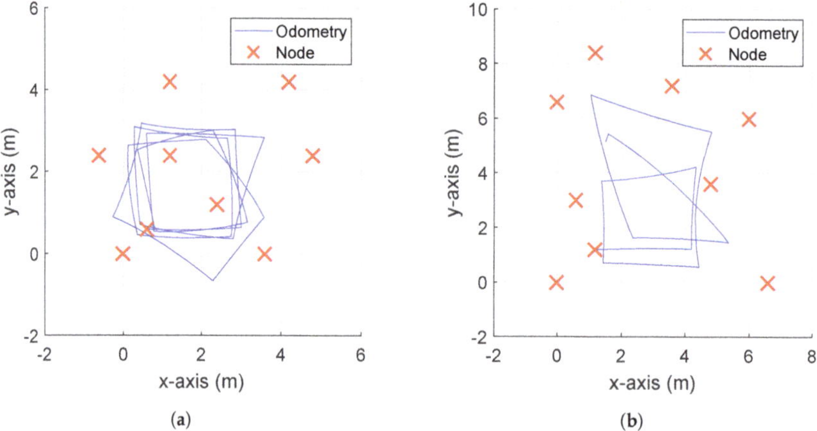

Figure 7. Deployment of nodes and odometry for (**a**) Experiment 1 and (**b**) Experiment 2.

4.1. Map Estimation Performance

The map estimation performance for the two experiments is illustrated in Figure 8, and the following results can be observed:

- The conventional algorithms exhibited a slower convergence rate and higher initial map estimation error than the proposed eCDRO-SLAM, since their convergence depends on both the movement of the robot and the rate of measurements. On the other hand, the convergence of the proposed eCDRO-SLAM depends only on the rate of measurements and is accelerated by employing the inter-node measurements, as discussed before.

- As shown in Figure 8, the proposed eCDRO-SLAM estimates the map even without the movement of a robot; on the other hand, the conventional algorithms can only estimate the map when a robot moves.

- During the refinement of Experiment 1, all algorithms achieved a similar RMSE of about 0.2 m. On the other hand, for Experiment 2, the RBPF-RO-SLAM algorithm (0.56 m) yielded a higher RMSE than the EPRO-SLAM algorithm (0.72 m), as expected. Furthermore, the proposed eCDRO-SLAM algorithm (0.21 m) obtained a better performance than the non-cooperative RBPF-RO-SLAM algorithm. The difference between the two experiments comes from the odometry error. The performance of the RBPF-RO-SLAM (also the EPRO-SLAM) suffered from not only the measurement error but also the odometry error, especially in the large area; on the other hand, the proposed eCDRO-SLAM was affected only by the measurement error, which led to a similar map estimation result for both experiments, regardless of the difference in size.

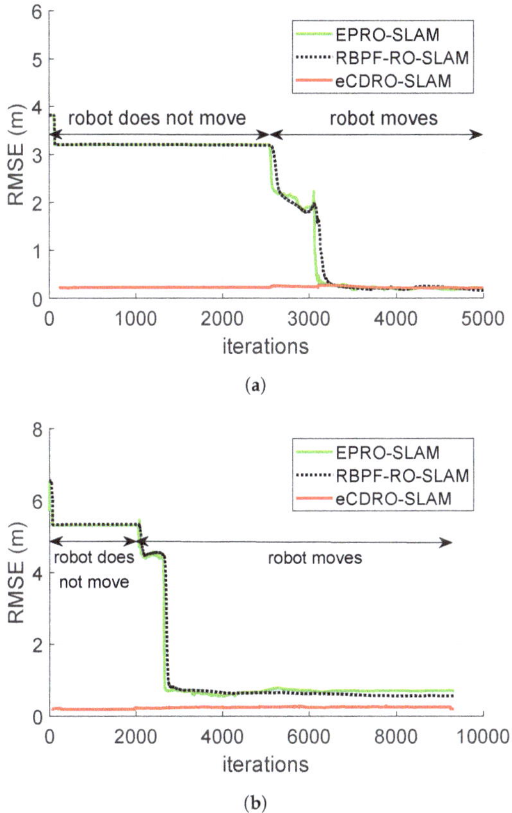

Figure 8. Map estimation performance obtained by the proposed eCDRO-SLAM and the conventional RO-SLAM algorithms [6,7] for (**a**) Experiment 1 and (**b**) Experiment 2.

4.2. Node Tracking Performance

The node tracking performance of the proposed eCDRO-SLAM was verified with Experiment 1. Each node moves continually in a square shape one-by-one. The tracking performance was measured after the initialization stage, and its result is illustrated in Figure 9. The average localization error was about 0.3688 m in the movement stage, somewhat larger than that of the initial stage; however, a reasonable tracking performance was achieved using the range sensor only. The tracking performance of the proposed eCDRO-SLAM can be further improved if a node moves in a stop-and-go way or if some sensors such as the inertial measurement unit (IMU) are fused. On the other hand, the most conventional RO-SLAM algorithms (including EPRO-SLAM [6] and RBPF-RO-SLAM [7]) cannot be applied to dynamic environments where nodes can have motion.

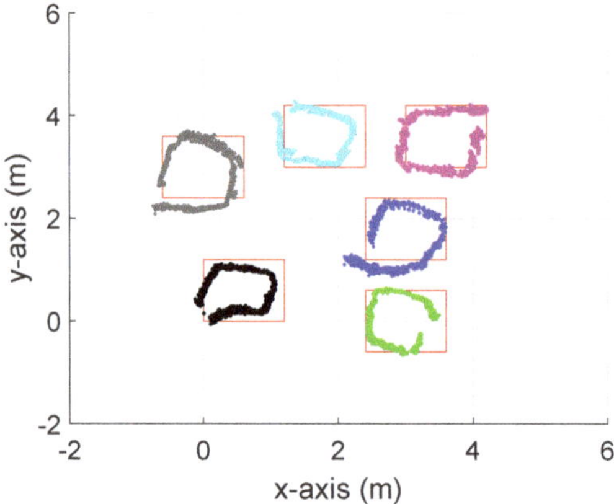

Figure 9. Node tracking performance of the eCDRO-SLAM for Experiment 1.

4.3. Computational Complexity

In Table 2, the computational complexity of the proposed algorithm and the conventional algorithms is summarized in terms of the average CPU time per sample for updating the weight w_{ij}^k and for executing the complete algorithm. The CPU time was measured through implementation in a MATLAB simulation with an Intel Core i7 processor of 3.3 GHz and 8 GB memory.

Table 2. Required CPU time for Experiment 1.

Algorithms	CPU Time (ms)	
	Weight	Total
EPRO-SLAM [6]	0.032	1.70
RBPF-RO-SLAM [7]	0.415	13.3
CDRO-SLAM	0.170	22.7
eCDRO-SLAM	0.101	9.20

The required CPU time for computing the weight shown in Table 2 has a similar tendency to that of the efficiency analysis shown in Section 3.4. When considering the total CPU time, it is assumed that all algorithms spend most of the CPU time in updating the weight and merging SoG variables (or particles) to simplify the analysis. From the total CPU time results, we can see the following:

- The total computational burden ratio of the proposed eCDRO-SLAM over the CDRO-SLAM ($9.20/22.7 \approx 0.405$) is smaller than the expected ratio of the weight ($0.101/0.170 \approx 0.594$), since the eCDRO-SLAM is efficiently implemented for both updating and merging SoG processes. As indicated in Section 3.4, the efficiency of the eCDRO-SLAM is about half that of the CDRO-SLAM.
- The eCDRO-SLAM requires $9.20 - 0.101 \times (9 \times 8) = 1.93$ ms to merge SoG variables between all inter-nodes, and the EPRO-SLAM needs $1.70 - 0.032 = 1.67$ ms to merge only one SoG distribution. This efficiency of the eCDRO-SLAM is obtained due to the fast convergence of the weight and because small enough weights are dropped in the merging process. Furthermore, the eCDRO-SLAM is computationally efficient in terms of achieving a similar ambiguity level, since the EPRO-SLAM may require $M = 9$ iterations (see Section 3.4) and, in this case, it needs $1.7 \text{ ms} \times 9 = 15.3$ ms in total.

- Despite the cooperative approach, the proposed eCDRO-SLAM was executed using a smaller total CPU time than the non-cooperative RBPF-RO-SLAM. Note that the RBPF-RO-SLAM spends most of the CPU time for transforming particles into a Gaussian filter.

5. Conclusions

In this paper, an efficient cooperative dynamic RO-SLAM (eCDRO-SLAM) based on the sum of Gaussian (SoG) filter integrating all inter-node measurements for the localization was proposed. A new framework to update the SoG variables using the inter-node measurements leads to faster map estimation with less error than conventional RO-SLAM algorithms using the direct measurements only. Also, several efficient implementation techniques such as the weight symmetry of the SoG distributions were introduced to alleviate the additional computational burden induced when calculating the SoG variables between inter-nodes.

Furthermore, the proposed eCDRO-SLAM was designed to track the movement of some nodes by resetting and updating the SoG variables. With the dynamic property and computational efficiency, its application can be extended to various dynamic environments that are full of movable objects such as unstable disaster sites, unstructured fields, or temporal deployment of robots/devices.

As a future work, a sensor fusion approach integrating the IMU, odometry, and range sensor will be considered to improve the localization and node tracking performance. Furthermore, we will apply the proposed algorithm to various RO-SLAM applications such as robot-assisted search and rescue application and ambient intelligence environments.

Author Contributions: J.-H.K. conducted the experiments and analyzed the data and wrote the manuscript; D.K. provided suggestions and guidance for the work and reviewed and edited the manuscript. All authors have read and agreed to the published version of the manuscript.

Funding: This work was supported by the Korea Institute of Science and Technology (KIST) Institutional Program (2E29460/2E30280).

References

1. Lee, T.J.; Kim, C.H.; Cho, D.I.D. A monocular vision sensor-based efficient SLAM method for indoor service robots. *IEEE Trans. Ind. Electron.* **2019**, *66*, 318–328. [CrossRef]
2. Yu, C.; Liu, Z.; Liu, X.; Xie, F.; Yang, Y.; Wei, Q.; Fei, Q. DS-SLAM: A semantic visual SLAM towards dynamic environments. In Proceedings of the IEEE/RSJ International Conference on Intelligent Robots and Systems, Madrid, Spain, 1–5 October 2018; pp. 1168–1174.
3. Bjrne, E.; Johansen, T.; Brekke, E. Redesign and analysis of globally asymptotically stable bearing only SLAM. In Proceedings of the IEEE International Conference on Multimedia and Expo, Hong Kong, China, 10–14 July 2017; pp. 1–8.
4. Menegatti, E.; Zanella, A.; Zilli, S.; Zorzi, F.; Pagello, E. Range-only SLAM with a mobile robot and a wireless sensor networks. In Proceedings of the IEEE International Conference on Robotics and Automation, Kobe, Japan, 12–17 May 2009; pp. 8–14.
5. Kantor, G.; Singh, S. Preliminary results in range-only localization and mapping. In Proceedings of the IEEE International Conference on Robotics and Automation, Washington, DC, USA, 11–15 May 2002; pp. 1818–1823.
6. Blanco, J.L.; Fernandez-Madrigal, J.A.; Gonzalez, J. Efficient probabilistic range-only SLAM. In Proceedings of the IEEE/RSJ International Conference on Intelligent Robots and Systems, Nice, France, 22–26 September 2008; pp. 1017–1022.
7. Blanco, J.L.; Gonzalez, J.; Fernandez-Madrigal, J.A. A pure probabilistic approach to range-only SLAM. In Proceedings of the IEEE International Conference on Robotics and Automation, Pasadena, CA, USA, 19–23 May 2008; pp. 1436–1441.

8. Djugash, J.; Singh, S.; Kantor, G.; Zhang, W. Range-only SLAM for robots operating cooperatively with sensor networks. In Proceedings of the IEEE International Conference on Robotics and Automation, Orlando, FL, USA, 15–19 May 2006; pp. 2078–2084.

9. Torres-Gonzalez, A.; de Dios, J.M.; Ollero, A. Efficient robot-sensor network distributed SEIF range-only SLAM. In Proceedings of the IEEE International Conference on Robotics and Automation, Hong Kong, China, 31 May–5 June 2014; pp. 1319–1326.

10. Wang, C.; Zhang, H.; Nguyen, T.M.; Xie, L. Ultra-wideband aided fast localization and mapping system. In Proceedings of the IEEE/RSJ International Conference on Intelligent Robots and Systems, Vancouver, BC, Canada, 24–28 September 2017; pp. 1602–1609.

11. Urzua, S.; Munguia, R.; Grau, A. Monocular SLAM system for MAVs aided with altitude and range measurements: A GPS-free approach. *J. Intell. Robot. Syst.* **2019**, *94*, 203–217. [CrossRef]

12. Torres-Gonzalez, A.; de Dios, J.M.; Ollero, A. An adaptive scheme for robot localization and mapping with dynamically configurable inter-beacon range measurements. *Sensors* **2014**, *14*, 7684–7710. [CrossRef] [PubMed]

13. Torres-Gonzalez, A.; de Dios, J.M.; Ollero, A. Range-only SLAM for robot-sensor network cooperation. *Auton. Robot.* **2018**, *42*, 649–663. [CrossRef]

14. Kim, J.; Kim, D. Cooperative range-only SLAM based on sum of Gaussian filter in dynamic environment. In Proceedings of the IEEE/RSJ International Conference on Intelligent Robots and Systems, Macau, China, 4–8 November 2019; pp. 2139–2144.

15. Vallicrosa, G.; Ridao, P.; Ribas, D.; Palomer, A. Active range-only beacon localization for AUV homing. In Proceedings of the IEEE/RSJ International Conference on Intelligent Robots and Systems, Chicago, IL, USA, 14–18 September 2014; pp. 2286–2291.

16. Youssef, A.; Agrawala, A. Accurate anchor-free node localization in wireless sensor networks. In Proceedings of the IEEE International Performance, Computing, and Communications Conference, Phoenix, AZ, USA, 7–9 April 2005; pp. 465–470.

17. Kwok, N.; Dissanayake, G.; Ha, Q. Bearing-only SLAM using a SPRT-based Gaussian sum filter. In Proceedings of the IEEE International Conference on Robotics and Automation, Barcelona, Spain, 18–22 April 2005; pp. 1109–1114.

18. Kwok, N.; Ha, Q.; Huang, S.; Dissanayake, G.; Fang, G. Mobile robot localization and mapping using a Gaussian sum filter. *Int. J. Control Autom. Syst.* **2007**, *5*, 251–268.

19. Moratuwage, D.; Vo, B.N.; Wang, D. Collaborative multi-vehicle SLAM with moving object tracking. In Proceedings of the IEEE International Conference on Robotics and Automation, Karlsruhe, Germany, 6–10 May 2013; pp. 5702–5708.

20. Torres-Gonzalez, A.; de Dios, J.M.; Ollero, A. Integrating internode measurements in sum of Gaussians range only SLAM. In *ROBOT2013: First Iberian Robotic Conference*; Springer: Cham, Switzerland, 2014; pp. 473–487.

21. Pozyx Labs, Pozyx Positioning System. Available online: https://www.pozyx.io/ (accessed on 27 April 2020).

Letter

LoRaWAN Geo-Tracking Using Map Matching and Compass Sensor Fusion [†]

Nico Podevijn [1], Jens Trogh [1], Michiel Aernouts [2], Rafael Berkvens [2], Luc Martens [1], Maarten Weyn [2], Wout Joseph [1] and David Plets [1,*]

[1] Department of Information Technology, Ghent University, imec-WAVES, 9052 Ghent, Belgium; nico.podevijn@ugent.be (N.P.); jens.trogh@ugent.be (J.T.); luc1.martens@ugent.be (L.M.); wout.joseph@ugent.be (W.J.)

[2] Faculty of Applied Engineering, University of Antwerp, imec-IDLAB, 2000 Antwerp, Belgium; Michiel.Aernouts@uantwerpen.be (M.A.); rafael.berkvens@uantwerpen.be (R.B.); maarten.weyn@uantwerpen.be (M.W.)

[*] Correspondence: david.plets@ugent.be

[†] This paper is an extended version of our paper published in Compass Aided TDoA Tracking in LoRaWAN networks. In Proceedings of the 2020 IEEE/ION Position, Location and Navigation Symposium (PLANS), Portland, OR, USA, 20–23 April 2020.

Received: 14 September 2020; Accepted: 5 October 2020; Published: 14 October 2020

Abstract: In contrast to accurate GPS-based localization, approaches to localize within LoRaWAN networks offer the advantages of being low power and low cost. This targets a very different set of use cases and applications on the market where accuracy is not the main considered metric. The localization is performed by the Time Difference of Arrival (TDoA) method and provides discrete position estimates on a map. An accurate "tracking-on-demand" mode for retrieving lost and stolen assets is important. To enable this mode, we propose deploying an e-compass in the mobile LoRa node, which frequently communicates directional information via the payload of the LoRaWAN uplink messages. Fusing this additional information with raw TDoA estimates in a map matching algorithm enables us to estimate the node location with a much increased accuracy. It is shown that this sensor fusion technique outperforms raw TDoA at the cost of only embedding a low-cost e-compass. For driving, cycling, and walking trajectories, we obtained minimal improvements of 65, 76, and 82% on the median errors which were reduced from 206 to 68 m, 197 to 47 m, and 175 to 31 m, respectively. The energy impact of adding an e-compass is limited: energy consumption increases by only 10% compared to traditional LoRa localization, resulting in a solution that is still 14 times more energy-efficient than a GPS-over-LoRa solution.

Keywords: LoRa; localization; positioning; LoRaWAN; TDoA; tracking; map matching; compass; sensor fusion

---

1. Introduction

The Internet of Things (IoT) allows connecting objects to the internet with the use of wireless sensors. Typical use cases are monitoring temperature, humidity, and soil moisture for smart farming applications [1,2], condition monitoring of air cargo [3], or monitoring vital signs of cows in rural areas [4]. Other examples are a bus positioning system [5,6] and asset tracking for logistics [7]. In nearly all cases, the sensor device transmits this information wirelessly to a gateway or access point which has a back-haul to the internet. In this manner, the data can be further processed and visualized. This paper considers the use case of asset tracking.

IoT devices are often powered by batteries, which will sooner or later need replacement, for example, nodes installed in equipment, pallets, or bikes. For economic reasons, a long-range wireless link is also required in order to have a large coverage area with the least amount of access points or gateways. The advent of Low-Power Wide Area Networks (LPWAN) and their deployment in many countries [8] brings benefits at the level of scalability (thousands of sensors per gateway), coverage range (more than 15 km), and power consumption (battery lifetimes up to more than 5 years). The only constraint is the fact that uplink should (in general) be infrequent and limited in the number of payload (information) bytes. Some examples of LPWAN standards are NB-IoT [9], SigFox [10], and LoRaWAN [11].

Many outdoor asset tracking implementations are using a Global Navigation Satellite System (GNSS) receiver to send GPS coordinates over LPWAN. In [5,6], these implementations are used to track city buses. In [3,12], similar GPS-based approaches are used to track air cargo and bikes. Another approach to perform localization is using the network itself to locate sensor nodes without GPS embedded in the node. The basic principle is that when a node send uplink data to the network, the incoming packets' meta-data such as Time of Arrival (ToA) and Received Signal Strength (RSS) is recorded on the different gateways. The meta-data and gateway locations are then forwarded to a so-called geo-location solver to estimate the location with a suitable algorithm. The output of the solver is then a (Latitude, Longitude) coordinate [13]. The provided location estimates are not as accurate as for GPS (order of a few meters), but an important advantage is that network-based localization (geo-locating) consumes less power on the mobile node compared to integrating a GPS receiver in the node. Another clear advantage of using LPWAN for localization is that only one technology is used for communication and localization, enabling the manufacturing of low cost and compact sensor nodes. Some use cases of geo-locating using LPWAN include tracking of valuable assets during transportation such as railway cars, truck trailers, and containers [7]. The main downside of geo-locating is the relatively low localization accuracy which might not be beneficial for some use cases. Therefore, this paper will focus on improving this performance metric.

In [14], the authors introduced the principle of using a compass on top of LoRaWAN raw location data and tested it for a single route. The goal of this paper is to realize and thoroughly analyze the LoRaWAN Geo-Tracking by the use of map matching and compass sensor fusion. A significant number of test routes is examined for different modes of transportation in different networks, not only in Belgium but also in the Netherlands. Furthermore, the difference between real-time localization and offline a posteriori localization is compared. The main idea is that we take into account the road infrastructure, maximum speed of the node, and (communicated) compass heading. Assuming the tracked item stays on the road network, heading info (e.g., item heading west) can be exploited to exclude other candidate trajectories (e.g., trajectory heading north). The basic setup for this is shown in Figure 1. The novelties of this paper are as follows.

(i)   A compass sensor fusion implementation for accurate LoRaWAN localization.
(ii)   Improved map matching technique to outperform current available LoRa geo-location possibilities, which works for all LoRa sensor nodes.
(iii) All algorithms are tested with experimental data obtained in different environments and using different modes of transportation (walking, cycling, and driving). The reproducibility of the results is also investigated.

The remainder of the paper is structured as follows. Section 2 presents an overview of related work on LoRa geo-localization methods and improvements proposed in literature. In Section 3, we describe the trajectories and the data collection method, the algorithm implementations, and the different investigated scenarios. Section 4 presents and discusses the results that are obtained using our techniques and compares the energy consumption of our geo-localization solution with GPS-based solutions. Finally, Section 5 summarizes the paper's findings and provides recommendations for future work.

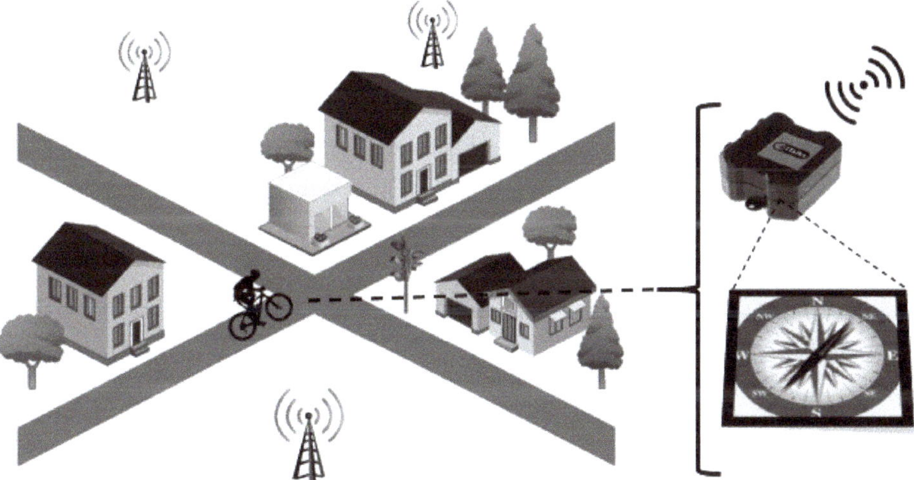

Figure 1. Measurement environment where a LoRaWAN node is attached to a bicycle. The node transmits the e-compass directional data to the network and the Time of Arrival is recorded on all gateways.

2. Related Work

The most accessible and available way to localize sensor nodes in a LoRaWAN network is by using the RSS metadata that is received by the gateways after an uplink. Such approaches have been studied in our previous works in [15,16]. In [15], the RSS received from up to three gateways was converted to a location using different algorithms, yielding a median accuracy between 1250 m and 2500 m. In case of frequent uplinks and when the mode of transportation is known, map matching further reduced the median error to 700 m. In [16], a mean accuracy around 400 m was obtained with an RSS fingerprinting method based on the collection of a large training database. Similar accuracies (around 350 m) were reported in [17] using other fingerprinting methods.

Current LoRaWAN gateways can accurately determine the Time of Arrival (ToA) of an incoming packet sent by a sensor node. Based on the observed time stamps and the known gateway locations, a Time-Difference-of-Arrival (TDoA) algorithm can estimate the location more accurately than RSS-based methods. Previous research [13,15,18] reported a median accuracy around 200 m when using this method in combination with a maximum likelihood (ML) algorithm. Similar accuracies of around 100 m were reported in [19] for the case of a privately deployed network with static nodes. Improving localization accuracy of static nodes is also investigated in [20], where multiple messages are merged to provide a more accurate estimate. The simulations in [20] showed that it was possible to achieve an average error of less than 100 m using this method. In [21], an improved TDoA algorithm is proposed and compared with a Least Squares (LS) approach. Ninety-five percent percentile values improved from 2200 m to 840 m in a simulated environment. Another approach, correcting the received timestamps of a mobile node by the use of machine learning in combination with stationary reference nodes, reported an accuracy around 61 m [22]. However, it is unclear how many reference nodes are needed, making it potentially unsuitable for large deployments from an economic point-of-view.

This paper will report results from experiments in real public LoRaWAN networks, without using GPS or any other additional infrastructure. Furthermore, it will consider non-stationary nodes and will compare the proposed method with available state-of-the-art TDoA algorithms.

3. Materials and Methods

3.1. Data Collection and Trajectories

The device for which we estimate its location and trajectory is a LoRaWAN sensor node (MCS iTalks 1608). This device is provisioned on either the Proximus (Belgium) or KPN (Netherlands) LoRa network. The device is configured to have the highest possible uplink frequency possible: every 5 s a 5-byte packet is transmitted on spreading factor (SF) 7. The airtime for this packet is 50 ms: this transmission configuration complies with the ETSI regulations of a 1 percent duty cycle [23]. The transmitted packets are received by the numerous gateways deployed in the outdoor area around the node, after which they are forwarded to the network server. The following data are available:

- A payload (5 bytes) from the device, containing the 5 compass values (also called heading or bearing) recorded during the last 5 s. Due to the unavailability of such a node, the compass values are emulated on a (Samsung Galaxy A20E) smartphone application (OSMTracker for Android), which was carried with the device.
- RSS value received at each gateway with its ID.
- Signal-to-Noise Ratio (SNR) of the received packet at each gateway with its ID.
- Timestamp of the received packet (nanosecond resolution) at each gateway denoted with its ID .

Furthermore, the network topology of all gateways is known for both Proximus and KPN networks (latitude/longitude coordinates for each gateway ID). The recorded data allow processing the data and converting them into the most likely locations and/or trajectory using a suitable algorithm. In order to estimate the localization accuracy, the ground truth (and time) was logged with a GPS application on the smartphone (OSMTracker for Android).

The LoRa device and the smartphone were carried in the front pocket of a jacket along 7 trajectories and using different transportation modes. The ground truth trajectories are shown in Figure 2 as a black trace. Each transportation mode (walking, cycling, and driving) was tested twice in a different area. For the second walking trajectory, we traveled the exact same trajectory (noted by "Walking 2A" and "Walking 2B"). Repeating the measurements in the same and/or different areas allows us to check the reproducibility of our results. The characteristics of the trajectories such as duration, velocity and distance are shown in Table 1. Nearly all measurements were performed in Ghent (Belgium) and made use of the Proximus network. The measurements for driving trajectory 1 is the only exception which was done in Eindhoven (Netherlands) and made use of the KPN network.

Table 1. Characteristics for the 7 mobile trajectories.

Trajectory	Duration	Distance	Avg. Velocity
Walking 1	40 min	3.1 km	4.6 km/h
Walking 2A	20 min	1.8 km	5.2 km/h
Walking 2B	19 min	1.8 km	5.4 km/h
Cycling 1	34 min	10.6 km	19 km/h
Cycling 2	21 min	6.5 km	19 km/h
Driving 1	18 min	16.2 km	55 km/h
Driving 2	19 min	8.5 km	26 km/h

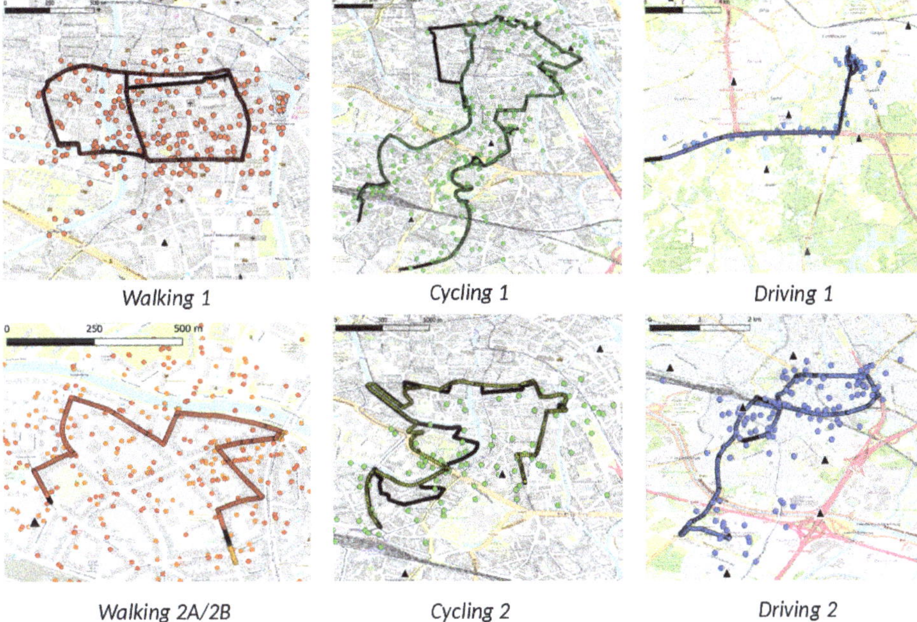

Figure 2. Trajectories with different mobility: left (walking), middle (cycling), and right (driving). The black trace denotes the ground truth. Localization by raw TDoA is shown as coloured dots. Gateways are shown as black triangles. The colored trace is the final result of combining TDoA and absolute compass heading in the map matching algorithm (Scenario OC-Abs. Heading).

3.2. Algorithm

The algorithms described hereafter were implemented in Python. Algorithm 1 shows the pseudocode of the map matching method with compass sensor fusion. The variables and different steps are discussed in the text below. The algorithm is initialized based on the first TDoA measurement (TM_0) for which the TDoA solver can calculate a location (L_0), i.e., if three gateways receive a packet. Next, a predefined number of other locations (MP) are selected around this location and their probability is initialized to one, e.g., the 1000 closest grid points to the current position. This ensures that the algorithm can recover from initially noisy data, e.g., 1000 grid points on a grid with a size of 10 m results in covered surfaces of around 50 hectares (the exact area depends on the density of the road network). This initialization phase forms the starting point of all possible paths that are kept in the memory of the location tracking algorithm (*paths*).

Algorithm 1: Map matching with compass data.

Input: compass data + TDoA measurements (TMS)
Result: map matched trajectory (MMT)
$MoT \leftarrow$ mode of transportation (known in advance);
$TM_0 \leftarrow$ first TDoA measurements;
$L_0 \leftarrow$ location based on TM_0;
$t_{prev} \leftarrow$ first timestamp;
$MP \leftarrow 1000$;
`// maximum number of paths in memory`
$L_R \leftarrow$ real time location
$paths \leftarrow$ list with MP grid points closest to L_0 initialized with probability 1;
`// iterate over all TDoA measurements`
for $TM \in TMS$ **do**
    $t \leftarrow$ current timestamp;
    $\Delta t \leftarrow t - t_{prev}$;
    $tmp \leftarrow$ empty list;
    for $path \in paths$ **do**
        $P \leftarrow$ current probability of $path$;
        $E \leftarrow$ current endpoint of $path$ (parent);
        $RGP \leftarrow$ reachable grid points along roads with MoT within time span Δt starting from
        E;
        $CD \leftarrow$ compass data between t_{prev} and t;
        `// update new path probabilities`
        for $CP \in RGP$ **do**
            `// CP: candidate position`
            $RS \leftarrow$ road segments between E and CP;
            `// compass probability`
            $P_{comp} \leftarrow$ probability of RS given CD;
            `// tdoa probability`
            $P_{tdoa} \leftarrow$ probability of CP given TM;
            $path_{new} \leftarrow path + RS + CP$;
            $P_{new} \leftarrow P \cdot P_{tdoa} \cdot P_{comp}$;
            add $(path_{new}, P_{new})$ to tmp
        end
    end
    $paths \leftarrow$ retain MP paths from tmp based on highest probability;
    $L_R \leftarrow$ endpoint of most likely path from tmp $t_{prev} \leftarrow t$;
end
$MMT \leftarrow$ reconstruct trajectory along $path$ with highest probability in $paths$

Then, for the subsequent TDoA measurements (TM), all reachable grid positions (RGP) starting from the path's current endpoint (E) are determined for all *paths* in memory, by using the surrounding road network; the elapsed time since last location update (Δt); the known mode of transportation (MoT); and OpenStreetMap metadata, i.e., maximum speed, road type, and one-way information. These reachable grid positions are the candidate positions for the next location update. Transitions between grid points are constrained by the road infrastructure. Each candidate position (CP) retains a link to its parent (i.e., the previous endpoint E), a list with visited road segments RS, and a probability that represents this new branch along the road network. This new path (branch) and updated probability are added to the temporary list (tmp) as a tuple $(path_{new}, P_{new})$. The updated probability is the product of the previous probability P with a TDoA and compass contribution (P_{tdoa} and P_{comp}).

P_{tdoa} is based on the probability of the TDoA measurements given CP, the gateway locations, and the standard deviation of LoRa TDoA measurements, e.g., 383 m in our experimental validation. P_{comp} is based on the probability of the compass data between t_{prev} and t given the bearing of the visited road segments RS and is calculated with a standard deviation of $60°$.

The MP paths with the highest probability are retained and serve as input for the next iteration. After all TDoA measurements are processed, the entire trajectory of the path with highest probability in memory is reconstructed. The trajectory is only estimated after all measurements are processed but it is also possible to provide real time location output by retaining the endpoint of the most likely path from tmp in each iteration.

3.3. Scenarios

Different scenarios are evaluated, all of which are applicable for different use cases. First, we distinguish between Real-Time (R) Tracking and Offline Tracing (O). In the case of Real-Time Tracking, the localization result is made available immediately after the packet was received. For the case of offline tracing, it is assumed one only requires the fully estimated trajectory at a later instance (this can provide a more accurate trajectory estimation thanks to more data being used). Second, we also differentiate between a device which does not contain a compass (A, for agnostic) and a device that does contain a compass sensor (C). To emulate a device without compass, we ignore the recorded compass data in our algorithms. We denote the agnostic or compass-enabled device by the letters "A" and "C", respectively. It is therefore possible to combine these considerations into 4 scenarios:

- RA: Real-Time Agnostic: Tracking is needed in real-time and no compass is available in the device
- RC: Real-Time Compass-enabled: Tracking is needed in real-time and compass data are available.
- OA: Offline Agnostic: Offline Tracing at a later instance is needed, the device has no compass data available.
- OC: Offline Compass: Offline Tracing at a later instance is needed with a compass-enabled device.

For each of the 4 scenarios we consider at least 2 methods/algorithms to perform the respective localization and evaluate them for the different trajectories. For the cases that use compass data, there are 2 possibilities. In the first implementation, it is assumed that the compass produces absolute headings. For example, 0 degrees means the asset moves towards the North. This implementation can be used for cases for which it is known how the compass was installed relative to the tracked device, e.g., for tracking bicycles. The second implementation does not rely on this assumption and uses only relative headings, e.g, a transition from a 90 to a 180 degree bearing or from a 30 to a 120 degree bearing means a turn (of 90 degrees) to the right. For example, this method can be used for tracking parcels in transit, where we do not know the absolute orientation of the installed compass node relative to the direction of movement of the tracker. Table 2 gives an overview of all the methods used in this paper.

Table 2. Overview of the different methods used to obtain the real-time estimated locations and/or reconstructed trajectory.

Scenario	Method	Output	Map Matched	Compass Used	Description
RA	Semtech	Real-Time location	no	no	TDoA solver from Semtech [24]
	Semtech mm	Real-Time location	yes	no	solver, next 'snap to road' [13]
	TDoA mm	Real-Time location	yes	no	algorithm 1 with Pcomp=1 and output L_R
OA	Semtech mm	Reconstructed Trajectory	yes	no	solver, next estimate trajectory [13]
	TDoA mm	Reconstructed Trajectory	yes	no	algorithm 1 with Pcomp=1 and output MMT
RC	Rel. Heading	Real-Time location	yes	yes-Relative	algorithm 1 with output L_R
	Abs. Heading	Real-Time location	yes	yes-Absolute	algorithm 1 with output L_R
OC	Rel. Heading	Reconstructed Trajectory	yes	yes-Relative	algorithm 1 with output MMT
	Abs. Heading	Reconstructed Trajectory	yes	yes-Absolute	algorithm 1 with output MMT

An illustrative example showing the difference between the use of relative and absolute compass headings is shown is Figure 3. The performance of all our implementation methods for the various scenarios is compared to a state-of-the-art commercial solver from Semtech [24], which takes the raw timestamps and gateway locations as inputs.

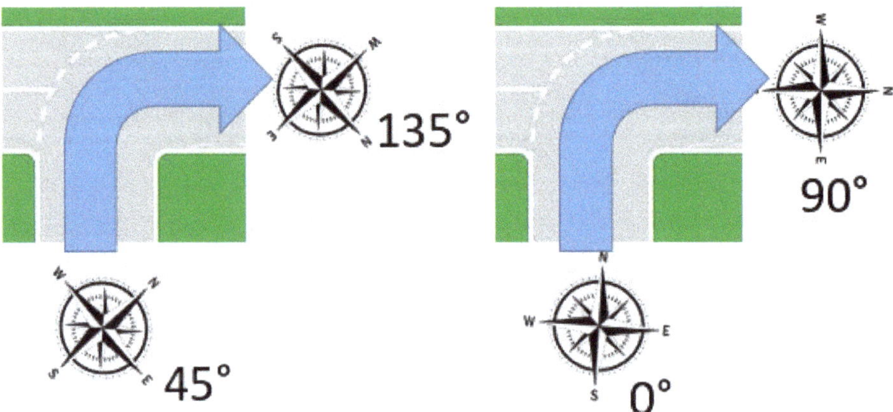

Figure 3. Left figure: compass is misaligned (parcel tracking use case) and heading goes from 45° to 135°. We only know that the relative heading is therefore +90°, e.g., a turn to the right. Right figure: compass is aligned (bicycle tracking use case) and heading goes from 0° to 90°. We now know the tracker went from South–North to East–West on the road map.

4. Results

4.1. Localization Accuracy

Figure 4a–d shows the accuracy CDF of the various scenarios, methods, and trajectories. The median (p50) and 90th percentile (p90) accuracy metrics are obtained from each CDF and are summarized in Table 3. The tracking results from method RA/Semtech for the walking, cycling, and driving mobility cases are displayed as dots on the different respective maps in Figure 2. The estimated trajectory of the OC/abs heading scenario is also shown as a colored trace on the maps. Throughout this paper each estimated result (localization on the map, accuracy CDF) is denoted by a different color according to the mobility scenario being evaluated: red corresponds with walking, while green and blue are assigned to cycling and driving. From the maps we can clearly see the improvement made: while the dots give a very rough estimate about the location in real-time, our tracing result nearly overlaps with the ground truth.

Relative improvements (in percentages) when comparing a method to the RA/Semtech traditional localization solver (no compass, no map matching) are shown in Table 4. This table shows that all our methods result in improvements ranging from 1 to 92 percent when compared to the Semtech localization solver. Large improvements are possible, especially when the mobility is low. According to all four CDFs the walking (red lines) did better then the cycling (green lines), which in turn did better than the driving (blue lines) mobility case. This is due to the fact fewer candidate positions are selected for slower modes of transportation (MoT). Regardless of the method used, a minimum improvement of 45 percent is possible for the walking/slow moving case. Therefore, our first recommendation is to let the user supply a mobility motion indicator to each asset to be tracked in case the asset is very likely to be constrained on the road infrastructure (e.g., a bicycle).

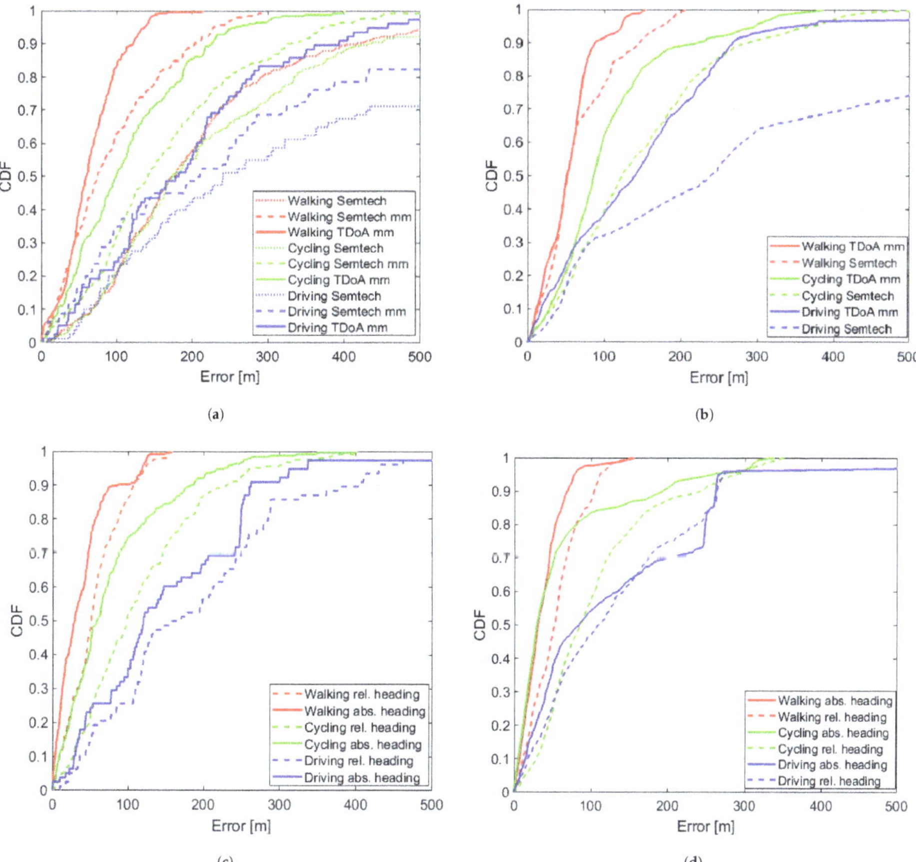

Figure 4. CDFs of the different scenarios evaluated for the walking 1, cycling 1, and driving 1 trajectories.
(**a**) Real-time Agnostic (**RA**) scenario: tracking in real-time and no compass is available in the device.
(**b**) Offline Agnostic (**OA**) scenario: offline tracing at a later instance, the device has no compass.
(**c**) Real-time Compass (**RC**) scenario: tracking is needed in real-time and a compass is available.
(**d**) Offline Compass (**OC**) scenario: offline tracing with a compass enabled device.

Table 3. Summary of the obtained p_{50}/p_{90} accuracies (in meters) for the various scenarios, methods, and trajectories.

Scenario	Method	Transportation Mode						
		WALK 1	WALK 2A	WALK 2B	CYCLE 1	CYCLE 2	DRIVE 1	DRIVE 2
RA	Semtech	175/413	187/494	226/501	175/433	197/704	242/1165	206/599
	Semtech mm	75/203	102/169	101/209	142/328	159/469	208/574	177/436
	TDoA mm	60/122	69/147	58/120	100/226	141/290	181/384	157/421
OA	TDoA_mm	52/89	39/63	68/120	87/222	147/417	140/268	126/291
	Semtech_mm	52/146	48/89	30/175	124/288	169/569	239/881	160/460
RC	Rel. Heading	52/107	45/149	49/149	99/230	149/319	164/397	154/421
	Abs. Heading	30/84	21/137	20/75	60/184	94/280	121/262	148/367
OC	Rel. Heading	52/104	22/42	41/137	86/241	156/412	109/264	118/359
	Abs. Heading	31/69	17/35	19/47	29/193	47/289	84/262	68/227

When considering the cases for a device without compass (scenarios RA and OA), we see from Table 4 that the best approach is to work with the raw timestamps instead of processed locations for the map matching algorithm: TDoA mm performance is always better than Semtech mm. This is our second recommendation and its impact is also visible in Figure 4a,b.

Table 4. Percentual reduction in median positioning error for the various scenarios, methods, and trajectories.

Scenario	Method	Transportation Mode							
		WALK 1	**WALK 2A**	**WALK 2B**	**CYCLE 1**	**CYCLE 2**	**DRIVE 1**	**DRIVE 2**	**AVERAGE**
RA	Semtech	Reference	Reference	Reference	Reference	Reference	Reference	Reference	**Reference**
	Semtech mm	57	45	55	19	19	14	14	**32**
	TDoA mm	66	63	74	43	28	25	24	**46**
OA	TDoA_mm	70	79	70	50	25	42	39	**54**
	Semtech_mm	70	74	87	29	14	1	22	**42**
RC	Rel. Heading	70	76	78	43	24	32	25	**50**
	Abs. Heading	83	89	91	66	52	50	28	**66**
OC	Rel. Heading	70	88	82	51	21	55	43	**59**
	Abs. Heading	82	91	92	83	76	65	67	**79**

Table 3 shows that offline tracing (scenarios OA and OC) gives better localization results than their tracking counterparts (scenarios RA and RC). This is due to the fact that when the route is reconstructed all historic data from the start till the end of the measurement are used. Our third recommendation is therefore to implement a feature for which the tracing history can be viewed along with its current most likely localization result.

Relative comparisons for the cases with a compass versus no compass are further shown in Table 5. This table shows that using sensor fusion with a compass improves the localization results: Depending on the mobility a real time median accuracy improvement of up to 70% was possible when the sensor was aligned (which provides an absolute heading = 0 degrees for a moving asset heading South–North). For the reconstructed tracing result, this improvement was up to 72% in the aligned case. When the sensor was not aligned (relative heading) the improvement was up to 35% in the real-time tracking case to 44% for the reconstructed tracing case. The final recommendation is therefore to embed a compass in the LoRa sensor node and align it with the movement direction (if possible)

Table 5. Relative improvement percentage when comparing the scenarios with compass to the absence of such compass.

Scenario	Method	Transportation Mode						
		WALK 1	**WALK 2A**	**WALK 2B**	**CYCLE 1**	**CYCLE 2**	**DRIVE 1**	**DRIVE 2**
RC vs RA	Rel. Heading	13	35	16	1	−6	9	2
	Abs. Heading	50	70	66	40	33	33	6
OC vs OA	Rel. Heading	0	44	40	1	−6	22	6
	Abs. Heading	40	56	72	67	68	40	46

4.2. Discussion

Table 4 shows that regardless of the method used, improvements (between 1% and 92%) are obtained for all seven trajectories. This clearly demonstrates that our methods increase accuracy of the raw output of the Semtech localization solver ("Semtech"), which here serves as reference method. Map matching these raw outputs ("Semtech mm") always produces better results, with improvements between 14% and 57%. Working with the raw timestamps (TDoA mm) instead of already processed locations in our map matching algorithm gives additional improvements. In Table 4, we note all percentages of TDoA mm are better than Semtech mm regardless if the scenario is offline tracing (OC) or real-time tracking (RA). When adding relative compass data to our TDoA mm algorithm, further improvement of accuracy is shown in Table 5 for nearly all (6/7) trajectories with only 1 exception for the second cycling trajectory (the degradation was only 6 %). When the compass is indicating 0 degrees for a movement from South to North and does not "rotate", while the tracker (and asset) are in transit (absolute compass) additional improvement is always guaranteed (Table 5) for all trajectories. Table 3 shows that our algorithm provides a median accuracy lower then 85 m while the Semtech solver median accuracy is always higher than 175 m. The same reasoning can be done for 90th percentile error which is always larger than 400 m, while our algorithm guarantees less then 290 m for this metric on all trajectories. The observations are summarized in Table 6.

Table 6. General observations.

Test Case	True For
All methods better than Semtech	7/7 trajects
Semtech mm better than Semtech	7/7 trajects
TDoA mm better than Semtech mm (RA and OA)	6/7 trajects
Adding relative compass improves results (RC vs RA and OC vs OA)	6/7 trajects
Adding absolute compass improves results (RC vs RA and OC vs OA)	7/7 trajects
p50 Semtech >175 m and p50 OC-Abs <85 m	7/7 trajects
p90 Semtech >400 m and p90 OC-Abs <290 m	7/7 trajects

When repeating the measurement along the exact same trajectory on a different day ("Walking 2A" and "Walking 2B"), we observe two different raw Semtech localization results (orange vs. red dots in Figure 2). After application of our OC/Abs algorithm, the same trajectory is estimated which in turn overlaps the ground truth trace. This clearly demonstrates the reproducibility of the obtained results.

The above analysis shows that the map matching algorithm delivers a first accuracy improvement over the raw data, as it is able to restrict to paths that are physically possible, given the road network topology and speed limits (with margin). Adding compass data further reduces localization errors. Absolute heading data gives information on the estimated travel direction, possibly excluding other candidate paths that remained after map matching. If absolute heading data is not available, relative heading data is still able to detect direction changes. When mapping these direction changes (e.g., a turn right of 60 degrees) to the road network, it is clear that certain trajectories will be more likely than others. Results confirm that the algorithm intelligently makes use of recorded location and heading data, by investigating to what extent these input data match with the different possible trajectories along the road network. It can be expected that results will be better for sparse road networks, as fewer alternatives are available. On the other hand, for grid street plans like in New York, the algorithm might have more trouble distinguishing between two parallel streets, as such grid plan contains only two street directions and allows only 90 degree turns.

4.3. Energy Consumption

In this section we briefly compare two common tracking implementations in terms of energy consumption. First, we calculate the energy consumption when a node sends its GPS coordinates using a GNSS receiver over LoRaWAN (GPS-over-LoRa). In the second implementation, we consider the proposed LoRaWAN tracking with compass. For both implementations, we consider a tracking update interval of 5 s and we quantify the energy consumption in mAh. For a node equipped with a GNSS module, the minimum on-time for a location fix is about 1 s [25]. During this acquisition period the receiver typically consumes 30 mA [25]. Therefore, the consumed acquisition energy per location update is

$$E_{GPS} = \frac{1}{3600} * 30 = 0.0083 \text{ mAh} \tag{1}$$

Transmission of 5 bytes on SF7 over LoRaWAN takes 0.05 s airtime while consuming 40 mA of current in a typical transceiver [26]. Therefore, the consumed energy per location update is

$$E_{LoRa} = \frac{0.05}{3600} * 40 = 0.00056 \, \text{mAh} \tag{2}$$

The energy consumption for continuous sampling with a (low-power) typical magnetometer sensor [27] is very low and and can be estimated as

$$E_{magneto} = \frac{5}{3600} * 0.040 = 0.000056 \, \text{mAh} \tag{3}$$

The energy reduction factor R for our proposed LoRa+compass solution versus a GPS over LoRa implementation is approximately given by

$$R = \frac{E_{GPS-over-LoRa}}{E_{LoRa+compass}} = \frac{E_{GPS} + E_{LoRa}}{E_{LoRa} + E_{magneto}} = 14 \tag{4}$$

Our approach is thus 14 times more energy efficient than a GPS-over-LoRa solution. Furthermore the price of integrating a magnetometer (e.g., LIS3MDL) is lower when compared to integrating a GNSS receiver (e.g., CAM-M8Q): 0.8€ compared to 13€ (for an order quantity of 500 samples). Although a GPS-based solution will still deliver superior accuracies, the lower cost and energy consumption will make our solution preferable for use cases where positioning accuracy requirements are not so strict (e.g., tracking of truck trailers). Although we only compared two tracking implementations, there are some other solutions such as Wi-Fi scanning and transmitting MAC addresses and signal strength levels of access points in the area over LPWAN, followed by a calculation in a dedicated server (e.g., Skyhook). The accuracy and energy consumption are in between the two earlier investigated approaches. A disadvantage of this method is maintaining an accurate and up to date database with respect to the locations of the access points. For more information on this implementation we refer to the work in [28].

5. Conclusions

In this paper, we proposed methods to track and trace assets building on a combination of LoRaWAN technology and a compass sensor. When comparing with a standard TDoA solver (Semtech), we obtained an improvement between 14% and 79% depending on the mobility scenario (walking, cycling, and driving) and whether the result should be immediately available (Real-Time-Tracking) or known at a later instance (Offline-Tracing). These improvements were possible thanks to the map matching algorithm proposed in this paper. Combined with a compass in the sensor node, the bearing information is transmitted periodically to give prevalence to specific paths for which the bearing matches the heading of the road segment. Using this sensor fusion approach, an additional improvement between 1% and 72% was obtained compared to TDoA map matching. The improvement depends on the mobility of the node/asset, usage of the compass node (fixed or rotatable) and whether the result should again be immediately available (Real-Time-Tracking) or known at a later instance (Offline-Tracing). Our best result was obtained for a walking scenario with a fixed (0 degrees means towards "North") compass for which we obtained a median accuracy of 17 m (an improvement of 91% versus Semtech). All results were obtained from experiments and the reproducibility was verified. Although our results are not as accurate as GPS, we clearly demonstrated that our implemented geo-tracking is 15 times more energy efficient and far less expensive to implement. Further research directions are using the same techniques for LoRaWAN networks which only have RSS values available and future networks which have also Angle of Arrival (AOA) capabilities. Other future work is to expand our implementations to other LPWAN technologies such as SigFox and Narrow-band IoT.

Sensors **2020**, *20*, 5815

Author Contributions: Conceptualization, D.P. and N.P.; methodology, D.P., N.P., and W.J.; software, N.P. and J.T.; validation, D.P., N.P., and W.J.; formal analysis, N.P. and J.T.; investigation, N.P.; resources, N.P. and R.B., M.A., and M.W.; data curation, N.P., R.B., and M.A.; writing—original draft preparation, N.P.; writing—review and editing, N.P., D.P., J.T., and W.J.; visualization, N.P.; supervision, D.P., W.J., and L.M. All authors have read and agreed to the published version of the manuscript.

Funding: This research received no external funding.

Conflicts of Interest: The authors declare no conflicts of interest.

Abbreviations

The following abbreviations are used in this manuscript.

GPS	Global Positioning System
LoRaWAN	Long-Range Wide Area Network
TDoA	Time Difference of Arrival
LoRa	Long Range
IoT	Internet of Things
LPWAN	Low-Power Wide Area Network
NB-IoT	Narrowband Internet of Things
GNSS	Global Navigation Satellite System
RSS	Received Signal Strength
ToA	Time of Arrival
ML	Maximum Likelihood
SF	Spreading Factor
SNR	Signal-to-Noise Ratio
CDF	Cumulative Distribution Function

References

1. Grunwald, A.; Schaarschmidt, M.; Westerkamp, C. LoRaWAN in a rural context: Use cases and opportunities for agricultural businesses. In Proceedings of the 24th ITG-Symposium on Mobile Communication-Technologies and Applications, Osnabrueck, Germany, 15–16 May 2019.
2. Ali, T.A.; Choksi, V.; Potdar, M. Precision Agriculture Monitoring System Using Green Internet of Things (G-IoT). In Proceedings of the 2018 2nd International Conference on Trends in Electronics and Informatics (ICOEI), Tirunelveli, India, 11–12 May 2018.
3. Poenicke, O.; Kirch, M.; Richter, K.; Schwarz, S. LoRaWAN for IoT Applications in Air Cargo-Development of a GSE Tracking System for DHL Air Cargo Hub Leipzig. In Proceedings of the Smart SysTech 2018, European Conference on Smart Objects, Systems and Technologies, Munich, Germany, 12–13 June 2018.
4. Li, Q.; Liu, Z.; Xiao, J. A Data Collection Collar for Vital Signs of Cows on the Grassland Based on LoRa. In Proceedings of the 2018 IEEE 15th International Conference on e-Business Engineering (ICEBE), Xi'an, China, 12–14 October 2018.
5. Boshita, T.; Suzuki, H.; Matsumoto, Y. Compression Method of Position Information for IoT-based Bus Location System Using LoRaWAN. In Proceedings of the 2018 Eleventh International Conference on Mobile Computing and Ubiquitous Network (ICMU), Auckland, New Zealand, 5–8 October 2018.
6. Hattarge, S.; Kekre, A.; Kothari, A. LoRaWAN based GPS Tracking of City-buses for Smart Public Transport System. In Proceedings of the 2018 First International Conference on Secure Cyber Computing and Communication (ICSCCC), Jalandhar, India, 15–17 December 2018.
7. Actility, IoT Geolocation Enables Asset Tracking for Logistics. Available online: https://www.actility.com/iot-geolocation-enables-asset-tracking-for-logistics/ (accessed on 13 October 2020).
8. Mekki, K.; Bajic, E.; Chaxel, F.; Meyer, F. A comparative study of LPWAN technologies for large-scale IoT deployment. *ICT Express* **2019**, *5*, 1–7. [CrossRef]

9. Ratasuk, R.; Vejlgaard, B.; Mangalvedhe, N.; Ghosh, A. NB-IoT System for M2M Communication. In Proceedings of the 2016 IEEE Wireless Communications and Networking Conference, Doha, Qatar, 3–6 April 2016.

10. Zuniga, J.C.; Ponsard, B. Sigfox System Description. *LPWAN@ IETF97 Nov. 14th* **2016**, *25*. Available online: https://www.ietf.org/proceedings/97/slides/slides-97-lpwan-25-sigfox-system-description-00 (accessed on 13 October 2020).

11. Poursafar, N.; Alahi, M.E.E.; Mukhopadhyay, S. Long-range Wireless Technologies for IoT Applications: A review. In Proceedings of the 2017 Eleventh International Conference on Sensing Technology (ICST), Sydney, NSW, Australia, 4–6 December 2017.

12. Croce, D.; Garlisi, D.; Giuliano, F.; Valvo, A.L.; Mangione, S.; Tinnirello, I. Performance of LoRa for Bike-Sharing Systems. In Proceedings of the 2019 AEIT International Conference of Electrical and Electronic Technologies for Automotive (AEIT AUTOMOTIVE), Torino, Italy, 2–4 July 2019.

13. Podevijn, N.; Plets, D.; Trogh, J.; Martens, L.; Suanet, P.; Hendrikse, K.; Joseph, W. TDoA-based outdoor positioning with tracking algorithm in a public LoRa network. *Wirel. Commun. Mob. Comput.* **2018**, *2018*. [CrossRef]

14. Podevijn, N.; Trogh, T.; Aernouts, M.; Berkvens, R.; Martens, L.; Weyn, M.; Joseph, W.; Plets, D. Compass Aided TDoA Tracking in LoRaWAN networks. In Proceedings of the 2020 IEEE/ION Position, Location and Navigation Symposium (PLANS), Portland, OR, USA, 20–23 April 2020.

15. Plets, D.; Podevijn, N.; Trogh, J.; Martens, L.; Joseph, W. Experimental Performance Evaluation of Outdoor TDoA and RSS Positioning in a Public LoRa Network. In Proceedings of the 2018 International Conference on Indoor Positioning and Indoor Navigation (IPIN), Nantes, France, 24–27 September 2018.

16. Aernouts, M.; Berkvens, R.; Van Vlaenderen, K.; Weyn, M. Sigfox and LoRaWAN Datasets for Fingerprint Localization in Large Urban and Rural Areas. *Data* **2018**, *3*, 13. [CrossRef]

17. Anagnostopoulos, G.G.; Kalousis, A. A Reproducible Comparison of RSSI Fingerprinting Localization Methods Using LoRaWAN. *arXiv* **2019**, arXiv:1908.05085.

18. Podevijn, N.; Trogh, J.; Karaagac, A.; Haxhibeqiri, J.; Hoebeke, J.; Martens, L.; Suanet, P.; Hendrikse, K.; Plets, D.; Joseph, W. TDoA-based Outdoor Positioning in a Public LoRa Network. In Proceedings of the 12th European Conference on Antennas and Propagation (EuCAP 2018), London, UK, 9–13 April 2018.

19. Fargas, B.C.; Petersen, M.N. GPS-free Geolocation Using LoRa in Low-power WANs. In Proceedings of the 2017 global internet of things summit (Giots), Geneva, Switzerland, 6–9 June 2017.

20. Bakkali, W.; Kieffer, M.; Lalam, M.; Lestable, T. Kalman Filter-based Localization for Internet of Things LoRaWAN™ End Points. In Proceedings of the 2017 IEEE 28th Annual International Symposium on Personal, Indoor, and Mobile Radio Communications (PIMRC), Montreal, QC, Canada, 8–13 October 2017.

21. Ghany, A.; Uguen, B.; Lemur, D. A Parametric TDoA Technique in the IoT Localization Context. In Proceedings of the 2019 16th Workshop on Positioning, Navigation and Communications (WPNC), Bremen, Germany, 23–24 October 2019.

22. Cho, J.; Hwang, D.; Kim, K.H. Improving TDoA Based Positioning Accuracy Using Machine Learning in a LoRaWan Environment. In Proceedings of the 2019 International Conference on Information Networking (ICOIN), Kuala Lumpur, Malaysia, 9–11 January 2019.

23. ETSI. ETSI TR 103 526; System Reference document (SRdoc); Technical Characteristics for Low Power Wide Area Networks Chirp Spread Spectrum (LPWAN-CSS) Operating in the UHF Spectrum Below 1 GHz. Technical Report. 2018. Available online: https://www.etsi.org (accessed on 13 October 2020).

24. Semtech. LoRaCloud Geolocation. Technical Report. 2020. Available online: https://www.loracloud.com/portal/geolocation (accessed on 13 October 2020).

25. ublox. CAM-M8 Datasheet. Technical Report. 2019. Available online: https://www.u-blox.com (accessed on 13 October 2020).

26. Microchip. RN2483 Datasheet. Technical Report. 2019. Available online: http://www.microchip.com (accessed on 13 October 2020).

27. ST. LIS3DML Datasheet. Technical Report. 2019. Available online: http://www.st.com (accessed on 13 October 2020).
28. Skyhook. Wi-Fi Scanning, Power Consumption and Bandwidth for LPWAN. Technical Report. 2020. Available online: https://www.skyhook.com (accessed on 13 October 2020).

 *sensors*

Article

A Robust Multi-Sensor Data Fusion Clustering Algorithm Based on Density Peaks

Jiande Fan, Weixin Xie * and Haocui Du

Automatic Target Recognition (ATR) Key Laboratory, Shenzhen University, Shenzhen 518060, China; jdfan@szu.edu.cn (J.F.); hcdu@szu.edu.cn (H.D.)
* Correspondence: wxxie@szu.edu.cn; Tel.: +86-0755-2673-2055

Received: 31 October 2019; Accepted: 24 December 2019; Published: 31 December 2019

Abstract: In this paper, a novel multi-sensor clustering algorithm, based on the density peaks clustering (DPC) algorithm, is proposed to address the multi-sensor data fusion (MSDF) problem. The MSDF problem is raised in the multi-sensor target detection (MSTD) context and corresponds to clustering observations of multiple sensors, without prior information on clutter. During the clustering process, the data points from the same sensor cannot be grouped into the same cluster, which is called the cannot link (CL) constraint; the size of each cluster should be within a certain range; and overlapping clusters (if any) must be divided into multiple clusters to satisfy the CL constraint. The simulation results confirm the validity and reliability of the proposed algorithm.

Keywords: clustering; data fusion; target detection

---

1. Introduction

As a powerful tool, clustering analysis is usually used in machine learning [1], image analysis [2], information retrieval [3] and data mining [4] to eliminate noise data-points and find hidden groups or patterns in a dataset. Due to the diversity/variability of the dataset to be processed, many clustering algorithms, such as density-based clustering [5,6], hierarchical clustering [7], and k-means clustering [8], have been developed to solve specific problems. It can be seen that, although there are many clustering algorithms, none of them can be applied in all cases.

Clustering is often taken as an unsupervised learning technique in many pre-processing processes, as no information is provided. Nevertheless, for many of the problems, including the MSDF clustering problem, to be solved in this paper, an amount of prior information can be obtained through additional data features [9,10], which can be employed to obtain better clustering results, namely, semi-supervised clustering.

Constraining the dataset during the clustering process to obtain specific clustering results is a hot issue in clustering research. In constrained clustering, "must-link" constraints (ML) and "cannot-link" constraints (CL) are two basic rules. An ML constraint is used to specify that the two instances should be associated with the same cluster, whereas a CL constraint is used to specify that the two instances should assigned to different clusters, allowing users to specify constraint rules to obtain the desired clustering results. Typical constrained clustering algorithms include the constrained k-means [11], pairwise constrained k-means [12], complete link [13], constrained hierarchical clustering algorithms [14].

In clustering research, the number of clusters and cluster center initialization have a great impact on the clustering convergence speed and clustering result. The research on the number of clusters mainly focuses on running the clustering algorithm multiple times, with different values of k, and the estimated k is chosen based on a specific criterion, such as the Bayesian information criterion [15], rate distortion theory [16], Akaike information criterion [17], etc. The research on cluster center initialization mainly focuses on how to maximize the distance of the initial cluster center through statistical information, such as the *k*-means++ [18].

In this paper, we address the problem of MSDF, which is involved in multi-sensor multi-target tracking [19], using the density peaks clustering (DPC) algorithm [5]. The DPC algorithm was published in the journal, Science, in 2014. The core idea of DPC is that cluster centers are characterized by a higher density and a relatively longer distance. The outstanding performance of DPC has attracted many scholars' attention, and many variants based on DPC have been proposed to address various clustering problems, such as BDDPC [20], and DPC-KNN [21]. In this paper, the original DPC algorithm is shown to be able to handle the type of dataset that contains observations of targets that obey a zero-Gaussian distribution well. Therefore, we use the original DPC algorithm to solve the MSDF problem. The purpose of clustering is to divide the observations of targets into multiple clusters; overlapping clusters (if any) must be divided into multiple sub-clusters to satisfy the CL constraint; the data points in each cluster correspond to the observations of targets; and the cluster center is the estimated target position.

In the past, the MSDF problem was solved using a model-based method [22,23], but in recent years, many scholars have begun to solve the MSDF problem using a clustering-based method. Tiancheng Li has produced a lot of groundbreaking work [24–27] on this issue. He not only solves the MSDF problem using density-based clustering, but also uses multi-sensor clustering to improve the performance of a model-based filter [28]. Tianxian Zhang uses the clustering method to solve the MSTD problem with a distributed radar network [29]. This paper made several improvements on the basis of Li's work: (1) A more accurate and robust clustering algorithm, based on DPC, is proposed; and (2) the proposed algorithm can accurately filter out clutter, and the performance of the algorithm does not vary with the change of the detection probability of the sensors, which has great advantages, given the low detection probability of sensors.

The rest of the paper is organized as follows. The problem model is discussed in Section 2. Section 3 presents details of the proposed clustering method. Section 4 discusses the experimental simulation results, which are summarized in Section 5.

2. Multi-Sensor Data Fusion

2.1. Multi-Sensor Data Fusion

The MSDF problem involves estimating the state of the unknown number and unknown motion mode targets in the presence of noise data, which has a wide application space in the remote sensing image fusion, oceanography and military fields. The general MSDF problem can be modeled by the following assumptions:

Assumption 1. *Each target evolves and generates observations/measurements independently from the others.*

Assumption 2. *The observations of targets obey a zero-Gaussian distribution. Both the noise and observations constitute the measurement dataset of each sensor.*

Assumption 3. *One target can generate no more than one measurement in each scan.*

Assumption 4. *The distribution density of the clutter is significantly lower than the density of the observations of targets.*

The goal of the MSDF is to distinguish the observations of each target from those of others using a clustering method, as shown in Figure 1.

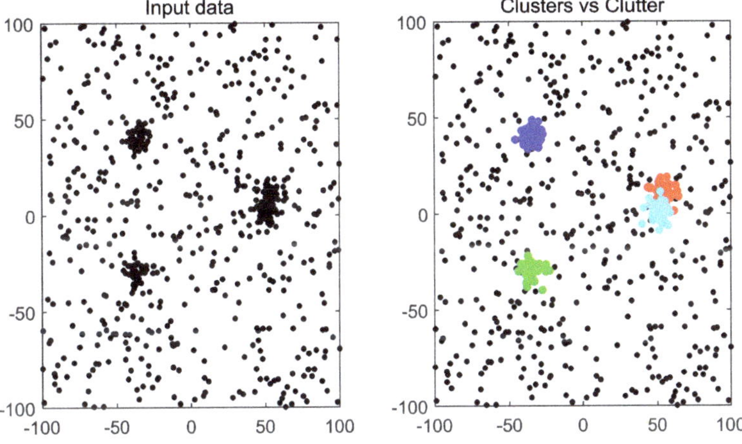

Figure 1. Multi-sensor i.i.d data points.

2.2. Problem Formulation

The above MSDF problem can be formulated as a CL-constrained clustering problem. Considering a dataset Z, which consists of observations from multiple sensors, z_i is included in dataset Z.

$$z_i \in P, i = 1, \cdots, N \tag{1}$$

where parameters N and P are the number of data points and the parameter space, respectively. In this paper, we define z_i as a point in a two-dimensional Cartesian coordinate system.

The dataset Z can be written in the form of a union of multi-sensor observations. We define the sth sensor as $S_s = \left\{ z_1^s, z_2^s, \cdots, z_{m_s}^s \right\}$, where m_s is the number of data-points, and all the data-points in Z can be written as:

$$Z := \{S_1, S_2, \cdots, S_n\} = \left\{ z_1^1, z_2^1, \cdots, z_{m_1}^1, z_1^2, z_2^2, \cdots, z_{m_2}^2, \cdots, z_1^n, z_2^n, \cdots, z_{m_n}^n \right\} \tag{2}$$

where n is the number of sensors. The MSDF problem requires that the dataset Z be divided into k clusters, namely, $C_1, C_2, \cdots, C_k$, and the CL constraint requires that:

$$c \neq (z_i^s, z_j^s), \forall i, j \in \{1, 2, \cdots, m_s\}, s \in \{1, 2, \cdots, n\} \tag{3}$$

where $c \neq (z_i^s, z_j^s)$ means z_i^s, z_j^s cannot be within the same cluster.

We define the set of noisy data points in dataset Z as C_0 and the observations of targets as C_T. The dataset Z can be defined as:

$$Z = C_0 \cup C_1 \cup C_2 \cup \ldots \cup C_k = C_0 \cup C_T, \, k \in T \tag{4}$$

At the same time, each cluster cannot have any intersection with the rest of the subsets.

$$C_i \cap C_j = \Phi, \forall i, j \in \{1, 2, \cdots k, 0\} \tag{5}$$

As mentioned above, the MSDF clustering problem can be described as: A dataset Z is divided into k clusters, the size of each cluster must satisfy the CL constraint (3), and each cluster cannot have any intersection with the others (5).

2.3. CL Constraint and the Size of Clusters

The CL constraint (3) limits the size of each cluster, which must be smaller or equal to the number of sensors n.

$$|C_i| \lesssim n, \forall i \in \{1, 2, \cdots k\} \tag{6}$$

where $|C_i|$ means the number of data points in cluster C_i, $\lesssim$ means smaller than or equal to. Denoting the detection probability of the sensor s on target i as $p_D^s(i) \leq 1$, to simplify the calculation, we simplify $p_D^s(i)$ as a constant p_D, then the size of a cluster can be calculated as:

$$E[|C_i|] = \sum_{s=1}^{n_i} p_D \leq n_i \tag{7}$$

Given p_D and the number of sensors n, $E[|C_i|]$ can be considered as a constant:

$$E[|C_i|] = r \tag{8}$$

The number of sub-clusters (targets) in each cluster C_i is:

$$k_i \approx \left\lceil \frac{|C_i|}{r} \right\rceil \tag{9}$$

3. Multi-Sensor Data Clustering Algorithm

3.1. Density Peaks Clustering Algorithm

In this paper, we use the DPC to calculate the local density. For each data point, we compute two quantities: its local density ρ_i and distance δ_i from points of higher density. Both these quantities depend only on the distances d_{ij} between data points [5]. The local density ρ_i is defined as:

$$\rho_i = \sum_j \chi(d_{ij} - d_c) \tag{10}$$

where d_{ij} means the distances between data points. $\chi(x) = 1$ if $x < 0$ and $\chi(x) = 0$; otherwise, d_c is a cutoff distance. ρ_i is equal to the number of data points within the cutoff distance to point i. The larger the ρ_i, the higher density of data point i, and the more likely are the observations of targets.

δ_i is measured by computing the minimum distance between point i and the other points with a higher density:

$$\delta_i = \min_{j:\rho_j > \rho_i} (d_{ij}) \tag{11}$$

The original DPC algorithm defines the data points of $\rho_i \geq 0.8 \times r$ and $\delta_i > 2d_c$ as cluster centers. Figure 2 shows the clustering results of the DPC algorithm of 50 i.i.d sensors. Clusters of different colors represent observations of different targets. The red "+" represents the true position of the targets, and the red "o" represents the clustering results. It can be seen, from Figure 2, that for non-overlapping clusters, the real position and estimated position of the targets are very close; for overlapping clusters, the clustering result has a large deviation from the target real position, and the target number is incorrect. In subsequent calculations, we need to re-cluster the overlapping clusters to obtain correct estimates.

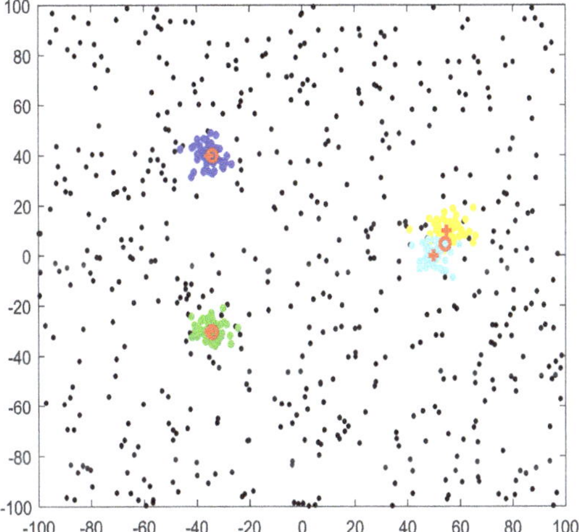

Figure 2. Clustering results of the density peak clustering algorithm for data from 50 i.i.d sensors.

From Figure 2, we can draw a conclusion: the cutoff distance in the clusters of $k_i \geq 2$ (overlapping clusters) is larger than that in the clusters of $k_i = 1$ (non-overlapping clusters). We define the cutoff distance in the non-overlapping clusters as d_c, and the cutoff distance in overlapping clusters is $d'_c = m d_c, m \in [1.1, 1.5]$. Assuming the cluster center of C_i is data point $i, i \in \{1, 2, \cdots, k\}$, the number of data points closer than i is equal to the size of the cluster C_i, that is, $\rho_i \approx |C_i|$. The data points of cluster C_i can be defined as:

$$C_i = \sum_{j \in \{i \mid d_{ij} < d_c\}} z_j \tag{12}$$

The number of targets in cluster C_i can be defined as:

$$k_i \approx \left\lceil \frac{\rho_i}{r} \right\rceil \tag{13}$$

The difference between Equations (9) and (13) is that Equation (13) can determine whether cluster C_i is an overlapping cluster using the ρ_i of cluster center i. The calculation of k_i is also an important step in the subsequent re-clustering process.

3.2. Target Observations Set and Target Number

The multi-source n-points algorithm searches for the number of data points within the cutoff distance of data point i to determine whether the union of point i and the data points within the cutoff distance is a cluster formed by observations of targets. The position of the data point i and detection probability p_D have a greater impact on the effect of the multi-source n-points algorithm. How to quickly and efficiently filter out noise and obtain the target observations is the key to designing MSDF clustering algorithms. Using the DPC algorithm, we find that data point i in C_T has a prior rule: ρ_i must be larger than a threshold. The data points in C_T can be defined as:

$$C_T = \sum_{j \in \{i \mid \rho_i \geq l \times n\}} z_j \tag{14}$$

where $l = 0.4$ is a reference and can be chosen roughly between 0.3~0.45, $\rho_i \geq l \times n$ means that the number of data points closer in data points i must be larger than or equal to $l \times n$, and the data point i in $\rho_i \geq l \times n$ is considered to be the target observations.

Based on the same dataset shown in Figure 2, the data points in C_T are circled with a red "o" in Figure 3. As shown in Figure 3, the observations of targets (color data points) are almost circled with a red "o", and only few data points are not circled. Considering the impact that noisy data points may have in clusters of C_T, Equation (14) is still very reliable.

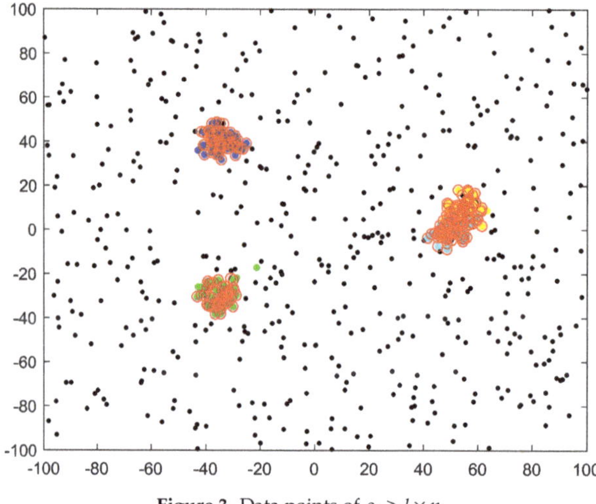

Figure 3. Data points of $\rho_i \geq l \times n$.

The CL constraint requires that C_T be divided into multiple clusters of roughly the same size, and the number of clusters/targets in dataset Z is:

$$\sum k_i = \left\lceil \frac{|C_T|}{r} \right\rceil \tag{15}$$

The number of targets in cluster C_i can be calculated using the ρ_i of the cluster center i through Equation (9), while the total number of targets in dataset Z can be calculated through Equation (15).

Given the number of clusters (targets) $\sum k_i$ and dataset C_T, the preferred choice is to use the k-means algorithm for clustering, as this saves the computing resources of δ_i; however, the k-means algorithm has difficulty handle cases where the local density differs greatly between clusters. During the experiment, we found that if the size of the clusters is roughly equal (no overlapping clusters), the k-means algorithm can obtain correct clustering results. Conversely, if there is at least one overlapping cluster contained in the dataset, the clustering result obtained by the k-means algorithm does not satisfy the CL constraint. In order to correctly cluster C_T using the k-means algorithm, we must first determine whether there are overlapping clusters in dataset Z.

The key to determining whether there are overlapping clusters in dataset Z is to compare $\max(\rho)$ and $1.1 \times r$. If $\max(\rho) < 1.1 \times r$, there are no overlapping clusters in dataset Z, and the number of targets in each cluster is 1, that is, $k_i = 1, i \in \{1, 2, \cdots, k\}$; otherwise, at least one overlapping cluster is contained in dataset Z. The reason for why we choose $1.1 \times r$, instead of the number of sensors n, is that the noisy data points may otherwise fall into cluster C_i.

$$\begin{cases} \max(\rho) < 1.1 \times r, k_i = 1, i \in \{1, 2, \cdots, k\} \\ \max(\rho) \geq 1.1 \times r, k_i \geq 2, i \in \{1, 2, \cdots, k\} \end{cases} \tag{16}$$

3.3. Proposed Clustering Method

The original DPC algorithm needs to calculate ρ_i and δ_i to find the cluster centers, which involves a computational burden that is too great in the case of no overlapping clusters in dataset Z, in this case, we can obtain the correct clustering results using the k-means clustering algorithm to cluster dataset C_T (a total of $\sum k_i$ targets), and dataset C_T can be obtained by a threshold rule.

While the DPC algorithm cannot correctly cluster overlapping clusters, and we have a fast and more efficient solution for the non-overlapping clusters. For the above reasons, we divide the dataset Z into two cases for processing: (1) Non-overlapping clusters in dataset Z (Algorithm 1); and (2) at least one overlapping cluster in dataset Z (Algorithm 2). The main difference between Algorithm 1 and Algorithm 2 is that Algorithm 2 requires an additional calculation of parameter δ_i and re-clustering of the cluster centers of overlapping clusters.

The proposed clustering Algorithm 1 includes 3 steps: (1) Calculate the ρ_i for each data point and determine whether there is an overlapping cluster in dataset Z according to (16); (2) filter out clutter and obtain dataset C_T and $\sum k_i$ for k-means clustering; and (3) revisit each cluster to make sure each cluster satisfies the CL constraint.

Algorithm 1 Clustering without any overlapping cluster in dataset Z

Input: dataset Z. **Output:** cluster C_i and its cluster center z_i, $i \in \{1, 2, \cdots, k\}$.
1.1: Calculate ρ_i according to (10) and determine whether there is any overlapping cluster in dataset Z according to (16). If there is no overlapping cluster, go to step 1.2; otherwise, see Algorithm 2.
1.2: Calculate C_T and $\sum k_i$ according to (14) and (15), then cluster C_T using the k-means algorithm.
1.3: Revisit each cluster C_i to make sure that the CL constraint was satisfied, then calculate the cluster center z_i of each cluster.

Algorithm 2 Clustering with at least one overlapping cluster in dataset Z

Input: dataset Z. **Output:** cluster C_i and its cluster center z_i, $i \in \{1, 2, \cdots, k\}$.
2.1: Calculate δ_i according to (11), and we can obtain estimated cluster centers $\widehat{z}_i$, $i \in \{1, 2, \cdots, k\}$ using the DPC algorithm.
2.2: According to (16), for cluster centers $\widehat{z}_i$ that are $\max(\rho_i) < 1.1 \times r$, the cluster center is $\widehat{z}_i$; for cluster centers $\widehat{z}_i$ that are $\max(\rho_i) \geq 1.1 \times r$, calculate C_i and k_i according to (12) and (13), then cluster C_i with the k-means algorithm (k_i clusters).
2.3: Repeat step 2.2, until all the overlapping clusters are all divided into sub-clusters.
2.4: Revisit each cluster C_i to make sure that the CL constraint was satisfied, then calculate the cluster center z_i of each cluster.

Remark 1. *The cutoff distance is the key to the proposed and the existing MSDF clustering algorithm. The C4F [24] algorithm selects two times the standard deviation of the observation noise as the cutoff distance, the multi-source n-points [26] calculates the cutoff distance using an online learning algorithm, and the proposed algorithm selects 2% of the sorted distances matrix d_{ij} (from small to large) as the cutoff distance. The multi-source n-points algorithm and the algorithm proposed in this paper can deal with unknown observation noise associated with the proposed clustering problem, whereas C4F can only deal with the case of known observation noise.*

Remark 2. *An indispensable step in the existing multi-sensor data fusion clustering algorithm is to calculate the point-to-point distance, which is also the most time-consuming part of the algorithm. The runtime complexity/storage space requirements of the proposed algorithm and the multi-source n-points algorithm are $O(N^2)/(N^2 - N)/2$, $O(N \log N)/O(N)$, respectively. Compared with the multi-source n-points algorithm, it can be seen that the proposed algorithm runs more slowly and requires more storage space.*

Remark 3. *For the multi-source n-points algorithm, the selection of sensor s is very critical. If one target in sensor s is lost, this target will not be detected during the subsequent clustering process, while the proposed algorithm can well deal with the case of some targets avoiding detection. This is the advantage of the proposed algorithm, which is more obvious when the sensor detection probability is lower.*

4. Simulation Results

In this section, we compare the proposed algorithm with the k-means algorithm [8], multi-source n-points algorithm [26], and typical DBSCAN [6] algorithm to obtain the performance of the various algorithms.

4.1. Given Cutoff Distance

The k-means clustering needs one parameter k (the number of clusters), and the DBSCAN algorithm needs two parameters ε (neighborhood radius) and m (minimum number of points). Both the multi-source n-points algorithm and the proposed algorithm need one parameter: the cutoff distance d_c. All the parameters used in the four algorithms are provided in Table 1. The number of sensors is set to n = {20, 50}, and the experimental results of n = {20, 50} are given in Figures 4 and 5, respectively.

Table 1. Parameters used in the four algorithms.

Algorithms	k-Means	DBSCAN	Multi-Source n-Points	Proposed Method
Parameters	$k = 4$	$\varepsilon = 8/m = 6$	$d_c = 8$	$d_c = 8$

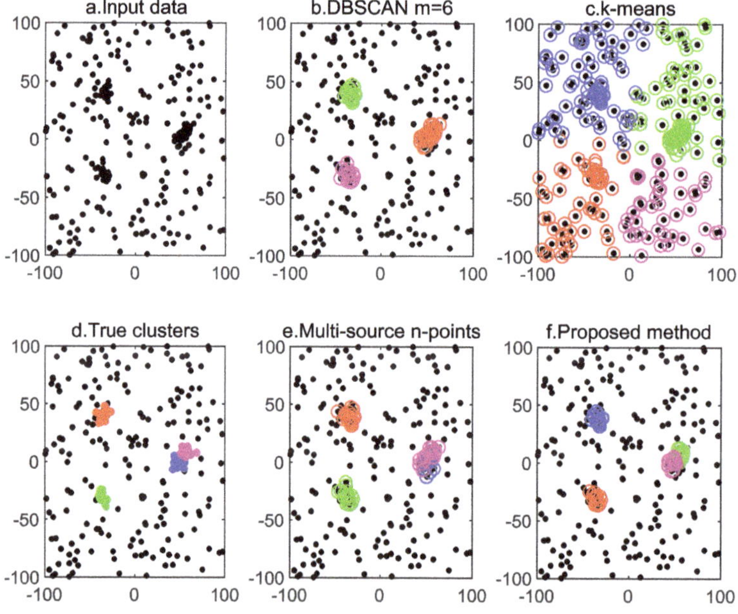

Figure 4. Outcomes of different clustering methods for data from 20 i.i.d sensors.

In each Monte Carlo simulation, the color of the circles is assigned randomly, and circles of the same color represent the same cluster. The clustering results show that both the proposed method and the multi-source n-points algorithm can solve the MSDF clustering problem, but the proposed method algorithm has a smaller variance. The k-means algorithm is unable to deal with clutter, and the clustering result is incorrect. The DBSCAN algorithm can detect observations of targets, but the overlapping cluster clustering result is incorrect.

Table 2 shows the average computing time of different algorithms for the 100 Monte Carlo simulations shown in Figures 4 and 5. It shows that the proposed method is slower than the other three algorithms. To speed up the multi-sensor data fusion clustering algorithm, a target motion model can be employed to determine the potential cluster centroid.

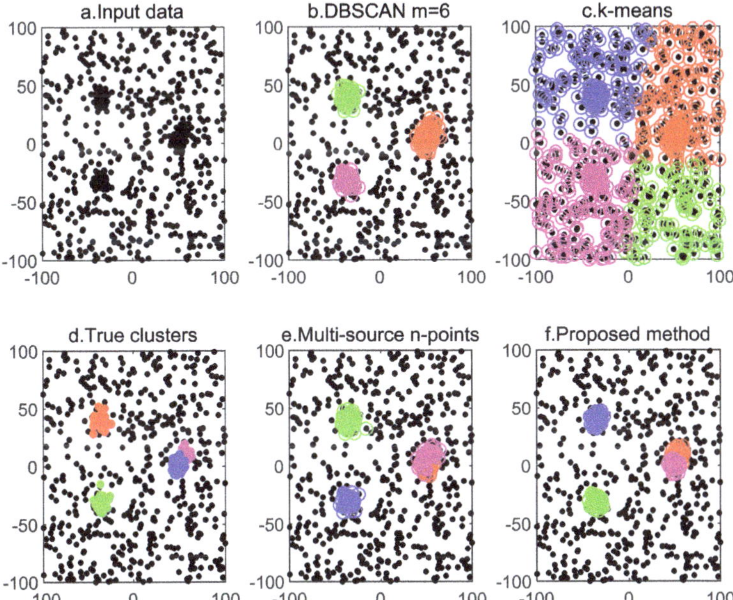

Figure 5. Outcomes of different clustering methods for data from 50 i.i.d.

Table 2. Computing time of different clustering methods (s).

Algorithms	*k*-Means	DBSCAN m = 6	Multi-Source n-Points	Proposed Method
Figure 4	0.0059	0.0021	0.0051	**0.0078**
Figure 5	0.0087	0.0145	0.0198	**0.0346**

4.2. Unknown Cutoff Distance

Based on the same dataset as that given in Figure 2, we assume the cutoff distance d_c is unknown and must be calculated from dataset using an algorithm, such as the DPC algorithm. The cutoff distances of the multi-source n-points and the proposed method are shown in Table 3. The clustering results of the proposed algorithm are given in Figure 6. Compared with the clustering results shown in Figures 4 and 5, the clustering results shown in Figure 6 are also good. This demonstrates that the cutoff distance calculation used in the DPC algorithm is effective.

Table 3. Cutoff distance of different clustering methods (m).

Algorithms	Multi-Source n-Points	Proposed Method
20 sensors	6.7815	**8.0932**
50 sensors	5.9779	**9.1440**

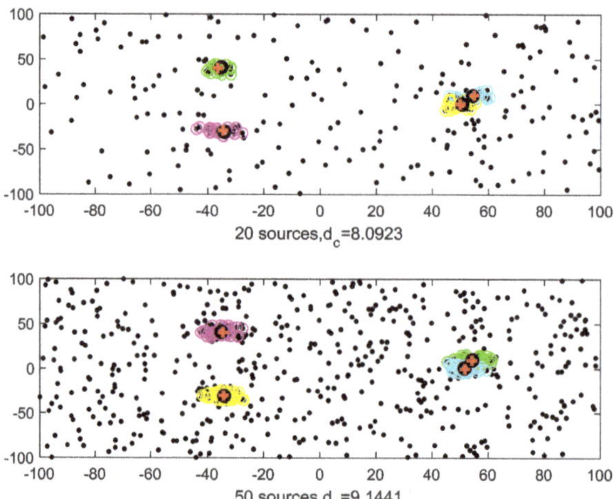

Figure 6. Clustering results with {20, 50} sensors (different color "O") and the. Cluster centers, estimated using proposed method (black "O") and multi-source n-points (red "+").

4.3. Clustering-Based Model

In this simulation, we compare our algorithm with the C4F and multi-source n-points algorithm for multiple target trajectories, provided in the excellent sample MATLAB code in [26]. Information on, for example, clutter and the target dynamic model, are unknown, and the only information that can be used is contained in the observations dataset (the data points in the two-dimensional Cartesian coordinate system) of multiple sensors. The surveillance area is $[-100,100] \times [-100,100]$ (m), and the start/end time and the initial position (green "□") of each target are recorded near the target trajectory, as shown in Figure 7. The average clutter rate per scan is 10, and the observation noise obeys a zero-mean Gaussian distribution, with a variance of 4.

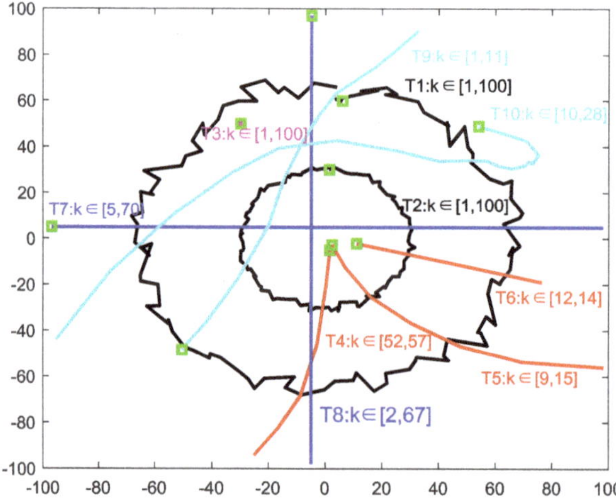

Figure 7. Trajectories of the targets with a fully unknown movement.

To test the clustering accuracy, we use the optimal sub-pattern assignment (OSPA) metric [29] to compare the proposed algorithm with the C4F and multi-source n-points algorithms. We set the cutoff parameter c = 100 and the ordered parameter p = 2.

First, we use 20 sensors. The clustering results of different algorithms for t = 16 are given in Figure 8. The average clustering target numbers and the average OSPA versus time over 100 Monte Carlo trails of different algorithms are given in Figure 9. The average OSPA of the proposed method is 3.7979, which is better than that of the C4F (5.9163) and the multi-source n-points (12.24).

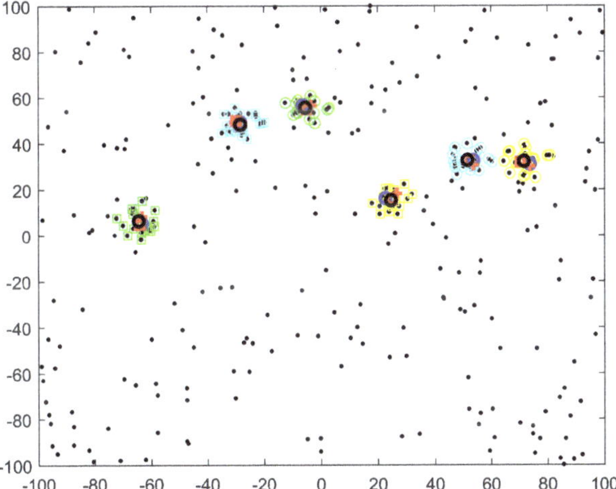

Figure 8. Clustering results of 20 sensors and different clusters (differently colored "o" and "□"), true target positions (red "□"), cluster centers of MS n-points (red "+"), C4F (blue "o") and proposed method (black "o").

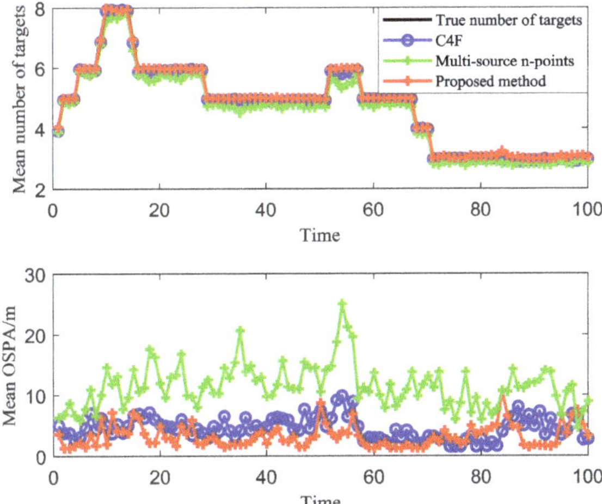

Figure 9. Mean estimated number of targets and mean OSPA of different algorithms over 20 MC trials.

Figure 10 shows the clustering results of different algorithms with 100 sensors for t = 16. The average clustering target number and the average OSPA comparison of different algorithms over 100 Monte Carlo trails are given in Figure 11. The average OSPA of the proposed method is 1.6717, which is better than that of the C4F (5.5241) and multi-source n-points (7.8788) algorithms. Compared with Figure 8, the clustering accuracy of the two algorithms increases with the increase of the sensor number.

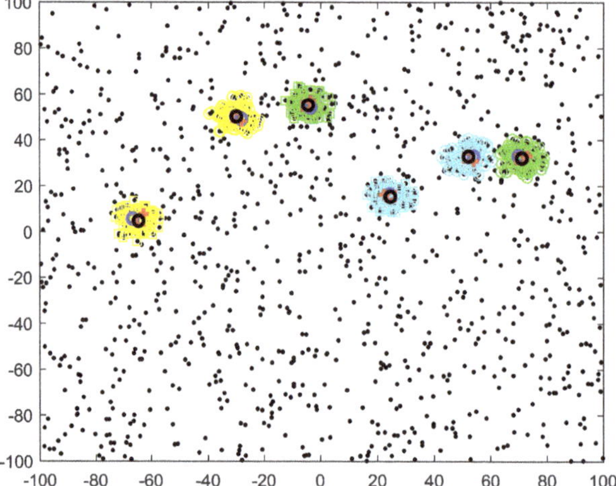

Figure 10. Clustering results of 100 sensors and different clusters (differently colored "o" and "□"), true target positions (red "□"), cluster centers of MS n-points (red "+"), C4F (blue "o"), and proposed method (black "o").

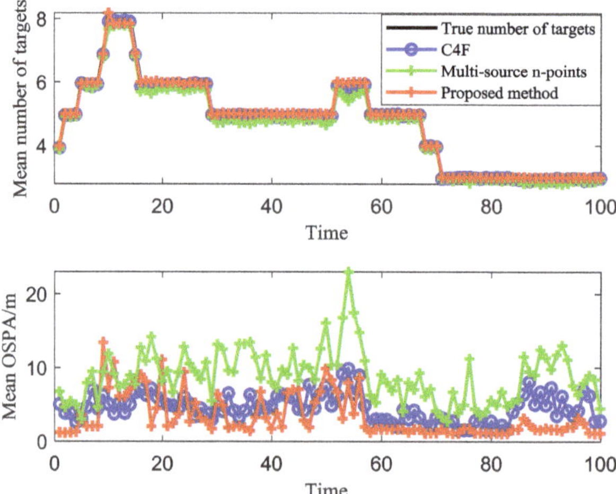

Figure 11. Mean estimated number of targets and mean OSPA of different algorithms over 100 MC trials.

Figure 12 gives the average time-consuming and average OSPA comparison of different algorithms versus different numbers of sensors over 100 Monte Carlo trials. It can be seen that the proposed method outperforms the C4F and multi-source n-points algorithm in the average OSPA. The clustering accuracy with 20 sensors using the proposed algorithm exceeds the clustering result with 100 sensors using the C4F and the multi-source n-points algorithms. As for the computing speed, the proposed method is slower than the C4F and multi-source n-points algorithms.

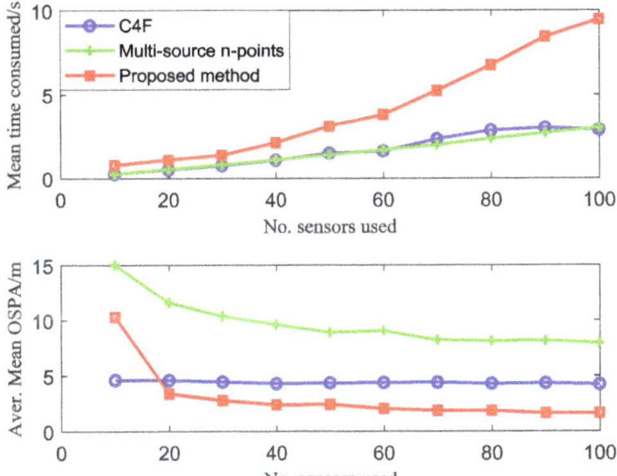

Figure 12. Mean OSPA and computing time of 100 steps × 100 MC runs, with different numbers of sensors.

5. Conclusions

We propose a robust multi-sensor clustering algorithm to solve the MSDF problem. The MSDF problem corresponds to the clustering dataset of the observations (containing a large amount of noise) of multiple sensors, forming k clusters, and each cluster must satisfy the CL constraint. Unlike other model-based multi-sensor data fusion algorithms, no prior information, like the noise and motion model of a target, is needed in the proposed algorithm. Compared with the existing multi-sensor data fusion clustering algorithm, the proposed algorithm is more robust, and the lower the detection probability of the sensors, the better the performance of the proposed algorithm.

Author Contributions: Investigation, J.F.; formal analysis, W.X. and H.D.; methodology, H.D.; resources, W.X.; simulation, W.X.; writing—original draft, J.F.; writing—review and editing, J.F. and H.D. All authors have read and agreed to the published version of the manuscript.

Funding: This project was funded by the Shenzhen Basic Research Project, grant number JCYJ20170818102503604, and National Natural Science Foundation of China, grant number NO.61271107 and No.61703280.

Acknowledgments: The first author acknowledges Tiancheng Li for his insightful discussions and encouragement on this series of work.

Conflicts of Interest: The authors declare no conflict of interest.

References

1. Belkin, M.; Niyogi, P. Laplacian Eigenmaps for dimensionality reduction and data representation. *Neural Comput.* **2003**, *15*, 1373–1396. [CrossRef]
2. Felzenszwalb, P.F.; Huttenlocher, D.P. Efficient graph-based image segmentation. *Int. J. Comput. Vis.* **2004**, *59*, 167–181. [CrossRef]
3. Li, J.; Wang, J.Z. Real-time computerized annotation of pictures. *IEEE Trans. Pattern Anal. Mach. Intell.* **2008**, *30*, 985–1002. [PubMed]
4. Liu, H.; Yu, L. Toward integrating feature selection algorithms for classification and clustering. *IEEE Trans. Knowl. Data Eng.* **2005**, *17*, 491–502.
5. Rodriguez, A.; Laio, A. Clustering by fast search and find of density peaks. *Science* **2014**, *344*, 1492–1496. [CrossRef]
6. Ester, M.; Kriegel, H.; Sander, J. A density-based algorithm for discovering clusters in large spatial Databases with Noise. In *Knowledge Discovery and Data Mining*; AAAI Press: Palo Alto, CA, USA, 1996; pp. 226–231.
7. Jain, A.K.; Murty, M.N.; Flynn, P.J. Data clustering: A review. *ACM Comput. Surv.* **1999**, *31*, 264–323. [CrossRef]
8. Wong, J.A.H.A. Algorithm AS 136: A k-means clustering algorithm. *J. R. Stat. Soc. Ser. C (Appl. Stat.)* **1979**, *28*, 100–108.

9. Shi, J.; Malik, J. Normalized cuts and image segmentation. *IEEE Trans. Pattern Anal. Mach. Intell.* **2000**, *22*, 888–905.

10. Navarro, J.F.; Frenk, C.S.; White, S.D. A Universal density profile from hierarchical clustering. *Astrophys. J.* **1997**, *490*, 493–508. [CrossRef]

11. Wagstaff, K.L.; Cardie, C.; Rogers, S. Constrained k-means clustering with background knowledge. In Proceedings of the International Conference on Machine Learning 2001, Williamstown, MA, USA, 28 June–1 July 2001; pp. 577–584.

12. Han, L.; Luo, S.; Wang, H. An intelligible risk stratification model based on pairwise and size constrained Kmeans. *IEEE J. Biomed. Health Inform.* **2017**, *21*, 1288–1296. [CrossRef]

13. Hansen, P.; Delattre, M. Complete-link cluster analysis by graph coloring. *J. Am. Stat. Assoc.* **1978**, *73*, 397–403. [CrossRef]

14. Miyamoto, S.; Terami, A. Constrained agglomerative hierarchical clustering algorithms with penalties. In Proceedings of the IEEE International Conference on Fuzzy Systems, Taipei, Taiwan, 27–30 June 2011; pp. 422–427.

15. Goode, A. X-means: Extending k-means with efficient estimation of the number of clusters. In *Intelligent Data Engineering and Automated Learning—IDEAL 2000. Data Mining, Financial Engineering, and Intelligent Agents*; Springer: Berlin/Heidelberg, Germany, 2000.

16. Akaike, H.T. A new look at the statistical model identification. *IEEE Trans. Autom. Control* **1974**, *19*, 716–723. [CrossRef]

17. Davisson, L. Rate distortion theory: A mathematical basis for data compression. *IEEE Trans. Commun.* **2003**, *20*, 1202. [CrossRef]

18. Arthur, D.; Vassilvitskii, S. K-Means++: The advantages of careful seeding. In Proceedings of the Eighteenth Annual ACM-SIAM Symposium on Discrete Algorithms, SODA 2007, New Orleans, LA, USA, 7–9 January 2007.

19. Coue, C.; Fraichard, T.; Bessiere, P. Using Bayesian Programming for multi-sensor multi-target tracking in automotive applications. In Proceedings of the International Conference on Robotics and Automation 2003, Taipei, Taiwan, 14–19 September 2003; pp. 2104–2109.

20. Qiao, D.; Liang, Y.; Jiao, L. Boundary detection-based density peaks clustering. *IEEE Access* **2019**, *19*, 755–765. [CrossRef]

21. Jiang, J.; Chen, Y.; Meng, X.; Wang, L.; Li, K. A novel density peaks clustering algorithm based on k nearest neighbors for improving assignment process. *Phys. A Stat. Mech. Appl.* **2019**, *523*, 702–713. [CrossRef]

22. Li, T.; Corchado, J.M.; Sun, S. Partial consensus and conservative fusion of Gaussian mixtures for distributed PHD fusion. *IEEE Trans. Aerosp. Electron. Syst.* **2018**, *55*, 2150–2163. [CrossRef]

23. Vo, B.; See, C.M.; Ma, N. Multi-sensor joint detection and tracking with the Bernoulli filter. *IEEE Trans. Aerosp. Electron. Syst.* **2012**, *48*, 1385–1402. [CrossRef]

24. Li, T.; Corchado, J.M.; Sun, S. Clustering for filtering: Multi-object detection and estimation using multiple/massive sensors. *Inf. Sci.* **2017**, *388–389*, 172–190. [CrossRef]

25. Li, T.; Corchado, J.M.; Chen, H. Distributed flooding-then-clustering: A lazy networking approach for distributed multiple target tracking. In Proceedings of the International Conference on Information Fusion 2018, Cambridge, UK, 10–13 July 2018; pp. 2415–2422.

26. Li, T.; Pintado, F.D.; Corchado, J.M. Multi-source homogeneous data clustering for multi-target detection from cluttered background with misdetection. *Appl. Soft Comput.* **2017**, *60*, 436–446. [CrossRef]

27. Shi, Q.; Zhang, T.; Cui, G.; Kong, L. Multi-target tracking algorithm based on multi-sensor clustering in distributed radar network. *Fusion* **2019**, in press.

28. Li, T.; Prieto, J.; Fan, H. A robust multi-sensor PHD filter based on multi-sensor measurement clustering. *IEEE Commun. Lett.* **2018**, *22*, 2064–2067. [CrossRef]

29. Schuhmacher, D.; Vo, B.T.; Vo, B.N. A consistent metric for performance evaluation of multi-object filters. *IEEE Trans. Signal Process.* **2008**, *56*, 3447–3457. [CrossRef]

Article

Extended Target Marginal Distribution Poisson Multi-Bernoulli Mixture Filter

Haocui Du and Weixin Xie *

Automatic Target Recognition (ATR) Key Laboratory, Shenzhen University, Shenzhen 518060, China; hcdu@szu.edu.cn

* Correspondence: wxxie@szu.edu.cn

Received: 3 August 2020; Accepted: 16 September 2020; Published: 20 September 2020

Abstract: The existence of clutter, unknown measurement sources, unknown number of targets, and undetected probability are problems for multi-extended target tracking, to address these problems; this paper proposes a gamma-Gaussian-inverse Wishart (GGIW) implementation of a marginal distribution Poisson multi-Bernoulli mixture (MD-PMBM) filter. Unlike existing multiple extended target tracking filters, the GGIW-MD-PMBM filter computes the marginal distribution (MD) and the existence probability of each target, which can shorten the computing time while maintaining good tracking results. The simulation results confirm the validity and reliability of the GGIW-MD-PMBM filter.

Keywords: extended target tracking; gamma-Gaussian-inverse Wishart; Poisson multi-Bernoulli mixture

---

1. Introduction

Multiple target tracking (MTT) involves estimating the state of an unknown number of targets in the presence of clutter [1–3]. The traditional MTT algorithm assumes each target is a point as one target generates at most one measurement in the sensor at each time step [3]. However, the rapid development of the modern high-resolution sensors makes the "point assumption" impractical because with such sensors, one target generates at least one measurement per time scan. The MTT problem with such sensors becomes an extended MTT problem [4,5]. Compared with the point target, the extended target can not only provide accurate target movement information, but also the target' shape and size information, making it useful in the fields of robot recognition and positioning, moving crowd tracking, and tracking of close cars or large ships using the high-resolution sensors or automotive radars. Due to its wide application, the research on extended target tracking has received considerable attention from many scholars, making it a growing research hotspot.

As one of the most used extended target measurement models, The Poisson Point Process (PPP) model assumes a Poisson distributed random number of measurements are spatially distributed around the target at each time step [4,5]. To efficiently and correctly represent the spatially distributed measurements, many shape models have been developed, such as the random matrix model (RMM) [6–15], random hypersurface model (RHM) [16–21] and Gaussian process model (GPM) [22]. RMM assumes that the shape of the measurements can be approximated by an ellipse, and the measurements surrounding the centroid of the target obey a Gaussian distribution. RHM uses a more general star-convex shape to approximate the target contour. The GPM automatically learns the shape of the target through a Gaussian process and can give an analytic expression of its contour for an arbitrary shape target. Among the above three measurement models, the RMM has the smallest amount of calculation and is easy to implement, and can meet the requirements of many practical situations, so we use the RMM method in this paper.

Both in the point target-tracking filters and the extended target-tracking filters, the number of targets and the number of measurements are unknown and time-varying; thus, how to represent the targets and the measurements is a challenge. The random finite sets (RFS) theory [23], which represents the targets and the measurements as a finite variable set, is a suitable choice. To solve the point MTT problem, in the early stage, many RFS-based filters have been proposed, such as the Probability Hypothesis Density (PHD) filter [24–26], the Cardinalized Probability Hypothesis Density (CPHD) filter [27–30] and a series of multi-Bernoulli (MB) filters [31–34]. In recent years, scholars have proposed many RFS-based filters to solve the extended MTT problem, such as PHD for extended target tracking (ETT-PHD) [9,35–37], ETT-CPHD [38–40], gamma-Gaussian-inverse Wishart-Poisson multi-Bernoulli mixture (GGIW-PMBM) [40,41], GGIW implementation of the Labelled Multi-Bernoulli (GGIW-LMB) [42,43], and so on [22,44].

In the above-listed RFS-based filters, many kinds of conjugate priors are widely used, to give a closed-form solution of the posterior density. The prior conjugate refers to a time series of random finite set that satisfies the conjugate prior property; if the prediction is a Gaussian (Multi-Bernoulli) distribution, the update is also a Gaussian (Multi-Bernoulli) distribution. Using the conjugate prior, given enough parameters, the posterior distribution can be approximated arbitrarily.

In recent years, developing conjugate prior-based MTT filters has been a significant trend; the two kinds of most used conjugate priors in various MTT filters are the PMBM conjugate prior and the δ-Generalized Labelled Multi-Bernoulli (δ-GLMB) conjugate prior. In the δ-GLMB conjugate prior, the state of each target is added with a unique label, which is helpful to find each target's trajectory. The PMBM prior conjugate divides the target set into two disjoint subsets, targets that have been detected and targets that have not yet been detected, and then processes the two subsets separately. This strategy can greatly reduce the target missed detection rate and improve tracking accuracy. In reference [40], it has been shown that the point target tracking filter based on PMBM conjugate prior outperforms the filter based on the GLMB conjugate prior in terms of tracking accuracy and computing time; thus, in this paper we used the PMBM conjugate prior.

In this paper, to solve the extended MTT problem and combine the PPP model and PMBM conjugate prior, we propose a GGIW implementation of a marginal distribution Poisson multi-Bernoulli mixture (MD-PMBM) filter on the basis of the GGIW-PMBM filter. The main difference between the GGIW-MD-PMBM filter and the GGIW-PMBM filter is that in the prediction step and the update step, instead of recursively propagating the joint state distribution of targets (GGIW-PMBM), the GGIW-MD-PMBM filter recursively propagates the marginal distribution and the existence probability of each target; such a strategy can effectively reduce the calculation cost while maintaining good tracking results.

The rest of the paper is organized as follows. Several model assumptions, such as the Bayesian model, PPP model, MBM model, motion model, measurement model, and PMBM conjugate prior, are discussed in Section 2. Section 3 presents the prediction step and the update step of the GGIW-MD-PMBM filter. Section 4 gives the four extended target filters' experimental simulation results, which are summarized in Section 5.

2. Background

In this section, we outline various model assumptions, such as the Bayesian model [45–47], PPP model, Multi-Bernoulli (MB) process model, motion and measurement model, and PMBM conjugate prior model, which were used in the prediction and update steps of the GGIW-MD-PMBM filter.

2.1. Bayesian Model

We define the state of the i-th extended target at time step k as $\xi_k^{(i)}$, and the state of the target set at time step k can be written as

$$X_k = \left\{ \xi_k^{(i)} \right\}_{i=1}^{N_{\xi,k}},$$

(1)

where $N_{\xi,k} = |X_k|$ is the number of the target, which is unknown and time-varying. The measurements (the union of target generated measurements and clutter) obtained from the sensor at time k are

$$Z_k = \{z_k^j\}_{j=1}^{N_{z,k}} \tag{2}$$

where $N_{z,k} = |Z_k|$ is the number of measurements. Let Z^k denote the union of measurements set from time step 1 to k, that is, $Z^k = Z_1 \cup Z_2 \cup \cdots \cup Z_k$.

In many point target-tracking filters and extended target-tracking filters, the main purpose of using the Bayesian model is to recursively propagate the multi-target density distribution. In a Bayesian model, the multi-target set density, multi-target transition density, and the multi-target measurement likelihood are represented by $f_{k|k}(X_k|Z^k)$, $f_{k|k-1}(X_k|X_{k-1})$, and $f_{k|k}(Z_k|X_k)$, respectively. Using the Chapman–Kolmogorov equation, the multi-target set density can be defined as

$$f_{k|k-1}(X_k|Z^{k-1}) = \int f_{k|k-1}(X_k \mid X_{k-1}) f_{k-1|k-1}(X_{k-1} \mid Z^{k-1}) \delta X_{k-1}, \tag{3}$$

and the Bayes update as

$$f_{k|k}(X_k \mid Z^k) = \frac{f_{k|k}(Z_k \mid X_k) f_{k|k-1}(X_k \mid Z^{k-1})}{\int f_{k|k}(Z_k \mid X_k) f_{k|k-1}(X_k \mid Z^{k-1}) \delta X_k}, \tag{4}$$

2.2. PPP Model and Multi-Bernoulli (MB) Process Model

The PPP model, proposed by Gilholm et al. [4] and Granstrom et al. [5], is widely used in point target tracking filters and extended target tracking filters. In a PPP model, each target generates a Poisson distributed number of measurements, and each measurement is independent and identically distributed (i.i.d.). The Poisson rate μ and the spatial distribution $f(\cdot)$ determine the PPP intensity.

We define the spatial distribution of the target state $\xi_k^{(i)}$ as $f(\xi_k^{(i)})$. The Poisson density distribution of X_k is

$$f(X_k) = e^{-\mu} \prod_{\xi_k^{(i)} \in X_k} \mu f(\xi_k^{(i)}). \tag{5}$$

The existence of a single target can be modeled by a Bernoulli RFS [14,31–34,40–42]. The cardinality of a Bernoulli RFS $X_k^{(i)}$ can either be 1 (with probability $r_k^{(i)}$) or empty (with probability $1 - r_k^{(i)}$), and the Bernoulli density distribution of $X_k^{(i)}$ can be defined as

$$f(X_k^{(i)}) = \begin{cases} 1 - r_k^{(i)}, & X_k^{(i)} = \phi \\ r_k^{(i)} f(\xi_k^{(i)}), & X_k^{(i)} = \{\xi_k^{(i)}\} \\ 0, & |X_k^{(i)}| \geq 2 \end{cases}. \tag{6}$$

An MB RFS X_k is the union of a limited number of independent Bernoulli RFS $X_k^{(i)}$, $X_k = \overset{N_{\xi,k}}{\underset{i=1}{\cup}} X_k^{(i)}$. The MB density distribution of X_k is

$$f(X_k) = (\prod_{i=1}^{N_{\xi,k}} (1 - r_k^{(i)})) \sum_{1 \leq i_1 \neq \cdots \neq i_n \leq N_{\xi,k}} \frac{r_k^{(i_1)} f(\xi_k^{(i_1)})}{1 - r_k^{(i_1)}} \cdots \frac{r_k^{(i_n)} f(\xi_k^{(i_n)})}{1 - r_k^{(i_n)}}. \tag{7}$$

In this paper, we divide the targets into the undetected target set and the detected target set. The PPP represents the distribution of undetected target measurements, and the MB mixture (MBM) represents the distribution of detected target measurements.

2.3. The Motion Model and Measurement Model

To simplify the calculation, at each time step, we assume each target evolves independently from the other targets, and the measurements generated by each target are independent of each other and independent of those generated by other targets.

We assume each birth target can be described by a PPP model, and its intensity is $D_b(x_k)$. All the existing targets either survive with a probability $P_D(x_k)$ or die with a probability $1 - P_D(x_k)$ from time step k to time step $k + 1$. Each extended target state consists of three variables, the measurement rate γ_k, kinematic state x_k, and extent state E_k, that is, $\xi_k^{(i)} = \left\{ \gamma_k^{(i)}, x_k^{(i)}, E_k^{(i)} \right\}$. Given γ_k, x_k and E_k, the Markov motion model can be defined as

$$\gamma_{k+1} = \gamma_k, \tag{8}$$

$$x_{k+1} = f(x_k) + w_k, \tag{9}$$

$$E_{k+1} = M(x_k)E_kM(x_k)^T, \tag{10}$$

where w_k is Gaussian process noise with zero mean and covariance Q, $f(\cdot)$ and $M(\cdot)$ are the transformation matrix of kinematic state and extent state, respectively.

The clutter measurements at time step k can be modeled as a PPP model with measurement rate λ and spatial Poisson distribution $c(z)$. The clutter PPP density function is $k(z) = \lambda\, c(z)$. The measurements generated by each target at time step k are a Poisson distribution with a measurement rate γ_k. The measurement model is

$$z_k = H_k x_k + v_k, \tag{11}$$

where H_k is the known observation model function, and v_k is the Gaussian observation noise with zero mean and covariance E_k.

Assuming the measurements generated by a single target follow a Gaussian distribution, that is

$$p(z_k|x_k, v_k) = N(z_k; H_k x_k, E_k). \tag{12}$$

The distribution of each target state can be expressed as

$$\begin{aligned}
p(\gamma_k, x_k, E_k|Z^k) &= p(\gamma_k|Z^k)p(x_k, E_k|Z^k)p(E_k|Z^k) \\
&= G(\gamma_k; \alpha_{k|k}, \beta_{k|k})N(x_k; m_{k|k}, P_{k|k} \otimes E_k)IW_d(E_k; v_{k|k}, V_{k|k}) \\
&= GGIW(\xi_k; \zeta_{k|k})
\end{aligned} \tag{13}$$

where $G(\gamma_k; \alpha_{k|k}, \beta_{k|k})$ is the Gamma distribution, $\alpha_{k|k} > 0$ and $\beta_{k|k} > 0$ are its shape parameter and inverse scale parameter; $N(x_k; m_{k|k}, P_{k|k} \otimes E_k)$ is the Gaussian distribution, $m_{k|k}$ and $P_{k|k} \otimes E_k$ are its mean and covariance; $IW_d(E_k; v_{k|k}, V_{k|k})$ is the inverse Wishart distribution, $v_{k|k}$ and $V_{k|k}$ are its degrees of freedom and shape matrix. Note that $\zeta_{k|k} = \left\{ \alpha_{k|k}, \beta_{k|k}, m_{k|k}, P_{k|k}, v_{k|k}, V_{k|k} \right\}$ is a shorthand for the set of GGIW density parameters. The detailed implementation of GGIW prediction and update is given in [21,23–26].

2.4. PMBM Conjugate Prior

Existing studies [40–42] show that developing PMBM conjugate priors-based filters is a significant trend, due to their excellent tracking performance. In a PMBM model, the extended target set X_k at time k is divided into two disjoint sets, undetected targets X_k^u and detected targets X_k^d, then the PPP model is used to describe the distribution of undetected targets X_k^u, and the MBM model to describe the distribution of targets X_k^d that have been detected at least once.

$$X_k^u, X_k^d : X_k = X_k^u \cup X_k^d, X_k^u \cap X_k^d = \varnothing. \tag{14}$$

The PMBM set density can be defined as

$$f_{k|k}(X_k|Z^k) = \sum_{X_k^u \uplus X_k^d = X_k} f_{k|k}^u(X_k^u|Z^k) \sum_{j \in \mathbb{J}} \omega_{k|k}^{d,j} f_{k|k}^{d,j}(X_k^d|Z^k), \tag{15}$$

$$f_{k|k}^u(X_k^u|Z^k) = e^{-\mu_{k|k}^u} \prod_{\xi \in X_k^u} \mu_{k|k}^u f_{k|k}^u(\xi), \tag{16}$$

$$f_{k|k}^{d,j}(X_k^d|Z^k) = \sum_{\substack{\uplus X_k^i = X_k^d \\ i \in \mathbb{I}^j}} \prod_{i \in \mathbb{I}^j} f_{k|k}^{d,j,i}(X_k^i), \tag{17}$$

where $f_{k|k}^{d,j,i}(\cdot)$ is the Bernoulli set density. The MBM has $|\mathbb{J}|$ components, and $\mathbb{J}$ is the index set (or MBM component) of the MB in the MBM. $\mathbb{I}^j$ is the Bernoulli index set of the j-th MB, that is, the j-th component has $|\mathbb{I}^j|$ Bernoulli components. $\omega^{d,j}$ is the probability of the j-th MB component. The conjugacy property determines the intensities of the birth PPP and the initial undetected PPP, which are expressed by both GGIW mixture density.

Given the PMBM conjugate prior assumption, the extended target tracking is equivalent to recursively propagate the PMBM density parameters in the prediction step and the update step, which is shown in Section 3.

3. The GGIW-MD-PMBM Filter

In this section, the GGIW-MD-PMBM filter is presented. Instead of recursively propagating the joint state distribution of targets (GGIW-PMBM filter), the proposed GGIW-MD-PMBM filter recursively propagates the marginal distributions and the existence probabilities of each extended target. The GGIW-MD-PMBM filter can be modeled by the following assumptions:

(1) Clutter is uniformly distributed in the surveillance area with a given Poisson rate λ and is independent of targets' distribution;

(2) Each extended target generates at least one measurement per time scan and evolves independently of the other targets;

(3) The birth PPP intensity is a GGIW mixture with a given Poisson rate μ^b, weight ω^b, and various density parameters ς^b of GGIW;

$$D_{k+1}^b = \mu_{k+1}^b \sum_{j=1}^{N_{k+1}^b} \omega_{k+1}^{(b,j)} GGIW(\xi_{k+1}; \varsigma_{k+1}^{(b,j)}). \tag{18}$$

(4) The initial undetected PPP intensity is also a GGIW mixture with a known Poisson rate μ_0^u, weight ω_0^u, and various density parameters ς_0^u of GGIW.

$$D_0^u = \mu_0^u \sum_{j=1}^{N_0^u} \omega_0^{(u,j)} GGIW(\xi_0; \varsigma_0^{(u,j)}). \tag{19}$$

(5) The initial PMBM parameter is empty.

$$\mu_{0|0}^u = 0 \text{ and } J_{0|0} = 0. \tag{20}$$

We assume the birth PPP and the initial undetected targets are all GGIW mixture intensities; according to the PMBM conjugate property, each target density in the PMBM filter is also a GGIW density.

3.1. Prediction

3.1.1. Detected Targets

According to the Chapman–Kolmogorov prediction (3) and the posterior PMBM density (15–16), the predicted distribution $f_{k|k-1}^{d,j}(\xi_k)$, $j = 1, \cdots, J_{d,k|k-1}$ and the existence probability $r_{k|k-1}^{d,j}$ of each detected target at time step k are

$$f_{k|k-1}^{d,j}(\xi_k) = GGIW(\xi_k; \zeta_{k|k-1}^{d,j}), j = 1, \cdots, J_{d,k|k-1}, \tag{21}$$

$$r_{k|k-1}^{d,j} = p_S r_{k-1|k-1}^{d,j}, j = 1, \cdots, J_{d,k|k-1}, \tag{22}$$

where p_S is the survival probability, and $\zeta_{k|k-1}^{d,j}$ are the predicted parameters. The weights and number of predicted MBM components are $\omega_{d,k|k-1}^{j} = \omega_{d,k-1|k-1}^{j}$ and $J_{d,k|k-1} = J_{d,k-1|k-1}$, respectively.

3.1.2. Undetected Targets

To avoid missed detection, we regard previously undetected targets and current birth targets as the undetected target set, that is, the undetected targets at time step k include the undetected targets at time step $k - 1$ and the birth targets at time step k (this paper does not consider the spawning targets). The predicted PPP for each undetected target at time step k is as follows:

$$f_{k}^{ub,j}(\xi_k) = GGIW(\xi_k; \zeta_{k}^{ub,j}), j = 1, \cdots, N_{k}^{ub}, \tag{23}$$

$$f_{k|k-1}^{uu,j}(\xi_k) = GGIW(\xi_k; \zeta_{k|k-1}^{uu,j}), j = 1, \cdots, N_{k|k-1}^{uu}. \tag{24}$$

The existence probability is $\omega_{k}^{ub,j}$ or $\omega_{k|k-1}^{uu,j}$.

$$\omega_{k}^{ub,j}, j = 1, \cdots, N_{k}^{ub}, \tag{25}$$

$$\omega_{k|k-1}^{uu,j} = \omega_{k-1|k-1}^{uu,j} p_S, j = 1, \cdots, N_{k|k-1}^{uu}, \tag{26}$$

where N_{k}^{ub} is the number of birth targets at time step k, and $N_{k|k-1}^{uu}$ is the number of undetected targets at time step $k - 1$. $\zeta_{k|k-1}^{u,j}$ are the parameters ($\zeta_{k}^{ub,j}$ or $\zeta_{k|k-1}^{uu,j}$) of undetected targets at time step k. The predicted PPP for the undetected targets has a Poisson rate

$$\mu_{k|k-1}^{u} = \mu_{k}^{ub} + P_{S}^{uu} \mu_{k-1|k-1}^{uu}, \tag{27}$$

where

$$P_{S}^{uu} = \sum_{j=1}^{N_{k-1|k-1}^{uu}} \omega_{k-1|k-1}^{uu,j} p_S. \tag{28}$$

The existence probability of undetected targets $\omega_{k}^{ub,j}$ and $\omega_{k|k-1}^{uu,j}$ can be normalized, as $\omega_{k}^{ub,j} = \frac{\mu_{k}^{ub}}{\mu_{k|k-1}^{u}} \omega_{k}^{ub,j}$, $\omega_{k|k-1}^{uu,j} = \frac{\mu_{k-1|k-1}^{uu}}{\mu_{k|k-1}^{u}} \omega_{k|k-1}^{uu,j}$, and $\sum_{j=1}^{N_{k}^{ub}} \omega_{k}^{ub,j} + \sum_{j=1}^{N_{k}^{uu}} \omega_{k|k-1}^{uu,j} = 1$. For simplicity, the existence probability of each undetected target at time k can be represented as $\omega_{k|k-1}^{u,j}$, which includes $\omega_{k}^{ub,j}$ and $\omega_{k|k-1}^{uu,j}$.

Through the above analysis, in the prediction step, all extended target distributions and existence probabilities can be described as

$$f_{k|k-1}^{j}(\xi_k), j = 1, \cdots, N_{k|k-1}, \tag{29}$$

$$\omega_{k|k-1}^{j}, j = 1, \cdots, N_{k|k-1},\tag{30}$$

where $N_{k|k-1} = J_{d,k|k-1} + N_{k|k-1}^{uu} + N_k^{ub}$. $f_{k|k-1}^{j}(\xi_k) = f_{k|k-1}^{d,j}(\xi_k)$ and $\omega_{k|k-1}^{j} = r_{k|k-1}^{d,j}$ for $1 \leq j \leq J_{d,k|k-1}$, $f_{k|k-1}^{j-J_{d,k|k-1}}(\xi_k) = f_{k|k-1}^{uu,j}(\xi_k)$ and $\omega_{k|k-1}^{j-J_{d,k|k-1}} = \omega_{k|k-1}^{uu,j}$ for $J_{d,k|k-1} < j \leq J_{d,k|k-1} + N_{k|k-1}^{uu}$, $f_{k|k-1}^{j-J_{d,k|k-1}-N_k^{uu}}(\xi_k) = f_{k|k-1}^{ub,j}(\xi_k)$ and $\omega_{k|k-1}^{j-J_{d,k|k-1}-N_k^{uu}} = \omega_k^{ub,j}$ for $j > J_{d,k|k-1} + N_{k|k-1}^{uu}$.

3.2. Update

If each target predicted density is a PMBM at time k, according to the conjugate prior property, the updated density is also a PMBM. The updated MBM is given by the formulas (31) and (32), which contain the MB components predicted by each target in the previous processing step and their related data associations.

$$f_{k|k}^{i}(\xi_k^{(i)}|Z^k) = \sum_{Y \subseteq Z} \sum_{P \angle Y} \sum_{C_C \cap \mathbb{I}^j \neq \varnothing} \omega_{k|k}^{i,P} f_{k|k}^{d,j}(\xi_k^{(i)}|Z^k), j = 1, \cdots, J_{d,k|k-1} + N_{k|k-1}^{u},\tag{31}$$

$$\omega_{k|k}^{i,P} = \frac{\omega_{k|k-1}^{j} L_P L_w}{\sum_{Y \subseteq Z} \sum_{P \angle Y} \sum_{C_C \cap \mathbb{I}^j \neq \varnothing} \omega_{k|k}^{i,P} L_P L_w}, j = 1, \cdots, J_{d,k|k-1} + N_{k|k-1}^{u}.\tag{32}$$

Here, the predicted likelihood L_P of partition and the predicted likelihood L_w of each MB are as follows:

$$L_P = \prod_{\substack{C_C \cap \mathbb{I}^j = \varnothing \\ |C_C| > 1}} 1_C \prod_{\substack{C_C \cap \mathbb{I}^j = \varnothing \\ |C_C| = 1}} (\kappa^{C_C} + 1_C),\tag{33}$$

$$L_w = \prod_{j} L_C.\tag{34}$$

In formulas (33) and (34), for each MB component in each partition cell, the data association can be divided into the following three types:

(1) The measurement set Z can be divided into the union of two disjoint sets, $Z = Y \cup (Z \backslash Y)$, where Y corresponds to the clutter and the previously undetected targets' measurements, and $Z \backslash Y$ corresponds to the previously detected targets' measurements.

(2) For set Y, form a partition of P of non-empty cells C; each cell contains clutter or target-generated measurements.

(3) For set $Z \backslash Y$, the union of a set of index subsets $\left\{ \mathbb{I}^j \right\}$ can be used to represent the measurements related to the j-th MB component of the detected targets.

In the update step, the Bernoulli parameters of each detected target and the PPP parameters of each undetected target are determined, as given below.

3.2.1. Detected Targets

In this paper, there are two types of detected targets at time k: The targets detected for the first time in the measurement set Y, and the previously detected targets are detected again in the measurement set $Z \backslash Y$. The first detected targets are processed by the PPP model, and the second detected target is estimated with the existing MB. According to Equations (9) and (10), the spatial distribution $f_{k|k}^{d,j}(\xi_k)$ and existence probability $r_{k|k}^{d,j}$ of each GGIW component of the detected target at time k are updated as follows:

$$f_{k|k}^{d,j}(\xi_k^{(j)}) = \sum_{\substack{\uplus \xi_k^{i} = X_k^{d} \\ i \in \mathbb{I}^j}} \prod_{i_C \cap \mathbb{I}^j} f_{k|k}^{d,j,i}(\xi_k^{(j)}), j = 1, \cdots, J_{d,k|k-1},\tag{35}$$

$$r_{k|k}^{d,j} = \frac{\omega_{k|k-1}^{j} \prod\limits_{C_C \cap \mathbb{I}^j} r^{j,C}}{\sum\limits_{j \in \mathbb{J}} \omega_{k|k-1}^{j} \prod\limits_{i_C \cap \mathbb{I}^j} r^{j,C}}, j = 1, \cdots, J_{d,k|k-1}. \tag{36}$$

For each partition cell, its predicted likelihood L_C is related to the predicted likelihood of each GGIW component in the spatial density, such as

$$L_C = \begin{cases} \kappa^{C_C} + l_C, & \text{if } C \cap \mathbb{I}^j = \varnothing, \; |C_C| = 1 \\ l_C, & \text{if } C \cap \mathbb{I}^j = \varnothing, \; |C_C| > 1 \\ 1 - r_{k|k-1}^{d,j} p_D + r_{k|k-1}^{d,j} p_D \left(\dfrac{\beta_{k|k-1}^{d,j,C}}{\beta_{k|k-1}^{d,j,C} + 1} \right)^{\alpha_{k|k-1}^{d,j,C}}, & \text{if } C \cap \mathbb{I}^j \neq \varnothing, \; C_C = \varnothing \\ r_{k|k-1}^{d,j} L_k^{u,j,C}, & \text{if } C \cap \mathbb{I}^j \neq \varnothing, \; C_C \neq \varnothing \end{cases} \tag{37}$$

The predicted likelihood L_C is determined by $C \cap \mathbb{I}^j$ and the measurement cell C_C, as follows:

(1) When $C \cap \mathbb{I}^j = \varnothing$, $|C_C| = 1$, the predicted likelihood L_C is an approximation of the predicted likelihood of the cell composed of the undetected target-generated measurements and the clutter measurements in the first detection process. Since this cell contains the clutter and undetected targets' measurements, L_C is related to κ^{C_C} and l_C, where κ^{C_C} is the predicted likelihood of clutter, and l_C is the predicted likelihood of undetected target-generated measurements.

(2) When $C \cap \mathbb{I}^j = \varnothing$, $|C_C| > 1$, it means the targets detected for the first time are divided into multiple cells. If there are undetected targets, they must be contained in the above cells, thus L_C is only related to l_C.

(3) When $C \cap \mathbb{I}^j \neq \varnothing$, $C_C = \varnothing$, it means L_C is the predicted likelihood when the j-th MB component of the previously detected target is an empty set.

(4) When $C \cap \mathbb{I}^j \neq \varnothing$, $C_C \neq \varnothing$, the predicted likelihood L_C is the approximation when the j-th MB component of the previously detected target is not empty. The calculation $L_k^{u,j,C}$ is presented in [39], Table 2.

The predicted likelihood l_C of undetected targets can be defined as

$$l_C = \mu_{k|k-1}^{u} \sum_{j=1}^{N_{k|k-1}^{u}} \omega_{k|k-1}^{u,j} L_k^{u,j,C}. \tag{38}$$

According to the different values of $C \cap \mathbb{I}^j$ and C_C, the likelihood probability $r^{j,C}$ can be defined as

$$r^{j,C} = \begin{cases} \dfrac{l_C}{\kappa^{C_C} + l_C}, & \text{if } C \cap \mathbb{I}^j = \varnothing, \; |C_C| = 1 \\ 1, & \text{if } C \cap \mathbb{I}^j = \varnothing, \; |C_C| > 1 \\ \dfrac{r_{k|k-1}^{d,j} q_D^{d,j,C}}{1 - r_{k|k-1}^{d,j} + r_{k|k-1}^{d,j} q_D^{j,C}}, & \text{if } C \cap \mathbb{I}^j \neq \varnothing, \; C_C = \varnothing \\ 1, & \text{if } C \cap \mathbb{I}^j \neq \varnothing, \; C_C \neq \varnothing \end{cases} \tag{39}$$

The effective probability of the missed detection can be defined as

$$q_D^{d,j,C} = 1 - p_D + p_D \left(\frac{\beta_{k|k-1}^{d,j,C}}{\beta_{k|k-1}^{d,j,C} + 1} \right)^{\alpha_{k|k-1}^{d,j,C}}. \tag{40}$$

The spatial distribution $f_{k|k}^{d,j,i}(\xi_k)$ of each measurement cell at time step k is as follows:

$$
f_{k|k}^{d,j,C}(\xi_k) = \begin{cases} \dfrac{\omega_{k|k-1}^{u,j} L_k^{u,j,C}}{\sum\limits_{j=1}^{N_{k|k-1}^u} \omega_{k|k-1}^{u,j} L_k^{u,j,C}} GGIW(\xi_k; \zeta_{k|k}^{d,j,C}), & if\ C_C \cap \mathbb{I}^j = \varnothing \\[6mm] \dfrac{1-p_D}{q_D^{d,j,C}} GGIW(\xi_k; \zeta_{k|k-1}^{d,j,C}) + \dfrac{p_D(\frac{\beta_{k|k-1}^{d,j,C}}{\beta_{k|k-1}^{d,j,C}+1})^{\alpha_{k|k-1}^{d,j,C}}}{q_D^{d,j,C}} G(\gamma_k; \alpha_{k|k-1}^{u,j,C}, \beta_{k|k-1}^{u,j,C}+1) \\[3mm] \quad \times N(x_k; m_{k|k-1}^{u,j,C}, P_{k|k-1}^{u,j,C}) IW_d(X_k; v_{k|k-1}^{u,j,C}, V_{k|k-1}^{u,j,C}), & if\ C \cap \mathbb{I}^j \neq \varnothing,\ C_C = \varnothing \\[3mm] GGIW(\xi_k; \zeta_{k|k}^{d,j,C}), & if\ C \cap \mathbb{I}^j \neq \varnothing,\ C_C \neq \varnothing \end{cases}
\tag{41}
$$

The second equation in (41) uses a gamma-mixture reduction to reduce its bi-modal GGIW distribution to a uni-modal GGIW distribution [44]. Detailed calculations of parameter $\zeta_{k|k}^{d,j,C}$ and prediction likelihood $L_k^{u,j,C}$ in the GGIW update are provided in [41].

3.2.2. Undetected Targets

The detection probability p_D is always less than 1, and the Poisson parameter γ_k may be zero; these two cases may cause missed detection of targets. Therefore, we need to consider the above two situations, and the updated spatial density is

$$
\begin{aligned}
f_{k|k}^{u,j}(\xi_k) = {} & \frac{(1-p_D)\omega_{k|k-1}^{u,j}}{\sum\limits_{j=1}^{N_{k|k-1}^u} q_D^{u,j}\omega_{k|k-1}^{u,j}} GGIW(\xi_k; \zeta_{k|k-1}^{u,j}) + \frac{p_D(\frac{\beta_{k|k-1}^{u,j}}{\beta_{k|k-1}^{u,j}+1})^{\alpha_{k|k-1}^{u,j}}\omega_{k|k-1}^{u,j}}{\sum\limits_{j=1}^{N_{k|k-1}^u} q_D^{u,j}\omega_{k|k-1}^{u,j}} G(\gamma_k; \alpha_{k|k-1}^{u,j}, \beta_{k|k-1}^{u,j}+1), \\
& \times N(x_k; m_{k|k-1}^{u,j}, P_{k|k-1}^{u,j}) IW_d(X_k; v_{k|k-1}^{u,j}, V_{k|k-1}^{u,j}), j = 1, \cdots, N_{k|k-1}^u
\end{aligned}
\tag{42}
$$

where $q_D^{u,j}$ is the effective probability of the undetected target, and $\mu_{k|k}^u$ is the PPP Poisson rate of the undetected targets,

$$
q_D^{u,j} = 1 - p_D + p_D\left(\frac{\beta_{k|k-1}^{u,j}}{\beta_{k|k-1}^{u,j}+1}\right)^{\alpha_{k|k-1}^{u,j}},
\tag{43}
$$

$$
\mu_{k|k}^u = \mu_{k|k-1}^u \sum_{j=1}^{N_{k|k-1}^u} q_D^{u,j},
\tag{44}
$$

The existence probability is

$$
\omega_{k|k}^{u,j} = \frac{(1-p_D)\omega_{k|k-1}^{u,j}}{\sum\limits_{j=1}^{N_{k|k-1}^u} q_D^{u,j}\omega_{k|k-1}^{u,j}} + \frac{p_D(\frac{\beta_{k|k-1}^{u,j}}{\beta_{k|k-1}^{u,j}+1})^{\alpha_{k|k-1}^{u,j}}\omega_{k|k-1}^{u,j}}{\sum\limits_{j=1}^{N_{k|k-1}^u} q_D^{u,j}\omega_{k|k-1}^{u,j}}, j = 1, \cdots, N_{k|k-1}^u.
\tag{45}
$$

Note: Most of the mathematical details of formulas (42) and (45) are given in the Appendix A. The Gaussian and inverse Wishart parameters in formula (42) are the same in both cases, but the gamma parameters are different. Therefore, formula (42) can be reduced by gamma mixing [44] to a single-peak GGIW distribution and the number of GGIW components remains unchanged.

The main contribution of the GGIW-MD-PMBM proposed in this paper is to recursively propagate the marginal distribution and existence probability of each target through formulas (21), (22), (29), (34),

(35), (42), and (45), which is different from the GGIW-PMBM in [40] that it recursively propagates the joint target state distribution and existence probability.

3.3. Complexity Reduction and Data Association

In the tracking process, as the number of targets increases, the number of unknown data associations, the number of components in the MBM, and the number of parameters in the PMBM greatly increase, which brings the problem of "combination explosion". It is necessary to approximate the target density through reduction methods. Possible reduction methods include gating, clustering, pruning, merging, and recycling. In this paper, we focused on developing the approximate method to recursively propagate the marginal distribution of targets in the prediction and update steps. The complexity reduction and the data association process used in this paper follow the same methods used in [40], Section V; hence the complexity reduction and the data association are briefly discussed.

4. Simulation

In this simulation, we compared the proposed GGIW-MD-PMBM filter with the GGIW-PMBM, GGIW-LMB, and GGIW-PHD filters for multi-target trajectories in the four classic scenarios, which are provided in the excellent sample MATLAB code [40–43]. Each extended target can be defined as $x_k = [p_k, v_k]^T \in \mathbb{R}^4$, where $p_k \in \mathbb{R}^2$ is each target's position and $v_k \in \mathbb{R}^2$ is the velocity. The motion model's parameters used in Formulas (8)–(10) are

$$f(x_k) = \begin{bmatrix} I_2 & T_s I_2 \\ 0 & I_2 \end{bmatrix} x_k, \quad Q = G\sigma_a^2 I_2 G^T, \quad G = \begin{bmatrix} \frac{T_s^2}{2} I_2 \\ T_s I_2 \end{bmatrix}, \quad H_k = [I_2\ 0_2], \tag{46}$$

where σ_a is the acceleration standard deviation and $T_s = 1s$ is the sampling time.

To evaluate the four algorithms' performance, we used the computing time and the generalized optimal sub-pattern assignment (GOSPA) metric [47], the cutoff parameter $c = 10$, and the ordered parameter $p = 1$, among which, for the distance measurement, we use the Gaussian Wasserstein Distance metric [48]. We divided the GOSPA metric into three categories: localization error, false detection error, and miss detection error. For GGIW-LMB filter and GGIW-PHD filter, we extracted the target states by taking the mean vector of all Bernoullis whose existence probability is larger than 0.5. For the GGIW-PMBM filter and GGIW-MD-PMBM filter, target state extraction was performed similarly, but only from the MB component with the highest weight.

In the four scenarios, the target detection probability p_D, target survival probability p_S, the clutter Poisson rate λ, and the measurement Poisson rate γ used in different scenarios are given in Table 1. In the first scenario, there are 100 time steps, and 27 highly time-varying targets are randomly generated in four positions; this scenario aims to compare the four algorithms' tracking performances with a high clutter density and high target number scenario. In the second scenario, there are 100 time steps, and two targets are born well separated, move close to each other, and then split; this scenario aims to compare the four algorithms' tracking performance when highly close to each other. In the third scenario, there are 10 time steps, and five targets are born closely at the same time; this scenario aims to compare the four algorithms' tracking performances of handling dense birth. In the fourth scenario, there are 300 time steps, and two targets first get close and then they maneuver closely before splitting; this scenario aims to compare the four algorithms' tracking performance of seriously handling the data association problem. The target trajectories of different scenarios are given in Figure 1.

Table 1. The parameters used in the four scenarios.

Scenario	p_D	p_S	λ	γ
Scenario 1	0.90	0.99	60	{7,8,9}
Scenario 2	0.98	0.99	10	{10,20}
Scenario 3	0.90	0.99	20	10
Scenario 4	0.98	0.99	10	{10,20}

(a) (b) (c) (d)

Figure 1. True target trajectories of four scenarios. (**a**) Twenty-seven targets, (**b**) separate/close/split, (**c**) dense birth, (**d**) nonlinear maneuver.

Figure 2 gives the average GOSPA error of the four algorithms over 100 Monte Carlo (MC) trials. Table 2 shows the estimation errors and cycling time of the four algorithms in the four scenarios. From the simulation results in Figure 2 and Table 2, we can see that the proposed GGIW-MD-PMBM obtains the general minimum average GOSPA error, which outperforms the other three algorithms; GGIW-PMBM is the second, and GGIW-PHD obtains the highest average GOSPA error. As for the time consumption of each MC run, the GGIW-PHD has the lowest computational cost, GGIW-LMB is the second, and GGIW-MD-PMBM is the third. Compared with other results of normalized location error, the number of missed targets, and the number of false detection, the proposed GGIW-MD-PMBM algorithm is generally better than the other three algorithms. In the third scenario, the GGIW-PMBM filter outperformed the GGIW-MD-PMBM filter in terms of the location error and the number of false detections; this is because the targets in this scenario are intensively generated at the same location and at the same time. Compared with the joint state distribution (GGIW-PMBM), the marginal distribution (GGIW-MD-PMBM) has a larger deviation, but as time increases, the targets gradually move away, and the average GOSPA error of the GGIW-MD-PMBM filter is still lower than the GGIW-PMBM filter. The tracking performance of each algorithm in the fourth scenario is worse than in the other three scenarios; this is because the data association in the fourth scenario is worse than in the other three scenarios.

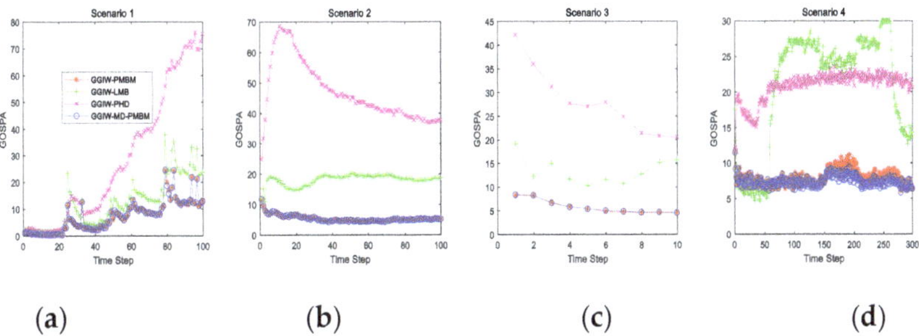

(a) (b) (c) (d)

Figure 2. Comparison of the simulation results of the four algorithms of GOSPA. (**a**) Twenty-seven targets, (**b**) separate/close/split, (**c**) dense birth, (**d**) nonlinear maneuver.

Table 2. Simulation results: the sum of estimation errors and cycling time; GO-GOSPA; LE-normalized location error; NM-number of missed targets; NF-number of false detection; CT-cycling time.

Scenarios		GGIW-PMBM	GGIW-LMB	GGIW-PHD	GGIW-MD-PMBM
	GO	732.8	1246.7	2873.2	729
	LE	56.5	76.6	60.7	56.3
Scenario 1	NM	61.3	481.0	2311.6	60.7
	NF	141.4	96.2	108.5	137
	CT	62.5	8.0	4.2	46.2
	GO	550.7	1818.2	4699.5	550.5
	LE	268.0	133.9	562.3	265.1
Scenario 2	NM	5.6	1479.5	997.5	5.0
	NF	16.9	193.8	3083.2	10.8
	CT	18.6	7.0	0.3	11.4
	GO	59.2	134.8	280.4	58.3
	LE	9.3	11.5	23.1	9.5
Scenario 3	NM	11.2	73.8	174.7	10.6
	NF	2.1	12.1	31.8	2.5
	CT	1.2	0.4	0.1	1.1
	GO	2835.6	6266.3	6257.8	2236.7
	LE	1011.2	869.3	176.8	1002.8
Scenario 4	NM	253.6	3770.6	5601.2	139.4
	NF	175.0	1445.2	470.8	106.2
	CT	49.7	6.5	2.2	40.6

5. Conclusions

In this paper, we propose an efficient filter to solve the extended target tracking problem. Unlike the existing GGIW-PMBM filter that recursively propagates the joint state distribution of targets, the proposed GGIW-MD-PMBM filter recursively propagates the marginal distributions and the existence probabilities of each extended target. Comparing the other three extended target filters in four classic scenarios, the experimental results show that the proposed filter in this paper improves tracking accuracy and reduces computing time.

Author Contributions: Investigation, H.D.; Formal analysis, W.X.; Methodology, H.D.; Resources, H.D.; Simulation, H.D.; Writing—original draft, H.D.; Writing—review and editing, H.D. All authors have read and agreed to the published version of the manuscript.

Funding: This project was funded by the Shenzhen Basic Research Project, grant number JCYJ20170818102503604, and National Natural Science Foundation of China, grant number NO.61271107 and No.61703280.

Appendix A

Let $p_{k,D}^{e}$ be the effective probability of target detection [35] (the set of target detections is nonempty with probability $1 - e^{-\gamma(j)}$), then the updated PPP intensity corresponding to previously existing targets that are not detected is

$$D_{k|k}^{u}(\xi_k) = \sum_{j=1}^{N_{k|k-1}^{u}} (1 - p_{k,D}^{e})\omega_{k|k}^{u,j} GGIW(\xi_k; \zeta_{k|k-1}^{u,j})$$

$$= \sum_{j=1}^{N_{k|k-1}^{u}} p^{(j)}(\gamma_k)N(x_k; m_{k|k-1}^{u,j}, P_{k|k-1}^{u,j})IW_d(X_k; v_{k|k-1}^{u,j}, V_{k|k-1}^{u,j})$$

$$(A1)$$

where $p^{(j)}(\gamma_k)$ is

$$p^{(j)}(\gamma_k) \;=\; (1-p_D)\omega_{k|k-1}^{u,j}G(\gamma_k;\alpha_{k|k-1}^{u,j},\beta_{k|k-1}^{u,j}) + p_D\left(\frac{\beta_{k|k-1}^{u,j}}{\beta_{k|k-1}^{u,j}+1}\right)^{\alpha_{k|k-1}^{u,j}}\omega_{k|k-1}^{u,j}G(\gamma_k;\alpha_{k|k-1}^{u,j},\beta_{k|k-1}^{u,j}+1). \tag{A2}$$

At this time, all existence probabilities of the undetected targets can be normalized, and the existence probability and spatial density of each target can be as follows:

$$\omega_{k|k}^{u,j} \;=\; \frac{(1-p_D)\omega_{k|k-1}^{u,j}}{\sum\limits_{j=1}^{N_{k|k-1}^u} q_D^{u,j}\omega_{k|k-1}^{u,j}} + \frac{p_D\left(\frac{\beta_{k|k-1}^{u,j}}{\beta_{k|k-1}^{u,j}+1}\right)^{\alpha_{k|k-1}^{u,j}}\omega_{k|k-1}^{u,j}}{\sum\limits_{j=1}^{N_{k|k-1}^u} q_D^{u,j}\omega_{k|k-1}^{u,j}}, j = 1,\cdots,N_{k|k-1}^u \tag{A3}$$

$$f_{k|k}^{u,j}(\xi_k) \;=\; f_{1,k|k}^{u,j}(\xi_k) + f_{2,k|k}^{u,j}(\xi_k), j = 1,\cdots,N_{k|k-1}^u \tag{A4}$$

where $f_{1,k|k}^{u,j}(\xi_k)$ represents the spatial density in the case of a missed detection caused by $p_D < 1$, and $f_{2,k|k}^{u,j}(\xi_k)$ represents the spatial density in the case of a missed detection caused by $\gamma_k = 0$.

$$f_{1,k|k}^{u,j}(\xi_k) \;=\; \frac{(1-p_D)\omega_{k|k-1}^{u,j}}{\sum\limits_{j=1}^{N_{k|k-1}^u} q_D^{u,j}\omega_{k|k-1}^{u,j}}GGIW(\xi_k;\zeta_{k|k-1}^{u,j}) \tag{A5}$$

$$f_{2,k|k}^{u,j}(\xi_k) = \frac{p_D\left(\frac{\beta_{k|k-1}^{u,j}}{\beta_{k|k-1}^{u,j}+1}\right)^{\alpha_{k|k-1}^{u,j}}\omega_{k|k-1}^{u,j}}{\sum\limits_{j=1}^{N_{k|k-1}^u} q_D^{u,j}\omega_{k|k-1}^{u,j}}G(\gamma_k;\alpha_{k|k-1}^{u,j},\beta_{k|k-1}^{u,j}+1)N(x_k;m_{k|k-1}^{u,j},P_{k|k-1}^{u,j})IW_d(X_k;v_{k|k-1}^{u,j},V_{k|k-1}^{u,j}) \tag{A6}$$

References

1. Vo, B.N.; Mallick, M.; Bar-shalom, Y.; Coraluppi, S.; Osborne, R., III; Mahler, R.; Vo, B.T. Multitarget tracking. *Wiley Encycl. Electr. Electron. Eng.* **2015**. [CrossRef]
2. Bar-Shalom, Y.; Li, X.R.; Kirubarajan, T. *Estimation with Applications to Tracking and Navigation: Theory Algorithms and Software*; John Wiley & Sons: Hoboken, NJ, USA, 2004.
3. Granstrom, K.; Baum, M. Extended Object Tracking: Introduction, Overview and Applications. *J. Adv. Inf. Fusion* **2017**, *12*, 139–174.
4. Gilholm, K.; Godsill, S.; Maskell, S.; Salmond, D. Poisson models for extended target and group tracking. In Proceedings of the Signal and Data Processing of Small Targets, San Diego, CA, USA, 31 July–4 August 2005; pp. 230–241.
5. Granstrom, K.; Salmond, D. Spatial distribution model for tracking extended objects. *IEEE Proc. Radarsonar Navig.* **2005**, *152*, 364–371.
6. Koch, J.W. Bayesian approach to extended object and cluster tracking using random matrices. *IEEE Trans. Aerosp. Electron. Syst.* **2008**, *44*, 1042–1059. [CrossRef]
7. Feldmann, M.; FraNken, D.; Koch, W. Tracking of Extended Objects and Group Targets Using Random Matrices. *IEEE Trans. Signal Process.* **2010**, *59*, 1409–1420. [CrossRef]
8. Feldmann, M.; Fränken, D. Advances on Tracking of Extended Objects and Group Targets using Random Matrices. In Proceedings of the 2009 12th International Conference on Information Fusion, Seattle, WA, USA, 6–9 July 2009; Volume 59, pp. 1029–1036.

9.   Granstrom, K.; Orguner, U. A phd Filter for Tracking Multiple Extended Targets Using Random Matrices. *IEEE Trans. Signal Process.* **2012**, *60*, 5657–5671. [CrossRef]
10. Orguner, U. A variational measurement update for extended target tracking with random matrices. *IEEE Trans. Signal Process.* **2012**, *60*, 3827–3834. [CrossRef]
11. Granstrom, K.; Orguner, U. A New Prediction Update for Extended Target Tracking with Random Matrices. *IEEE Trans. Aerosp. Electron. Syst.* **2014**, *2*, 324–336.
12. Liu, J.; Guo, G. A Random Matrix Approach for Extended Target Tracking Using Distributed Measurements. *IEEE Sens. J.* **2019**, *99*, 1–12.
13. Lan, J.; Li, X.R. Extended-Object or Group-Target Tracking Using Random Matrix with Nonlinear Measurements. *IEEE Trans. Signal Process.* **2019**, *99*, 1–15. [CrossRef]
14. Wang, L.; Huang, Y.; Zhan, R.; Zhang, J. Joint Tracking and Classification of Extended Targets Using Random Matrix and Bernoulli Filter for Time-Varying Scenarios. *IEEE Access* **2019**, *7*, 129584–129603. [CrossRef]
15. Lu, Z.; Hu, W.; Liu, Y.; Kirubarajan, T. Seamless Group Target Tracking Using Random Finite Sets. *Signal Process.* **2020**, *176*, 1–13. [CrossRef]
16. Gao, L.; Jing, Z.; Li, M.; Pan, H. Bayesian approach to multiple extended targets tracking with random hypersurface models. *IET Radar Sonar Navig.* **2019**, *13*, 601–611. [CrossRef]
17. Wang, X.; Li, H.G.; Kong, Y.B.; Pu, L.; Fan, P.F. SMC-PHD filter for extended target tracking based on star-convex random hypersurface models. *Appl. Res. Comput.* **2017**, *34*, 2144–2147.
18. Liu, Z.P.; Liu, Y.J. Extended Target Tracking Algorithm Based on Star-Convex Random Hypersurface Models. *Electron. Opt. Control* **2017**, *24*, 72–76, 82.
19. Li, Y.W. Extended Target Tracking Algorithm Based on Random Hypersurface Model with Glint Noise. *Chem. Eng. Trans.* **2017**, *59*, 685–690.
20. Baum, M.; Hanebeck, U.D. Random Hypersurface Models for extended object tracking. In Proceedings of the IEEE International Symposium on Signal Processing and Information Technology, Ajman, UAE, 14–17 December 2009; IEEE: Piscataway, NJ, USA, 2010; pp. 178–183.
21. Baum, M.; Hanebeck, U.D. Extended Object Tracking with Random Hypersurface Models. *IEEE Trans. Aerosp. Electron. Syst.* **2014**, *50*, 149–159. [CrossRef]
22. Wahlström, N.; Özkan, E. Extended Target Tracking Using Gaussian Processes. *IEEE Trans. Signal Process.* **2015**, *63*, 4165–4178. [CrossRef]
23. Mahler, R. *Statistical Multisource-Multitarget Information Fusion*; Artech House: Norwood, MA, USA, 2007; pp. 1–48.
24. Lian, F.; Han, C.; Liu, W.; Chen, H. Joint spatial registration and multi-target tracking using an extended probability hypothesis density filter. *IET Radar Sonar Navig.* **2011**, *5*, 441–448. [CrossRef]
25. Li, G.; Zhang, H.Q.; Wang, Y. An Efficient Multi-Target Tracking Algorithm Using Gaussian Mixture Probability Hypothesis Density Filter. In Proceedings of the 2018 IEEE CSAA Guidance, Navigation and Control Conference (CGNCC), Xiamen, China, 10–12 August 2018; IEEE: Piscataway, NJ, USA, 2020.
26. Liu, L.; Ji, H.; Zhang, W.; Liao, G. Multi-Sensor Multi-Target Tracking Using Probability Hypothesis Density Filter. *IEEE Access* **2019**, *7*, 67745–67760. [CrossRef]
27. Li, C.; Wang, R.; Hu, Y.; Wang, J. Cardinalised probability hypothesis density tracking algorithm for extended objects with glint noise. *IET Sci. Meas. Technol.* **2016**, *10*, 528–536. [CrossRef]
28. Vo, B.T.; Vo, B.N.; Cantoni, A. Analytic Implementations of the Cardinalized Probability Hypothesis Density Filter. *IEEE Trans. Signal Process.* **2007**, *55*, 3553–3567. [CrossRef]
29. Li, B.; Yi, H.W.; Li, X.H. Innovative unscented transform–based particle cardinalized probability hypothesis density filter for multi-target tracking. *Meas. Control Lond. Inst. Meas. Control* **2019**, *52*, 1567–1578. [CrossRef]
30. Chen, J.G.; Wang, X.H.; Ma, L.L.; Zhang, X.D.; Gong, L.M. Cardinalized Probability Hypothesis Density Smoother Using Piece-wise RTS. *Comput. Eng. Appl.* **2019**, *55*, 50–55, 95.
31. Gostar, A.K.; Hoseinnezhad, R.; Bab-Hadiashar, A. Robust Multi-Bernoulli Sensor Selection for Multi-Target Tracking in Sensor Networks. *IEEE Signal Process. Lett.* **2013**, *20*, 1167–1170. [CrossRef]
32. Park, W.J.; Park, C.G. Multi-target Tracking Based on Gaussian Mixture Labeled Multi-Bernoulli Filter with Adaptive Gating. In Proceedings of the 2019 First International Symposium on Instrumentation, Control, Artificial Intelligence, and Robotics (ICA-SYMP), Bangkok, Thailand, 16–18 January 2019; pp. 226–229.
33. Cament, L.; Correa, J.; Adams, M.; Pérez, C. The Histogram Poisson, Labeled Multi-Bernoulli Multi-Target Tracking Filter. *Signal Process.* **2020**, *176*, 837–851. [CrossRef]

34. Zhu, Y.; Wang, J.; Liang, S. Multi-Objective Optimization Based Multi-Bernoulli Sensor Selection for Multi-Target Tracking. *Sensors* **2019**, *19*, 1884–2021. [CrossRef]
35. Mahler, R. PHD filters for nonstandard targets, I: Extended targets. In Proceedings of the 2009 12th International Conference on Information Fusion, Seattle, WA, USA, 6–9 July 2009; pp. 915–921.
36. Granstrom, K.; Lundquist, C.; Orguner, U. A Gaussian mixture PHD filter for extended target tracking. In Proceedings of the International Conference on Information Fusion, Edinburgh, UK, 26–29 July 2011.
37. Han, Y.; Zhu, H.; Han, C.Z. A Gaussian-mixture PHD filter based on random hypersurface model for multiple extended targets. In Proceedings of the 16th International Conference on Information Fusion, Istanbul, Turkey, 9–12 July 2013.
38. Lundquist, C.; Granstrom, K.; Orguner, U. An extended target CPHD filter and a gamma Gaussian inverse Wishart implementation. *J. Sel. Top. Signal Process.* **2013**, *7*, 472–483. [CrossRef]
39. Orguner, U.; Lundquist, C.; Granstrom, K. Extended target tracking with a cardinalized probability hypothesis density filter. In Proceedings of the 14th International Conference on Information Fusion, Chicago, IL, USA, 5–8 July 2011.
40. Granstrom, K.; Fatemi, M.; Svensson, L. Poisson multi-Bernoulli conjugate prior for multiple extended object estimation. *IEEE Trans. Aerosp. Electron. Syst.* **2018**, *3*, 30–45.
41. Granstrom, K.; Fatemi, M.; Svensson, L. Gamma Gaussian inverse-Wishart Poisson multi-Bernoulli filter for extended target tracking. In Proceedings of the 19th International Conference on Information Fusion, Heidelberg, Germany, 5–8 July 2016.
42. Beard, M.; Reuter, S.; Granström, K.; Vo, B.T.; Vo, B.N.; Scheel, A. A Generalised Labelled Multi-Bernoulli Filter for Extended Multi-target Tracking. In Proceedings of the 18th International Conference on Information Fusion, Washington, DC, USA, 6–9 July 2015.
43. Beard, M.; Reuter, S.; Granström, K.; Vo, B.T.; Vo, B.N.; Scheel, A. Multiple extended target tracking with labelled random finite sets. *IEEE Trans. Signal Process.* **2016**, *64*, 1638–1653. [CrossRef]
44. Granstrom, K.; Orguner, U. Estimation and maintenance of measurement rates for multiple extended target tracking. In Proceedings of the 15th International Conference on Information Fusion, Singapore, 9–12 July 2012; pp. 2170–2176.
45. Liu, Z.X.; Xie, W.X. Multi-target Bayesian filter for propagating marginal distribution. *Signal Process.* **2014**, *105*, 328–337. [CrossRef]
46. Bar-Shalom, Y.; Willett, P.K.; Tian, X. *Tracking and Data Fusion: A Handbook of Algorithms*; YBS Publishing: Storrs, CT, USA, 2011.
47. Rahmathullah, A.S.; García-Fernández, Á.F.; Svensson, L. Generalized optimal sub-pattern assignment metric. In Proceedings of the 20th International Conference on Information Fusion (Fusion), Xi'an, China, 10–13 July 2017; pp. 1–8.
48. Yang, S.S.; Baum, M.; Granstrom, K. Metrics for performance evaluation of elliptic extended object tracking methods. In Proceedings of the International Conference on Multisensor Fusion and Integration for Intelligent Systems (MFI), Baden-Baden, Germany, 19–21 September 2016; pp. 523–528.

Article

A Multi-Core Object Detection Coprocessor for Multi-Scale/Type Classification Applicable to IoT Devices

Peng Xu [1,†], **Zhihua Xiao** [1,†], **Xianglong Wang** [1], **Lei Chen** [2], **Chao Wang** [3,4] **and Fengwei An** [1,2,5,*]

[1] School of Microelectronics, Southern University of Science and Technology, Shenzhen 518055, China; 11710124@mail.sustech.edu.cn (P.X.); 11713016@mail.sustech.edu.cn (Z.X.); 12031015@mail.sustech.edu.cn (X.W.)

[2] Pengcheng Laboratory, Shenzhen 518055, China; chenl03@pcl.ac.cn

[3] Department of Integrated Circuit Engineering, Huazhong University of Science and Technology, Wuhan 430074, China; chao_wang_me@hust.edu.cn

[4] Wuhan National Laboratory for Optoelectronics, Wuhan 430074, China

[5] Engineering Research Center of Integrated Circuits for Next-Generation Communications, Ministry of Education, Southern University of Science and Technology, Shenzhen 518055, China

* Correspondence: anfw@sustech.edu.cn

† These authors contributed equally to this paper.

Received: 20 September 2020; Accepted: 30 October 2020; Published: 31 October 2020

Abstract: Power efficiency is becoming a critical aspect of IoT devices. In this paper, we present a compact object-detection coprocessor with multiple cores for multi-scale/type classification. This coprocessor is capable to process scalable block size for multi-shape detection-window and can be compatible with the frame-image sizes up to 2048×2048 for multi-scale classification. A memory-reuse strategy that requires only one dual-port SRAM for storing the feature-vector of one-row blocks is developed to save memory usage. Eventually, a prototype platform is implemented on the Intel DE4 development board with the Stratix IV device. The power consumption of each core in FPGA is only 80.98 mW.

Keywords: power efficiency; object-detection coprocessor; histogram of oriented gradient; support vector machine; block-level once sliding detection window; multi-shape detection-window

---

1. Introduction

Real-time processing ability is required by multiple tasks such as auto-drive, Internet of Things (IoT) systems, security systems, and so on. Even though the CPU (Central Processing Unit) and GPU (Graphics Processing Unit)-based solutions are flexible and can be easily used on multiple devices for different tasks, its low processing speed and high-power consumption make it inefficient for edge computation devices. Typically, the edge devices using low-power processors find it hard to handle complex tasks or have to compromise on precision and speed. Meanwhile, hardware-friendly VLSI (Very Large-Scale Integration) becomes attractive for satisfying the speed and power need of edge devices. Even the VLSI has an undesirable design complexity and flexibility loss. The high performance and low power consumption of VLSI implementation are suitable to the edge device which needs to handle real-time tasks.

1.1. Related Work

Vehicle detection is extremely important for driverless cars or advanced driver assistance systems (ADAS). With the development of Convolutional Neuron Networks (CNN), many researchers are working on using CNN for object detection (e.g., YOLO [1], RFCN [2], VGG [3]), but only a small subset of papers discuss the running time in any detail. For example, the design in [4,5] have a considerable accuracy in detection tasks but incur high computation costs to execute their CNN models. Furthermore, these papers only claim the frame rate they achieve but do not give a full picture of the speed-power-accuracy trade-off [6]. Though the speed performance of CNN on the GPU platform has improved a lot, the high memory usage and computation complexity still make it inapplicable for time-critical and low-power-consumption devices.

Nakahara et al. [7] proposed a YOLO-based object detection architecture on FPGA. To make the complex CNN network more hardware friendly, they chose to use a light weighted YOLOv2 and further simplified it to binary weight. According to their paper, for an input image size of 224 × 224, the detection frame rate is 40 fps. Though their FPGA (Field Programmable Gate Array) implementation shows a great higher efficiency compared with ARM CPU and Pascal GPU, the detection speed is still slow and the memory is high when considering the small input resolution.

Feature descriptors like Histogram of Orientated Gradient (HOG), Local Binary Pattern (LBP), and Haar-like features are also proved to be efficient for vehicle detection [8,9]. As a consequence, those algorithms require much lower computational resources compared to CNN applications.

For object detection, a multi-scale detector always requires much more memory [10,11]. Usually, as the size of detection windows is fixed, to detect objects in different scales, a raw image needs to be buffered into different sizes and this process enhances the usage of storage (e.g., Static Random-Access Memory (SRAM)).

A sliding window is a common approach for object detection [12]. Classifiers will determine the similarity between the feature vector of the window and samples. A detection window is divided into some local regions (blocks and cells) to calculate feature vectors, while the window is shifting on the image, many cells and blocks are overlapped. One possible solution to avoid overlapping processing is to calculate each cell and use the results to construct the feature vector of blocks and construct a detection window. Meanwhile, feature-vector normalization among a block usually requires dividers. Traditional digital division methods cannot meet the speed requirement. A common way for the fast division is to create a look-up table to store the reciprocal of divisors and convert the division into a multiplication and table-lookup problem.

The hardware implementation of the HOG plus SVM (support vector machine) has been discussed to improve the speed-power performance [13]. Peng [14] studied the HOG feature extractor circuit for real-time detection and discussed how the sliding window size and cell size can affect the performance. To increase the robustness, the histograms of cells are combined and then normalized in L2 form to construct a descriptor of a block.

1.2. Contribution

In this paper, we propose a multi-core object detection coprocessor for multi-scale/type classification considering the speed-power-accuracy tradeoff within the HOG and SVM framework as shown in Figure 1. The contribution of this paper can be summarized into three aspects as follows:

Firstly, differing from our previous work [15], a scalable size of a block is implemented in this work, which enables the flexibility to detect objects in different shapes and scales. For instance, a vertical rectangle detection window (DW) is often used in pedestrian detection but a horizontal rectangle DW is suitable for vehicle detection.

Then, a sliding detection window mechanism with a scalable block size enables parallel partial SVM classifications of all DWs that contain the being-processed block. This can avoid repeated computations of the overlapped blocks in different DWs and large memory for buffering DWs. Compared to the previous FPGA-implemented HOG-SVM classifiers [15–18] with only a dedicated scale, we provide multi-scale detection for vehicle detection in this work.

Finally, an approximation divider with the multiplication of the reciprocal using Taylor expansion solves the critical path of the normalization circuitry so that it can synchronize to the working frequency of the image sensor for low dynamic power consumption. In contrast to the general implementation of a divider, not only less hardware-resource usage but also the max working frequency of this approximation divider can be significantly improved from 28.66 MHz to 162.81 MHz with 0.682% accuracy loss.

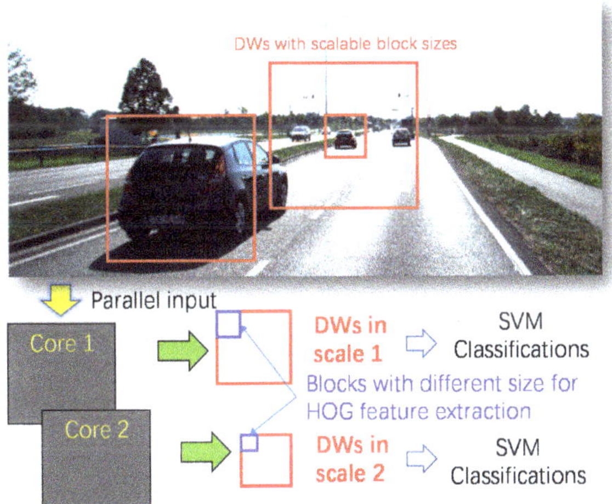

Figure 1. The overview of the proposed multi-core object detection coprocessor for multi-scale/type classification with the HOG and SVM framework.

1.3. Structure

The remaining of the paper is organized as follows. Section 2 introduces the framework with the HOG feature and SVM classifier. Section 3 presents the VLSI-oriented algorithm for the proposed framework. The experimental results are discussed in Section 3. Finally, we conclude the paper in Section 4.

2. VLSI-Oriented Hardware Algorithm for Feature Extraction

2.1. Block-Level Feature Extraction and Normalization Circuitry

The HOG feature extraction starts with dividing the input image into non-overlapping subcomponents, the so-called cells with the size of w_{cell} and h_{cell}. The gradients are calculated for each pixel within each cell. The gradient orientation of all pixels in a cell is mapped to B bins for each cell. The histogram normalized in a block with $C \times C$ cells to increase robustness to the variations, e.g., texture and illumination, results in a final $(B \times C)$-dimensional feature.

Suppose that the size of the detection window is w_{DW} and h_{DW}, since the blocks are overlapped by a cell, each DW has $m \times n$ blocks where $m = (w_{DW}/w_{cell} - 1)$ in horizontal and $n = (h_{DW}/h_{cell} - 1)$ in the vertical. The final dimensionality of the feature vector (FV) of a DW is $(B \times C^2) \times m \times n$.

A linear SVM classifier in conjunction with the HOG feature is a popular solution for object detection except for deep neural networks [4]. The SVM classifier for a DW with a $(B \times C^2) \times m \times n$-dimensional FV is trained off-line by training samples with the same size as a DW, i.e., $w_{DW} \times h_{DW}$, so that the SVM weight has the same dimensionality as a DW, i.e., $(B \times C^2) \times m \times n$.

It can be observed that the dimensionality keeps the same when the size of cells, i.e., w_{cell} and h_{cell} increases or decreases in the same ratio to the size of the DW, i.e., w_{DW} and h_{DW}. One main factor differencing from the previous works [15–18] is that the size of the DW in this work can be scaled. This is the reason that, rather than pyramid images, the designed HOG feature extraction module can produce different sizes of DWs with the same dimensionality and thus achieve the multi-scaled detection.

For each block, a normalization step is applied to adjust the descriptors of cells in the block during the construction of the block's local FV. Supposing that each block has 2×2 cells in Figure 2, block 1 (B1) and block 2 (B2) have two overlapped cells. It is observed that these two overlapped cells are reused two times in B1 and B2. Furthermore, the presence of, e.g., cell (2, 2) in four blocks (B1, B2, B4, and B5) leads to 4-fold reuse of this cell. Eventually, as shown in Figure 2, a cell-reuse map (CRM) within an image shows that corner cells, edge cells, and inter cells overlap one time, two times, and four times respectively.

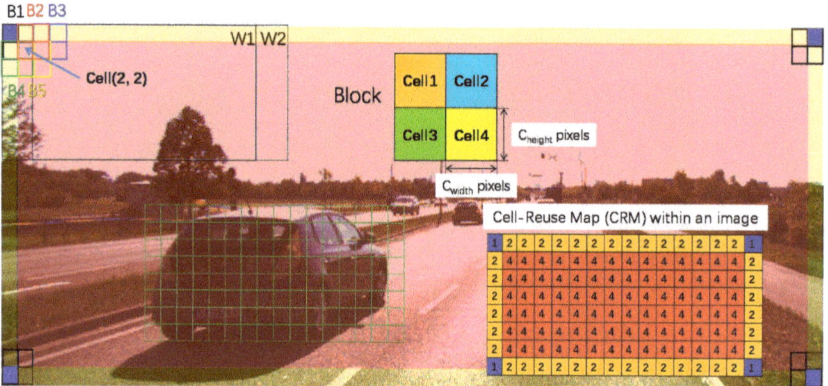

Figure 2. Illustration of the Cell-Reuse map based on the 128×64 DW and 2×2 cell blocks within an image.

Additionally, according to the CRM, the 2-D address of the cell in the image can derive the blocks which contain it and the locations (top-left, top-right, bottom-left, and bottom-right as shown in Figure 2) in the blocks. Then, the extracted cell-descriptor is added to the block descriptor memory for the block-level normalization. Since the blocks are overlapped by one cell, there are $w/w_{cell} - 1$ blocks. Once the Cell (1, 1) has been extracted, the accumulation of Block 0 is then completed. At the same moment, the Cell (0, 0) stored in the cell line-buffer is read out for the normalization of Block 0. Accordingly, besides the first cell line, Cell (1, 0) and Cell (1, 1) must be stored in the cell line buffer. In the normalization module, the memory usage including the intermediate block memory and the cell line buffer is $w/w_{cell} \times \text{WL}_{block} \times \text{Bin} + (w/w_{cell} + 2) \times \text{WL}_{cell} \times \text{Bin}$ bits. Here, WL_{block} is the word length of the intermediate value of the cell accumulation for block normalization so that WL_{block} is equal to $\text{WL}_{cell} + 2$.

In Figure 3, the HOG feature extraction unit has been implemented in our previous works [19] with the raster scan manner of the image sensor. It extracts the feature vectors of the cells from the image and the extracted cell vectors are then transferred into both the intermediate block memory and the cell line buffer. The cell position points to the location of the intermediate block memory to store the sum of the cells in their block. Once the sum of the cell vectors within a block is calculated, the feature vectors of the cells stored in the cell line buffer then divide this sum for normalization. Subsequentially, the normalized block feature is passed to the next module for partial SVM classification.

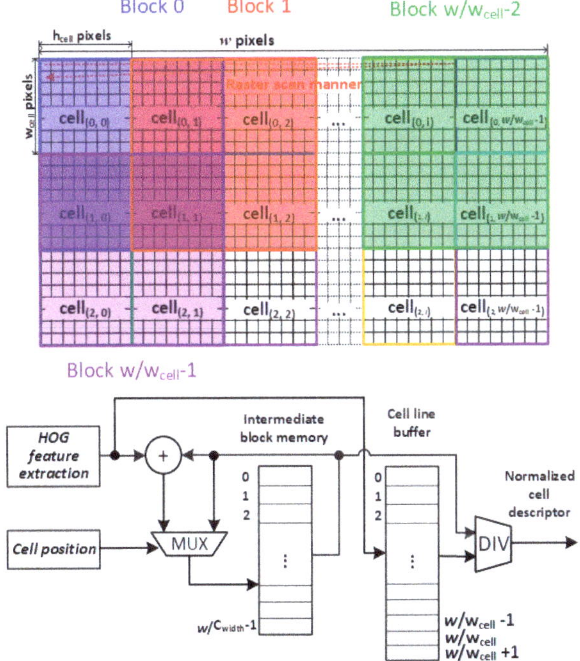

Figure 3. Block-level normalization. Cell position in a block and image is given by CRM.

The critical path in the block normalization module is the divider. In this work, we consider the division as the multiplication of the dividend and the reciprocal of the divisor as Equation (1) where the reciprocal uses Taylor expansion. First, the divisor X is normalized into $[1, 2)$ as illustrated in Equation (2). Then we apply the first two series of Taylor expansion to estimate x as in Equations (3–5).

$$\frac{Y}{X} = Y \times \frac{1}{X} \tag{1}$$

$$X = x \times 2^j \qquad x \in [1, 2) \tag{2}$$

$$f(x) = \frac{1}{x} \approx f_i(x_{i0}) + f_i'(x_{i0} - x) \tag{3}$$

$$f_i(x_{i0}) = \frac{1}{x_{i0}} \quad f_i'(x_{i0}) = \frac{1}{x_{i0}^2} \tag{4}$$

To minimize the look-up-table of the $f_i(x_{i0})$ and $f_i'(x_{i0})$, domain $[1, 2)$ is divided into n parts and we use the middle point x_{i0} of each part to represent the value of the part in Equation (6). To make up for the loss of accuracy, Newton iteration [20] is used to increase the accuracy in a very efficient way. Finally, the division value can be given by Equation (7).

$$x_{i0} = 1 + i \times 2^{-n} - 2^{-n-1} \tag{5}$$

$$f_2(x) \approx f(x) \times (2 - x \times f(x)) \tag{6}$$

$$\frac{Y}{X} = Y \times f_2(x) \times 2^{-j} \tag{7}$$

As shown in Figure 4, the normalization unit is to map the dividend into a number within $[1, 2)$ after a bit-shifting. In particular, the look-up-table, in which the number within $[1, 2)$ is divided into n parts, determines the accuracy precision of the result. Of course, this is a tradeoff of the result accuracy and the hardware resource.

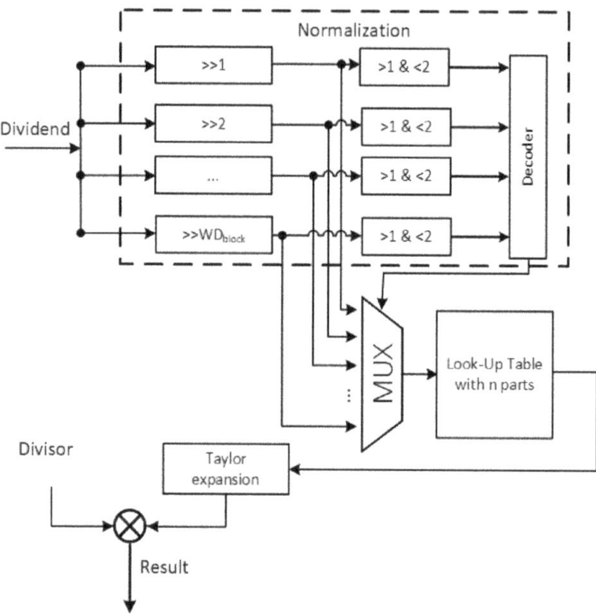

Figure 4. Optimized divider. The dividend is normalized into $(1, 2)$ and a lookup table is then used to estimate the Taylor expansion of the reciprocal of dividend.

2.2. Scalable-Block-Size Based Sliding-Detection-Window (SBSSDW) Mechanism

Compared to our previous works in [19], in this work, the block-level normalization requires a larger memory but brings a smaller index look-up table and a lesser memory address calculation in the partial classification stage. For instance, a 128×64-pixel DW contains 420 cells but only 105 blocks so that the index and address number assignment of the overlapped DWs can be reduced to four times less than the cell-based method.

In particular, this work develops a scalable block size for multi-shape objects instead of the fixed cell/block size. For detecting objects in the input image, usually, the object will be detected by a sliding window with the overlapping of a block. Instead of the repeated computation of the overlapped cells and a large memory for buffering the window and the entire image, we develop a scalable-block-size-based sliding-detection-window (SBSSDW) mechanism in this work. For every pixel (m, n) and $0 \le m < w, 0 \le n < h$, it should be included in a block (i, j), where $m / w_{block} = i$, $n / h_{block} = j$. Consequently, we can model the index of the first window which contains the block (i, j) as shown in (10) to construct the window vectors. Here, each DW with the index (i_{DW}) is a 1-dimensional vector to represent the number of DWs overlapping by a block, i.e., $(w - w_{DW} / w_{block} + 1) \times (h - h_{DW} / h_{block} + 1)$. For example, a VGA image has 832 DWs while each DW has 128×64 pixels and each block contain 2×2 cells with 8×8 pixels. Hence, the index starts from 0 to 831.

$$i_{FW} = \left| j - \frac{h_{DW}}{h_{block}} \right| \times \left(\frac{w - w_{DW}}{w_{block}} + 1 \right) + \left| i - \frac{w_{DW}}{w_{block}} \right| \tag{8}$$

$$\left(0 \le i \le \frac{w - w_{DW}}{w_{block}}, \ 0 \le j \le \frac{h - h_{DW}}{h_{block}} \right) \tag{9}$$

Furthermore, the number of DWs that contains the block (i, j) can be calculated to decide the numbers of replications of this block, i.e., $N_{hor} \times N_{ver}$ where N_{hor} in horizontal and N_{ver} in vertical by Equations (10) and (11) respectively.

$$N_{hor} = \begin{cases} i+1, \ 0 \le i < \frac{w_{DW}}{w_{block}} \\ \frac{w_{DW}}{w_{block}}, \ \frac{w_{DW}}{w_{block}} \le i < \frac{w}{w_{block}} - \frac{w_{DW}}{w_{block}} \\ \frac{w_{DW}}{w_{block}} - i, \ \frac{w}{w_{block}} - \frac{w_{DW}}{w_{block}} < i \ge \frac{w}{w_{block}} \end{cases} \tag{10}$$

$$N_{ver} = \begin{cases} j+1, \ 0 \le j < \frac{h_{DW}}{h_{block}} \\ \frac{h_{DW}}{h_{block}}, \ \frac{h_{DW}}{h_{block}} \le j < \frac{h}{h_{block}} - \frac{h_{DW}}{h_{block}} \\ \frac{h_{DW}}{h_{block}} - j, \ \frac{h}{h_{block}} - \frac{h_{DW}}{h_{block}} < j \ge \frac{h}{h_{block}} \end{cases} \tag{11}$$

As Figure 2 shows, a block can be used in different DWs. After the block-level normalization, the FV of each block is used to construct the partial FV of the overlapped DWs.

In the case of Block 3 in Figure 2, its index (i_{FW} =0) of the first window and the number of DWs ($N_{ver} \times N_{hor} = 2 \times 2 = 1$) containing Block (1, 1) can be calculated at the same moment based on the coordinate. The classification of the DWs is partially computed and the intermediate results are buffered in memory. This can significantly reduce memory usage and certainly reduce the delay in the classification of DWs. Once the FV of the last block of a DW0 is extracted, the classification for DW0 with a linear support vector machine (SVM) [8] can be completed with a short delay.

The linear SVM approach aims to construct a classifier which can be mathematically represented like $y(\vec{v}) = \text{sgn}[f(\vec{v}, \vec{w})]$. Here, $\vec{v}$ and $\vec{w}$ represent the FV of DWs and weight vector, $f(\vec{v}, \vec{w})$ is the kernel function and sgn($\cdot$) is the sign function. The FVs of DWs were mapped into two classes, positive and negative by the classifier $y(\vec{v})$ and then the cost function would be used to evaluate the cost between the label of the $\vec{v}$ and $y(\vec{v})$.

The SVM prediction part uses the parameters from the off-line trained model to classify the class of a DW. The weight vector $\vec{w}$ and the bias b are the two essential components from the trained linear SVM model, which is initially stored in the dual-port memory (DPM) in advance. Then the FV of a DW multiply-accumulated the weight vector and finally adds the bias b to predict the positive or negative class.

In the hardware architecture (Figure 5) for SBSSDW, the pixel coordinates are converted to the position of the processed cells and blocks in a frame for calculating the corresponding index of the first window, N_{hor} and N_{ver}. In the beginning, the index of the first window is set to i_{FW} and the index then increases after handling the components of block-level FVs from i_{FW} to $i_{FW} + N_{hor}$ horizontally. When the processing reached the end of a row, the index is reassigned to $i_{FW} + x \times n$ where n = $(w - w_{DW})/W_{block}$ representing the maximum overlapped sliding window in the horizontal direction of a frame and x is the row number of the DW. Finally, the classification of a block ends until the index reaches $i_{FW} + N_{hor} + (N_{ver} - 1) \times n$.

For a block, its corresponding components of FVs of a DW are readout. Meanwhile, the classification intermediate values of the DW stored in the SRAM are firstly read out to continually compute the SVM classification. The classification will be completed if this block is the final pitch of the DW. The above operation is then repeated until all DWs contain this block. As a consequence, the block should be stored in a FIFO.

It can be observed that the classification processing starts at the last pixel-row of a frame and need to be completed in $w \times h_{block}$ clocks. The feature-extraction and block-level normalization unit are synchronized with the working frequency of the image sensor. Whereas, a buffer for block FVs, i.e., FIFO, has to be used to store the unprocessed data. Hence, this causes a tradeoff between the buffer size and the power dissipation. The buffer size can be significantly reduced when the SVM classification module adopts a higher working clock frequency but with high power dissipation. On the other hand, the working frequency can synchronize to the HOG feature and normalization module when the classification DWs for a block under processing is parallelized. In the case of the maximum parallelism with $h_{DW}/h_{block} \times w_{DW}/w_{block}$, although the FIFO does not need to be set, the memory of the weight for SVM requires $h_{DW}/h_{block} \times w_{DW}/w_{block}$ ports. As a consequence, parallelism defines the complexity of the hardware resource in the SVM module.

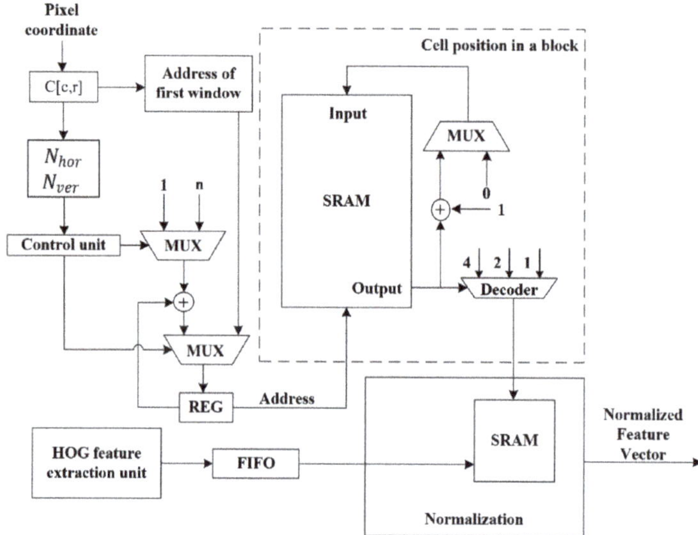

Figure 5. Hardware architecture of SBSSDW, the pixel index, N_{hor} and N_{ver} is calculated by this module.

2.3. Multi-Scale Detection with Multi-Core Implementation

Usually, an image pyramid is generated for each frame for supporting multi-scale detection. In the case of a single core, besides the large buffer for scaled images, a high working frequency must be applied to process these additional scales.

In this work, a multi-core architecture can avoid the frame buffer memory and a high working frequency. Each core detects objects with a scaled DW by adjusting the width and height of the cell in the HOG feature extraction module as described in Section 3.1 and Figure 2. For each core, the key parameters vary because of the different sizes of sliding windows. In other words, each core will construct their map of reuse times of each cell, the index of each cell, and the feature vectors. The number of cores is equivalent to the number of scaled images to detect objects in different sizes while each core has the same on the hardware structure.

3. Experiment Results and Discussion

3.1. Hardware Implementation and Performance Analysis

To estimate the performance, we deployed this work on the Intel DE4 development board (Stratix IV GX EP4SGX230 [21]) with an STC-MC83PCL [22] camera-input XGA signal (1024×768 @ 60 Hz), as shown in Figure 6. In this work, the image size is constrained by two factors: w the width of the line buffer in the Sobel filter module and $DP_{cellmem}$, the depth of the memory for storing the intermediate FVs of cells in a row in Figure 2, i.e., $DP_{cellmem} = w/w_{cell}$. As illustrated in Figure 5, synchronized with the pixel-based HOG feature extraction, the pixel coordinates are also converted to the position of the processed cell in the image frame for the simultaneous calculation of the corresponding index of the first window i_{FW}, N_{hor} and N_{ver}. The weight of the offline trained SVM classifier is initialized in the DPM.

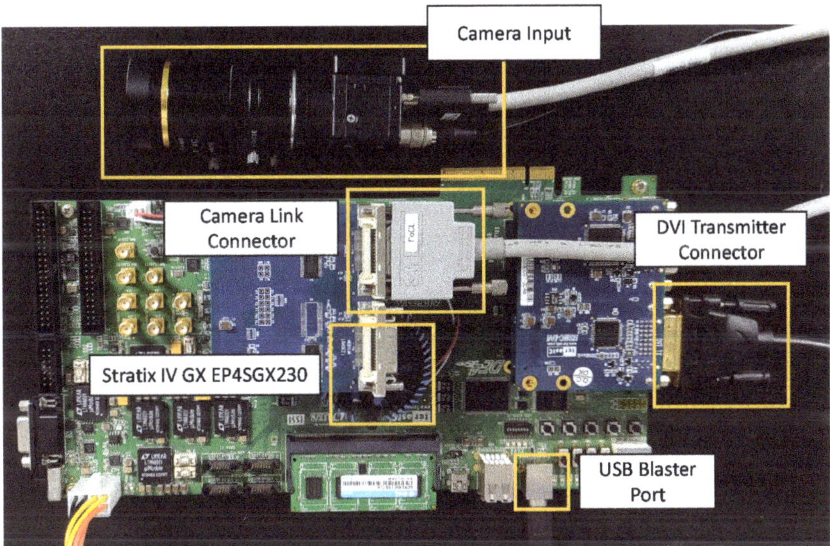

Figure 6. FPGA-based prototype system deployed on the Intel DE4 development board (Stratix IV GX EP4SGX230).

3.2. Discussion and Comparison

We define the width of the line buffer w = 2048 and the depth of the cell FV memory $DP_{cellmem} = 256$. Table 1 list the resources, working frequency, and power dissipation for different modules. Most of the intense computing process is the normalization and SVM module. To avoid data overflow between modules and large FIFO due to different processing speeds, the working frequency of the SVM module must be two times faster than the Sobel filter, HOG feature, and the normalization module. The working frequency of Sobel, HOG, and normalization can be synchronized with the pixel clock, which is 65 MHz for XGA. The max working frequency of the Sobel filter and the HOG feature can easily satisfy this speed requirement. In particular, the maximum working frequency of the Quartus LPM divider IP core, e.g., the 18-bit divider with 28.66 MHz, is hard to meet the speed for real-time processing. In this work, we thus improve the critical path caused by the divider in the normalization module using Taylor expansion. The accuracy loss of the divider is only 0.682% and while this is small enough, it has a limited effect on the SVM prediction. Meanwhile, the max working frequency of the normalization module has been improved to 162.81 MHz.

The power consumption of each module and the whole design estimations are shown in Table 1 where it is estimated using Quartus Power Estimator 18.1 with their max working frequency. In the case of the same working frequency, the powers of the Sobel Filter, HOG Feature, and normalization parts increase gradually because of the increased resource usage of each module. This can explain why the SVM part consumes the highest power, i.e., its highest clock frequency and most resource usage.

Table 1. Hardware resources of each module *.

	Sobel Filter	HOG Feature	Normalization	SVM	Total
Combinational ALUTs	442	549	4678	3017	8686
Dedicated Logic Registers	616	587	1694	5319	8216
Block Memory Bits	18,432	49,152	39,616	174,080	281,280
Max Working Frequency	356.63 MHz	323.31 MHz	162.81 MHz	136.43 MHz	
Actual Working Frequency	65 MHz	65 MHz	65 MHz	130 MHz	
Power Consumption	2.84 mW	3.51 mW	9.47 mW	65.17 mW	80.98 mW

* All the resources are measured by Quartus Premium 18.1.

By using the SBSSDW, only a small part of the intermediate FVs need to be stored for SVM prediction and finally, we reduced the total SRAM usage to 281 Kbit, in which 73.7 Kbit is used to store the weight for SVM classification. Consider that the max working frequency of SVM, 136.43 MHz, and the detection speed is 60 fps for an XGA (1024 × 768 pixels) video input or 30 fps for a 2K (2048 × 2048) video. Compared with other related work as shown in Table 2, our work uses significantly less memory than [16]. Additionally, the hardware resource with the throughput of 60 fps XGA video of this work is only one-tenth that of [17] with a throughput of 60 fps VGA.

Table 2. Resource comparison.

	[16]	[17]	[18]	[15]	This Study
Combinational ALUTs	85,837	87,306	6551	7652	8686
Dedicated Logic Registers	406,978	77,726	4375	4503	8216
Block Memory Bits (Mbit)	2.55	2.97	NA	0.13	0.28
Max Working Frequency (MHz)	135	108.19	50	* NA	130
Frame rate (fps)	68.18	60	25	60	60
Resolution	640 × 480	640 × 480	640 × 480	1920 × 1080	1024 × 768
FPGA platform	Virtex-6	Cyclone IV		Stratix IV	

* NA means this value is not listed in the reference paper.

We are using the KITTI object detection suite [23] as our database, which contains 7481 training images and 7518 test images, comprising a total of 39,595 labeled objects (including cars, pedestrians, and cyclists). As we focused on detecting cars, we picked out 6996 samples from all the 33,259 labeled cars as our positive training samples and randomly extracted 40,000 carless samples from the large picture as our negative training samples. Classification results are simulated by the software implementation and the results are illustrated in Figure 7. To quantify the performance of our implementation, we plot the precision-recall curve (Figure 7) with the measurement criteria discussed in [23], which is the formal evaluation mechanism of the KITTI dataset and the average precision of our result is 52.07%, where the error rate indicates that the detected bounding boxes are correct only if they overlap by at least 50% with a ground truth bounding box. This average precision ranks at 316th in the car precision list of KITTI except for the speed performance due to the hardware implementation.

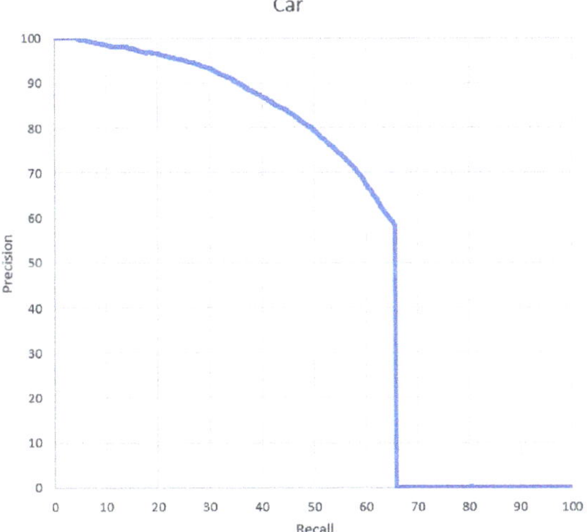

Figure 7. The precision-recall curve of our vehicle detection classifier. For a required recall, if our result can reach the recall, the corresponding precision can be calculated according to the measurement criteria in [23].

This result implies that it is difficult to detect vehicles in a relatively complex environment because of the inherent property of the HOG feature, which is sensitive to the complex edge characteristic information. As shown in Figure 8, the classifier not only detects the real vehicles but also treats some buildings and backgrounds as the vehicle. The complex environment means that the scenario with edge information of buildings and background appear as vehicle-like outlines. Further work needs to be conducted to detect vehicles in a complex environment with complex feature combinations.

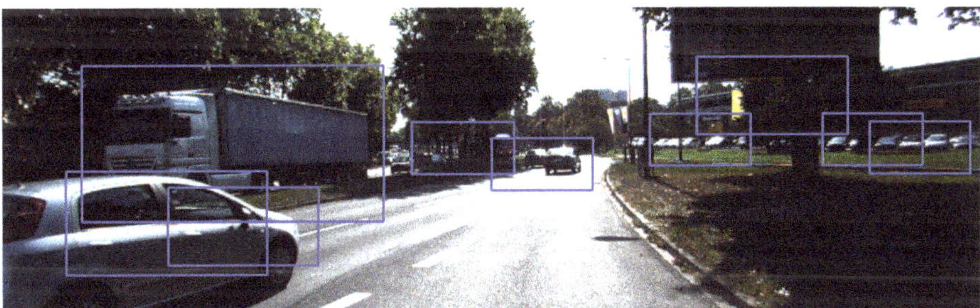

Figure 8. Detection example in the scenario with complex edge information of buildings and background-appearing vehicle-like outlines.

Since the size of the DW is scalable, in this work, we compare the classification result of both the 64×128 (1:2) and the 128×64 (2:1) detection windows. Normally, the SVM classifier is trained by a 1:2 scaled window for pedestrian detection, but in our design, we found that the shape of the detection window can significantly influence the result of classification since the orientation of cars is different compared with pedestrians. Finally, we chose to use a 2:1 scaled window with different sizes to detect car objects.

4. Conclusions

This paper presents a hardware-friendly architecture on FPGA that implements a cell-based Histogram of Oriented Gradient (HOG) feature extraction circuitry, a block-level normalization unit, and a partial Support Vector Machine (SVM) classifier module within a scalable-block-size-based sliding-detection-window (SBSSDW) mechanism. Within the SBSSDW method, the feature vector of each detection window (DW) can be constructed according to the location of the block in a window without a large buffer. Additionally, the proposed architecture is capable of the scalable size of the DW from 32 × 32 up to 2048 × 2048 pixels. The experimental results show that the coprocessor attains an average precision of 52.07% with the power dissipation of 80.98 mW. The accuracy performance implies that the complex edge characteristic information significantly affects the HOG feature. Thus, a feature combination is to be conducted to detect objects in a complex environment in our further work.

Author Contributions: F.A. provided the original idea for this study. X.W. and P.X. improved the original idea and provided guidance. P.X. handled the design and development of the proposed algorithm. Z.X. focused on the experimental task (i.e., setting up the system, the design and realization of the experiments). F.A. supervised the research and contributed to the article's organization. C.W. and L.C. edited the manuscript, which was revised by all authors. All authors have read and agreed to the published version of the manuscript.

Funding: This research received no external funding.

Acknowledgments: This research presented in the paper was partially supported by the China National Key Research & Development Program (2019YFB1310001), Fundamental Research Funds of the Central Universities (2019KFYXJJS049).

Conflicts of Interest: The authors declare no conflict of interest.

References

1. Redmon, J.; Farhadi, A. Yolov3: An incremental improvement. *arXiv* **2018**, arXiv:1804.02767.
2. Aziz, L.; Salam, S.B.H.; Ayub, S. Exploring deep learning-based architecture, strategies, applications and current trends in generic object detection: A comprehensive review. *IEEE Access* **2020**, *8*, 170461–170495. [CrossRef]
3. Haque, M.F.; Lim, H.; Kang, D. Object Detection Based on VGG with ResNet Network. In Proceedings of the 2019 International Conference on Electronics, Information, and Communication (ICEIC), Auckland, New Zealand, 22–25 January 2019; pp. 1–3.
4. Ren, J.; Chen, X.; Liu, J.; Sun, W.; Pang, J.; Yan, Q.; Tai, Y.; Xu, L. Accurate single stage detector using recurrent rolling convolution. In Proceedings of the IEEE Conference on Computer Vision and Pattern Recognition (CVPR), Honolulu, HI, USA, 21–26 July 2017; pp. 5420–5428.
5. Nayebi, A.; Bear, D.; Kubilius, J.; Kar, K.; Ganguli, S.; Sussillo, D.; DiCarlo, J.; Yamins, D. Task-driven convolutional recurrent models of the visual system. In Proceedings of the 32nd International Conference on Neural Information Processing Systems (NIPS'18), Montréal, QC, Canada, 3–8 December 2018; pp. 5290–5301.
6. Fukagai, T.; Maeda, K.; Tanabe, S.; Shirahata, K.; Tomita, Y.; Ike, A.; Nakagawa, A. Speed-Up of Object Detection Neural Network with GPU. In Proceedings of the 2018 25th IEEE International Conference on Image Processing (ICIP), Athens, Greece, 7–10 October 2018; pp. 301–305.
7. Nakahara, H. A Lightweight YOLOv2. In Proceedings of the 2018 ACM/SIGDA International Symposium on Field-Programmable Gate Arrays (FPGA), Monterey, CA, USA, 25–27 February 2018.
8. Arunmozhi, A.; Park, J. Comparison of HOG, LBP and Haar-Like Features for On-Road Vehicle Detection. In Proceedings of the 2018 IEEE International Conference on Electro/Information Technology (EIT), Rochester, MI, USA, 3–5 May 2018.
9. Shobha, B.S.; Deepu, R. A Review on Video Based Vehicle Detection, Recognition and Tracking. In Proceedings of the 2018 3rd International Conference on Computational Systems and Information Technology for Sustainable Solutions (CSITSS), Bengaluru, India, 20–22 December 2018; pp. 183–186.
10. Kawamoto, R.; Taichi, M.; Kabuto, M.; Watanabe, D.; Izumi, S.; Yoshimoto, M.; Kawaguchi, H. A 1.15-TOPS 6.57-TOPS/W DNN Processor for Multi-Scale Object Detection. In Proceedings of the 2020 2nd IEEE International Conference on Artificial Intelligence Circuits and Systems (AICAS), Genova, Italy, 31 August–2 September 2020; pp. 203–207.

11. Kawamoto, R.; Taichi, M.; Kabuto, M.; Watanabe, D.; Izumi, S.; Yoshimoto, M.; Kawaguchi, H.; Matsukawa, G.; Goto, T.; Kojima, M. A 1.15-TOPS 6.57-TOPS/W Neural Network Processor for Multi-Scale Object Detection With Reduced Convolutional Operations. *IEEE J. Sel. Top. Signal Process.* **2020**, *14*, 634–645.

12. Rashid, M.M.; Kamruzzaman, J.; Wasimi, S. Sliding Window-based Regularly Frequent Patterns Mining Over Sensor Data Streams. In Proceedings of the 2019 IEEE Asia-Pacific Conference on Computer Science and Data Engineering (CSDE), Melbourne, Australia, 9–11 December 2019; pp. 1–6.

13. Ghaffari, S.; Soleimani, P.; Li, K.F.; Capson, D.W. Analysis and Comparison of FPGA-Based Histogram of Oriented Gradients Implementations. *IEEE Access* **2020**, *8*, 79920–79934. [CrossRef]

14. Song, P.; Zhu, Y.; Zhang, Z.; Zhang, J. Subsampling-based HOG for Multi-scale real-time Pedestrian Detection. In Proceedings of the 2019 IEEE International Conference on Cybernetics and Intelligent Systems (CIS) and IEEE Conference on Robotics, Automation and Mechatronics (RAM), Bangkok, Thailand, 18–20 November 2019; pp. 24–29.

15. An, F.; Xu, P.; Xiao, Z.; Wang, C. FPGA-based object detection processor with HOG feature and SVM classifier. In Proceedings of the 2019 32nd IEEE International System-on-Chip Conference (SOCC), Singapore, 3–6 September 2019; pp. 187–190.

16. Ma, X.; Najjar, W.A.; Roy-Chowdhury, A.K. Evaluation and Acceleration of High-Throughput Fixed-Point Object Detection on FPGAs. *IEEE Trans. Circuits Syst. Video Technol.* **2015**, *25*, 1051–1062.

17. Wang, M.; Zhang, Z. FPGA implementation of HOG based multi-scale pedestrian detection. In Proceedings of the 2018 IEEE International Conference on Applied System Invention (ICASI), Chiba, Japan, 13–17 April 2018; pp. 1099–1102.

18. Bilal, M.; Khan, A.; Khan, M.U.K.; Kyung, C. A Low-Complexity Pedestrian Detection Framework for Smart Video Surveillance Systems, in IEEE Transactions on Circuits and Systems for Video Technology. *IEEE Trans. Circuits Syst. Video Technol.* **2017**, *27*, 2260–2273. [CrossRef]

19. An, F.; Zhang, X.; Luo, A.; Chen, L.; Mattausch, H.J. A Hardware Architecture for Cell-Based Feature-Extraction and Classification Using Dual-Feature Space. *IEEE Trans. Circuits Syst. Video Technol.* **2018**, *28*, 3086–3098. [CrossRef]

20. Wang, L.; Lin, Y.; Jiang, W. Implementation of Fast and High-precision Division Algorithm on FPGA. *Comput. Eng.* **2011**, *37*, 240–242.

21. Stratix IV GX FPGA DevelopmLent Kit. March 2014. Available online: https://www.intel.com/content/dam/www/programmable/us/en/pdfs/literature/ug/ug_sivgx_fpga_dev_kit.pdf (accessed on 21 October 2020).

22. STC-MC_MB83PCL_Spec_Manual. July 2017. Available online: https://aegis-elec.com/mwdownloads/download/link/id/2289/ (accessed on 21 October 2020).

23. Geiger, A.; Lenz, P.; Urtasun, R. Are we ready for autonomous driving? The kitti vision benchmark suite. In Proceedings of the 2012 IEEE Conference on Computer Vision and Pattern Recognition (CVPR), Providence, RI, USA, 16–21 June 2012.

Publisher's Note: MDPI stays neutral with regard to jurisdictional claims in published maps and institutional affiliations.

Letter

Leveraging Uncertainties in Softmax Decision-Making Models for Low-Power IoT Devices

Chiwoo Cho [1], Wooyeol Choi [2] and Taewoon Kim [3],*

[1]   Hallym Institute for Data Science and Artificial Intelligence, Hallym University, Chuncheon 24252, Korea; cwcho@hallym.ac.kr
[2]   Department of Computer Engineering, Chosun University, Gwangju 61452, Korea; wyc@chosun.ac.kr
[3]   School of Software, Hallym University, Chuncheon 24252, Korea
*   Correspondence: taewoon@hallym.ac.kr; Tel.: +82-33-248-2335

Received: 21 July 2020; Accepted: 14 August 2020; Published: 16 August 2020

Abstract: Internet of Things (IoT) devices bring us rich sensor data, such as images capturing the environment. One prominent approach to understanding and utilizing such data is image classification which can be effectively solved by deep learning (DL). Combined with cross-entropy loss, softmax has been widely used for classification problems, despite its limitations. Many efforts have been made to enhance the performance of softmax decision-making models. However, they require complex computations and/or re-training the model, which is computationally prohibited on low-power IoT devices. In this paper, we propose a light-weight framework to enhance the performance of softmax decision-making models for DL. The proposed framework operates with a pre-trained DL model using softmax, without requiring any modification to the model. First, it computes the level of uncertainty as to the model's prediction, with which misclassified samples are detected. Then, it makes a probabilistic control decision to enhance the decision performance of the given model. We validated the proposed framework by conducting an experiment for IoT car control. The proposed model successfully reduced the control decision errors by up to 96.77% compared to the given DL model, and that suggests the feasibility of building DL-based IoT applications with high accuracy and low complexity.

Keywords: deep learning; softmax; decision-making; classification; sensor data; Internet of Things

---

1. Introduction

The unprecedented success of the Internet of Things (IoT) paradigm has changed and reshaped our daily lives. We start a day by asking smart speakers, such as Amazon Echo or Google Home, about the weather, remotely control the home appliances and door lock, and switch off the light from mobile devices before going to bed. As was inevitable, IoT is now playing an important role in many other fields, such as industry [1,2], healthcare [3,4], energy [5,6], transportation [7,8] and environment monitoring [9], to name a few. The increasing number of pervasive and widespread, Internet-connected IoT devices capture the environment and generate an enormous amount of data, which is becoming one of the major sources of information nowadays. To understand such massive sensor data, and thus, to draw meaningful information out of it in an autonomous manner, various approaches have been applied, including deep learning.

Deep learning is a type of or a class of techniques in machine learning [10,11] that surpasses the capacity of machine learning in many applications such as computer vision and pattern recognition. Deep learning is a representation learning technique, and is capable of learning a proper representation for the given task, such as classification or detection, from the sensor data. Deep learning is now one of the most actively studied areas, and it is expected to contribute much to the success of many IoT applications. As a result, applying different deep learning techniques to IoT applications [12,13] is also gaining much attention nowadays.

Among those areas deep learning excels in, we focus on a classification task in this paper. Classification is a popular supervised learning task in deep learning [10], wherein a trained model is asked to predict which of the classes or categories the unseen input belongs to. Deep learning is carried out by using a neural network model that learns features from the input data. For clarification, a deep neural network with a multi-layer, fully connected perceptron is referred to as artificial neural network (ANN), whereas a network with convolution and pooling layers is a convolutional neural network (CNN). For the output classifier of ANN, CNN and other deep neural network models, the softmax function is the most widely used [10] combined with cross-entropy loss, thereby called, softmax loss.

Although the primitive or basic deep neural network models have shown surprisingly excellent classification performances in many applications, many research communities have been striving for better performance. Such efforts can be roughly divided into two research directions. One is to design sophisticated network models [14–16], and the other is to replace the softmax loss with an advanced equivalent [17–19]. Among these, we briefly review the latter which are closer to what we focus on in this paper.

In general, to achieve a better image recognition performance for face identification and verification, for example, many researchers have focused on how to effectively classify a given face, i.e., mapping a face to a known identity, and determining if the two given faces are identical, respectively [20,21]. For such tasks, the chosen features from the decision boundary can immensely affect the classification or recognition performance. Therefore, extracting ideal features having small intra-class and large inter-class distances under a given metric space is the most important, yet challenging, task. Among the previous work tackling the issue with softmax loss, angular softmax loss for CNN [17] is proposed to adjust the angular margin when the model determines decision boundaries for the feature distribution, and it helps CNN learn angularly discriminative features. In [19], the authors introduced another softmax loss that contributes to incorporating the margin in more instinctively and interpretable manners, while minimizing the intra-class variance, which was considered to be the main drawback of the widely-used softmax loss. Liu et al. [18] proposed a large-margin softmax loss function, called L-softmax, which effectively enhances both intra-class compactness and inter-class separability between the already-learned features. Furthermore, L-softmax can adjust the desired margin, while preventing the network model from being overfitted. With respect to the hardware-centric research, Wang et al. [22] proposed optimizing the softmax function by mitigating its complexity by means of an advanced hardware implementation. The authors showed that by reducing the total number of operations in the constant multiplication, the adjusted softmax function architecture enables multiple algorithm reduction, fast addition and less memory consumption at the expense of a negligible accuracy loss. These approaches successfully enhanced the classification performance in the domains of their concerns. However, many of such advanced techniques may not be viable or practical when dealing with low-power IoT devices.

In this paper, we assume a situation wherein a trained model from a simple neural network architecture is given to low-power IoT devices. In general, IoT devices are limited in computing power, energy and memory capacity. Thus, running sophisticated and complex neural networks or advanced loss models are challenging [13], and even impractical for real-time IoT applications, such as self-driving cars. Additionally, changing the existing network model to a different one for better performance may take a significant amount of time, and it will delay the deployment stage. In general, shifting to a different network model requires several iterations of training (from a vast amount of data), testing and hyper-parameter tuning tasks. Thus, it is necessary to devise a low-complexity method to enhance the performance of deep learning models for IoT.

On the other hand, in the cases without IoT devices, there exist some sophisticated deep learning models that achieve a close-to-perfect accuracy. Nevertheless, it is impossible for a model not to make any mistakes, and such subtle errors may result in severe damage in mission-critical systems, such as battlefields and hospitals [23]. One may propose an application-specific way to further reduce errors, but it cannot be applied to general applications. Thus, it is necessary to devise a one-size-fits-all approach to assisting a deep learning model to achieve a better accuracy that can be used with general deep learning models.

To enhance the performance of a given deep learning model without incurring any additional significant or time-consuming computation, we propose a low-complexity novel framework which operates as an add-on to the general deep learning models without requiring any modification on the model's side. The basic idea of the proposed framework is straightforward. For the given input $\mathbf{x}$, the softmax output $\mathbf{y}$ is a vector of y_i's, where $y_i = P(i|\mathbf{x})$ is the posterior probability of $\mathbf{x}$ belonging to class $i \in \{1, 2, \cdots, N_c\}$ and N_c is the number of classes/categories. Although the following arg max operation takes the most likely class, it does not care about how close the corresponding probability is to 1. Furthermore, if the largest probability $y_j = \max\{y_i | y_i \in \mathbf{y}\}$ was not so much different from the second-largest $y_k = \max\{y_i | y_i \in \mathbf{y} \backslash \{y_j\}\}$, the model might be, what we call, *uncertain* about its prediction or decision. We propose to measure such *decision uncertainty* in a single quantity by using the well-known Jain's fairness index [24], which has been widely used in the computer network domain [25]. In this paper, the computed fairness score of the softmax output is referred to as the *uncertainty score*, and is used to measure the level of uncertainty as to the model's prediction.

In this paper, we propose a light-weight, uncertainty-score-based framework that effectively identifies incorrect decisions made by softmax decision-making models. We also propose a novel way to make mixed control decisions to enhance the target performance when the given deep learning model makes an incorrect decision. Additionally, the proposed framework does not make any change to the given trained model, but it simply puts an additional low-complexity function on top of the softmax classifier. The specific contributions we make in this work are summarized as follows:

- We propose a novel framework for the widely-used softmax decision-making models to enhance the performance of the given deep learning task without making any modification to the given trained model. Therefore, the proposed framework can be used with any neural network models using softmax loss.
- We propose to use an uncertainty score to gauge the level of uncertainty as to the model's prediction. In a nutshell, the similarity among the softmax output is interpreted as how sure the model is about the current decision. To this end, we developed a practical method to effectively detect incorrect decisions to be made by the given deep learning model.
- We propose an effective way to enhance the performance of a deep leaning control system by making a mixed control decision. When the given model is believed to be yielding an incorrect decision/prediction, the proposed model replaces the model's output with the probabilistic mixture of the available actions in order not to deviate much from the correct decision.
- We propose a low-complexity yet effective method to enhance the performance of the softmax decision-making models for low-power IoT devices. By using the time complexity terms, we show that the proposed framework does not incur any significant load from the given decision-making model, and thus, it can be used for online tasks.
- We show by an empirical study how the proposed framework effectively enhances the performance of the softmax decision-making tasks. To be specific, we carried out an experiment for IoT car control; we designed a control decision system that utilizes the softmax output to make a mixed, probabilistic car control decision when the model prediction is of low certainty.

Our work presented in this paper is innovative in that it suggests a new and systematic way of enhancing the performances of deep learning models. The proposed method treats the trained model as a *black box*, and thus, it can be applied to general deep learning models with little overhead. Additionally, it takes advantage of the *entire* softmax output to generate a decision when the model fails. The proposed approach is different from the previous studies focusing on either revising the deep neural networks or loss models. Additionally, by statistical and evaluation studies we show that not only the largest softmax output to be taken by the arg max operator, but also the actual values in the entire softmax output can be utilized to enhance the performances of deep learning models in the low-power IoT device control domain.

The rest of this paper is organized as follows. Section 2 introduces a brief overview on image classification and softmax loss. In the following Section 3, we describe the proposed framework to enhance the performances of deep neural networks with softmax loss. Section 4 presents experiment results, and the following Section 5 includes some discussion along with some notes as to the proposed framework. Finally, Section 6 concludes the paper.

2. Background and Motivation

In this section, we briefly review the classification problems on deep neural networks with softmax loss, and Jain's fairness index, which plays a key role in the proposed framework as an uncertainty score. Then, we introduce our findings and understanding that motivated this work.

2.1. Background: Softmax and Uncertainty Score

As aforementioned, we consider ANN and CNN models with softmax loss for classification tasks. The lower layers in such networks can be seen as a feature extractor, and the last fully connected one as a classifier. As suggested in [18], we denote the combination of the cross-entropy loss and a softmax function of the last fully connected layer in a neural network by softmax loss. The given neural network is fed with the input data and trained using back-propagation by minimizing the loss which acts as an error signal. Therefore, which loss model to use is important for neural networks to effectively and efficiently train the network.

The classification problem, which frequently arises in deep learning tasks, is defined as follows: train the neural network model, $f : \mathbb{R}^d \to \{1, \cdots, N_c\}$ so that for the given input $\mathbf{x} \in \mathbb{R}^d$; let $i = f(\mathbf{x})$ be the correct prediction on to which class does the input belong to. The last network layer outputs a vector $\mathbf{y}$ which is the set of probabilities that the given input belongs to each class. Taking the arg max operator on $\mathbf{y}$ yields the class indicator $i \in \mathbb{Z}_{++}$, where the arg max operator finds the argument that yields the maximum value from a given function. Please note that depending on the beginning index of classes, we may have $i \in \mathbb{Z}_+$ be the case. The softmax loss is defined as below, following similar notation to that used in [26]:

$$L = -\frac{1}{M} \sum_{i=1}^{M} \log \frac{e^{\mathbf{W}_{y_i}^T \mathbf{x}_i + b_{y_i}}}{\sum_{j=1}^{N_c} e^{\mathbf{W}_j^T \mathbf{x}_i + b_j}} \tag{1}$$

where M is the size of the mini-batch, $\mathbf{x}_i \in \mathbb{R}^l$ is the learned feature that belongs to the y_i class, l is the feature dimension, $\mathbf{W}_j$ is the j-th column of the weight matrix $\mathbf{W}$ in the last fully connected layer and $\mathbf{b}$ is the bias.

One key component in the proposed framework is to use a fairness index or score to measure the level of uncertainty as to the model's prediction, called the uncertainty score. We use the well-known Jain's fairness index to compute the uncertainty score, which is defined as:

$$\mathcal{J}(y_1, y_2, \cdots, y_n) = \frac{\left(\sum_{i=1}^{n} y_i\right)^2}{n \cdot \sum_{i=1}^{n} y_i^2} \tag{2}$$

where $y_i \in \mathbf{y}$ and $\mathbf{y} \in \mathbb{R}^n$. The fairness index ranges from $1/n$, representing the most unfair values among y_i's, to 1 for being perfectly fair. Suppose a binary classification task, mapping each input to either of two classes. The softmax output vector $\mathbf{y}$ has two real values $y_i \in [0, 1], \forall i = \{1, 2\}$, and $\sum_i y_i = 1$. For a given input $\mathbf{x}_1$, if the softmax output is [0.0, 1.0], the model predicts the given input to belong to the second class. The corresponding fairness score is 0.5, having the worst fairness. Having a low fairness score implies that the model was certain about its decision, and thus, a high probability was given to the most-likely class. This is when there is little uncertainty in the model's decision. For another input x_2, if the softmax output is [0.49, 0.51], the model predicts the given input to belong to the second class. The corresponding fairness score is 0.9996, achieving almost the perfect fairness. Having a high fairness score indicates that the model was uncertain about its decision, and thus, high probabilities were given to both classes. This is when there is a high uncertainty in the model's decision.

For this reason, we interpret the fairness score among the softmax output as the uncertainty score which measures how uncertain the model is about its decision/prediction. The proposed framework relies much on the uncertainty score, and thus, the proposed framework is called UFrame hereafter.

2.2. Motivation: Uncertain Model Prediction

To address what motivated our study, we first take the classification task for the MNIST handwritten digit database [27] as an example. We modeled a simple ANN, as shown in Figure 1, which yielded about 95% accuracy on the test data. Please note that we consider running the trained model on a low-power IoT device, and thus, we chose an extremely simple ANN model. We also obtained the 2D embedding result of the learned features, as shown in Figure 2, in which the misclassified digits are colored in red.

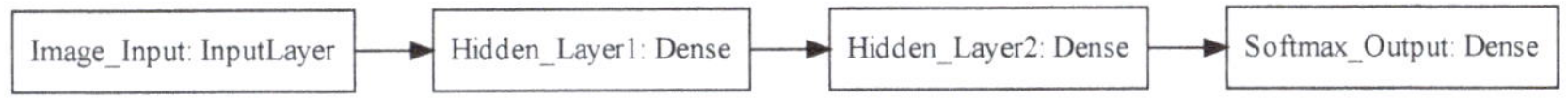

Figure 1. The simple ANN structure we used for the MNIST handwritten digit classification task. The second hidden layer is added to embed the learned features into 2D.

As can be seen from Figure 2, samples of the same labels form clusters, implying that different classes have different statistical characteristics. The misclassified samples tend to be apart from the corresponding cluster center. Said finding inspired many studies aiming at enlarging the inter-class gaps by using sophisticated loss models [17–19]. Since we assume a setting where re-training or adding additional complexity to the neural network model is challenging for the limited resources of IoT devices, we rather focus on discovering a *signal* of misclassification by means of the uncertainty score.

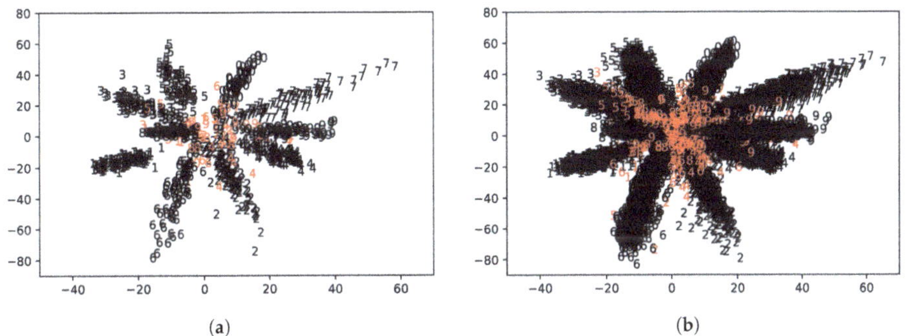

(a) (b)

Figure 2. 2D embedding of the learned features for the ANN is displayed. For better visibility, the left figure shows only the first 1000 samples in the test dataset. Correctly classified digits are colored in black, whereas misclassified ones are in red. (**a**) First 1000 samples in the test dataset; (**b**) entire sample of the test dataset.

Figure 3 shows the uncertainty scores computed from the test dataset. The two upper images, Figure 3a,b, show the uncertainty score histogram of the softmax output of each test sample. On the other hand, the lower two images, Figure 3c,d, show the uncertainty score histogram of the two largest softmax outputs for each test sample. It is clear from both upper and lower figures that correctly classified samples have concentrated uncertainty distributions centering to the left, whereas misclassified samples have widespread distributions comparatively centering to the right.

That is, most of the correct classifications are made with almost certainty; in other words, the softmax output of the likely digit is almost 1, and the rest are almost 0, resulting in a low uncertainty score. On the contrary, for the misclassified samples, the model prediction was uncertain, yielding high probabilities for many classes. In addition, by comparing Figure 3b–d, we can see that taking the two maximum softmax outputs to compute the uncertainty score, i.e., Figure 3d, results in a more distinct distribution pattern compared to its counterpart, Figure 3c.

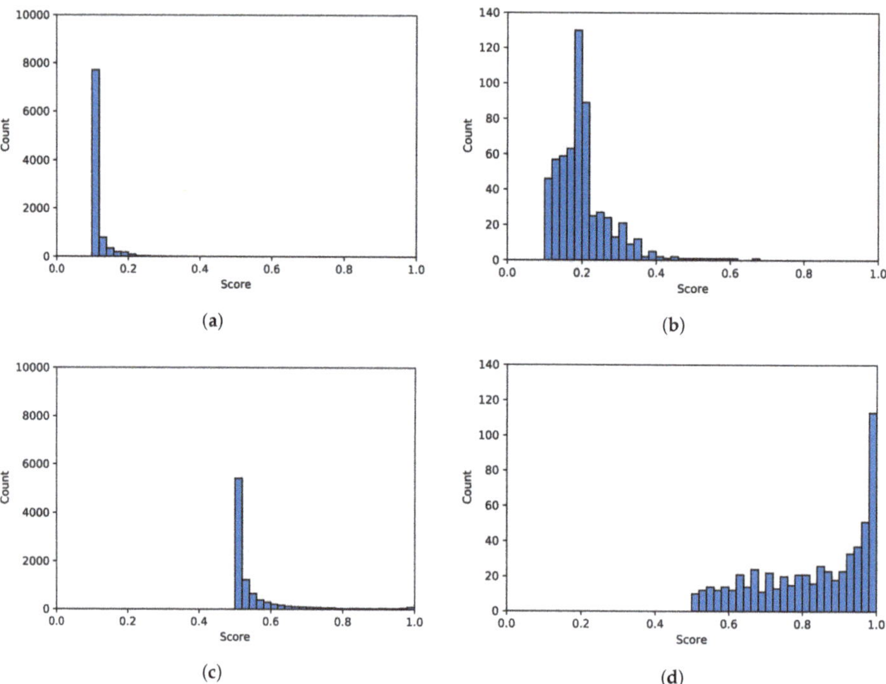

Figure 3. Histograms of the uncertainty scores with the interval of 0.02. Figures in the left column show the uncertainty score histograms of the correctly classified test samples, and the right column is for the misclassified test samples. (**a**) Uncertainty score histogram of the entire softmax output of the correctly classified samples. (**b**) Uncertainty score histogram of the entire softmax output of the misclassified samples. (**c**) Uncertainty score histogram of the two largest softmax outputs of the correctly classified samples. (**d**) Uncertainty score histogram of the two largest softmax outputs of the misclassified samples.

The proposed UFrame is based off of the aforementioned findings that: (1) the uncertainty scores of the correctly classified samples have a distinct pattern compared to that of the misclassified ones; (2) the uncertainty scores of the two largest softmax outputs are more informative because the distribution of the correctly classified digits is concentrated to the left and that of the misclassified one is to the right, and the difference between those two is noticeable; and (3) different classes have different statistical characteristics. Such findings led to the proposed UFrame, which is explained in detail in the following section.

3. Proposed Method

Based on the findings in Section 2, we propose a novel framework, called UFrame, that effectively identifies misclassification and makes a mixed decision to enhance the performance of the given control task. By using a widely-used MNIST dataset, we validated `UFrame` as to whether it can effectively identify misclassified samples.

3.1. Algorithm Description

Figure 4 illustrates the proposed `UFrame`. Those boxes with black solid lines are the regular (deep) neural network workflow, and the ones with red dash lines belong to `UFrame`. The proposed `UFrame` runs as follows. First, by using the model output from the validation dataset, assuming it is available, it learns the error detection threshold which is to be explained in detail later. This step is carried out offline without disturbing the operation of the regular deep learning workflow. The computed threshold values get stored in a table for constant-time access. Then, for each new datum given to the model, `UFrame` takes the two largest values in the softmax output, and then computes the uncertainty score. Said score is referred to as the *max2 uncertainty* score hereafter. The error detector module compares the max2 uncertainty score to the threshold to determine whether or not the upcoming decision by the $\arg\max$ function is likely to be incorrect. If the max2 uncertainty score exceeds the threshold, the error flag is set to true, 1 or on; otherwise, it sets the flag to false, 0 or off. If the error flag is false, the control decision module uses the model's prediction as it is. If the error flag is true, on the other hand, the control decision module makes a mixed control decision as follows.

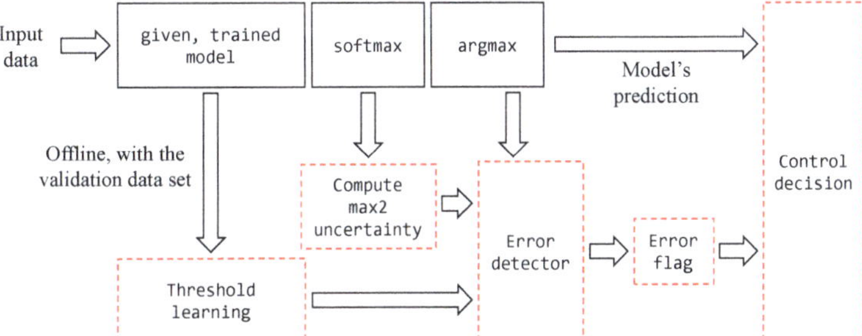

Figure 4. The proposed `UFrame` operates as a wrapper module for the given trained neural network model. `UFrame` includes the components inside the red dotted box, whereas the rest of the boxes in black solid lines belong the regular workflow of a general neural network model using softmax loss.

Let i be a class index, indicating one of the possible actions the IoT device can make. It can be one of the directions an IoT car can move towards, or a robot arm movement. In the regular deep learning workflow, for an input datum, the deep learning model predicts which action to take, and passes the decision to the IoT device as a control signal. Having N_c number of classes implies there are the same number of actions available. Let each viable action be u_i with i being the class index. The deep learning model predicts to which class i the input belongs to, and the corresponding control decision becomes u_i, if the error flag is off. That is, the model's prediction is passed to the IoT device as it is when the error flag is off. However, if the error flag is on, the model's prediction is uncertain, and thus, the control decision module makes a mixed control decision in the following manner. It first transforms all possible actions $u_i, \forall i$ to N_c number of different unit vectors in an $\mathbb{R}^q$ space, where q is task-specific. Given the softmax output $\mathbf{y}$, the mixed control decision becomes $\sum_{i=1}^{N_c} y_i \cdot u_i$, where $y_i \in \mathbf{y}$.

The error detection threshold plays a key role in `UFrame`; it is used to determine whether or not the upcoming decision for the input data is likely to be incorrect. The threshold learning algorithm (TLA) is carried out offline as follows. Given the validation dataset, TLA feeds the validation dataset to the trained model to retrieve the softmax output of each sample therein. Here, it does not matter whether or not the validation dataset is the same dataset that was used when training the model. At the same time, for each sample TLA sets a binary flag indicating whether the corresponding sample has been correctly identified, and stores the true label indicating the class to which each sample belongs. With the acquired records of (softmax output vector, binary flag, true label) for each sample,

TLA computes and collects the max2 uncertainty scores among the correctly classified samples for each class. Please note that having a different threshold for each class has shown better performance from our empirical study (see Section 3.2). Assuming the distribution of max2 uncertainty score follows a (one-sided) normal distribution within each class (see Figure 3), UFrame computes the mean and standard deviation of the max2 uncertainty scores among the correctly classified samples.

As for the last step, TLA takes in a design parameter $\alpha \in \mathbb{Z}_{++}$ so that the detection threshold for class i is $th_i = m_i + \alpha \cdot stdev_i$, where m_i is the mean of the max2 uncertainty scores of class i and $stdev_i$ is the standard deviation. The detection threshold for class i is referred to as th_i hereafter. The parameter α controls how *rigid* the error detector is to be, which will be explained in Section 3.2.

3.2. Validation: Identifying Misclassified Samples

To validate the performance of UFrame as to whether it can correctly identify the misclassified samples, we have applied UFrame to the MNIST digit recognition task. The MNIST database includes 60,000 samples for training, and 10,000 samples for testing. We further divided the training dataset into 50,000 for training and 10,000 for validation. We have used a similar ANN as in Figure 1, except the 2D embedding layer, which was omitted here. After tuning hyper-parameters, the neural network achieved 97% accuracy on the test dataset.

The per-class mean and standard deviation of max2 uncertainty score are shown in Table 1, and they indeed are different from each other. As aforementioned, difference classes have different characteristics, and having a different threshold for each class has shown better detection performance. Given the per-class mean and standard deviation values, we can compute the error detection threshold for class i as $th_i = m_i + \alpha \cdot stdev_i$, where $i \in \{0, 1, \cdots, 9\}$ for the given 10-digit recognition task. Please note that the threshold is computed from the correctly identified samples in the validation dataset, while the performance of UFrame is measured on the test dataset. When the neural network model processes an input datum x, UFrame intercepts the softmax output and computes the max2 uncertainty score. The error detector then compares the max2 uncertainty score to th_i, where i is the model's prediction. If the max2 uncertainty score exceeds th_i, the error flag is set to true, and false if not.

Table 1. Per-class mean and standard deviation of the max2 uncertainty scores among the correctly classified samples, acquired from feeding the validation dataset.

Digit (i.e., Class)	0	1	2	3	4	5	6	7	8	9
mean	0.506	0.521	0.519	0.527	0.520	0.523	0.514	0.516	0.515	0.519
standard deviation	0.039	0.064	0.068	0.080	0.068	0.073	0.055	0.057	0.060	0.066

Figure 5 shows the performance of UFrame in terms of error detection, and the exact values are reported in Table 2. In the figure, the x-axis is the value of α for the threshold, and the y-axis is the proportion of the corresponding samples. Regardless of α, the number of misclassified test samples from the ANN model remains the same. Among those misclassified samples by the ANN model, the proportion of the ones that are identified and reported by UFrame is shown in blue bars in Figure 5a. A smaller value of α makes the error detector more rigid or strict, and thus, UFrame frequently encounters samples violating the threshold. As a result, with $\alpha = 1$, UFrame successfully identified 218 cases of misclassification out of 271 in the test dataset, which is about 80% of the misclassified samples. On the other hand, a larger α increases the threshold for each class. Thus, UFrame becomes more tolerant to the samples with high max2 uncertainty score, resulting in a smaller number of cases with the error flag being true. As a reminder, high max2 uncertainty score means the first and second larges values in the softmax output are similar to each other. That is, the model is uncertain about its prediction, and thus, it is likely to be incorrect.

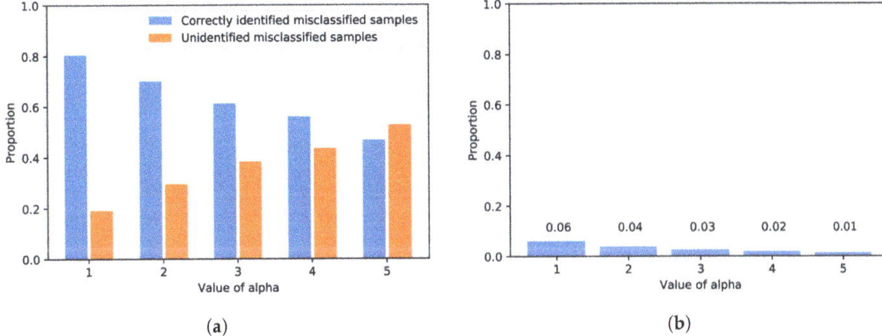

Figure 5. The detection performance of `UFrame` as to the MNIST test dataset is shown. (**a**) Among the samples misclassified by the ANN model, the proportions of the correctly identified samples (i.e., error flag was true) and those that were not (i.e., error flag was false) are shown in blue and orange, respectively. (**b**) Among the correctly identified samples by the ANN model, the proportion of the incorrectly identified samples (i.e., error flag was true) is shown with the corresponding value.

Table 2. The exact values reported in Figure 5 when there are 10,000 test samples.

Value of α	1	2	3	4	5
Misclassification by ANN	271	271	271	271	271
Misclassification with error flag on	218	190	166	152	127
Misclassification with error flag off	52	80	104	118	143
Correct classification with error flag on	589	376	265	196	138

However, the value of α has a negative effect at the same time. Figure 5b shows the proportion of the samples detected by the error detector among the correctly identified samples. As aforementioned, since α determines how strict or tolerant the error detector will become, a smaller α results in more *false positives*, i.e., an error flag is on even for the correctly classified samples. However, such cases amount to only a little portion—at most 6%. Additionally, such false positives indeed have a negligible effect on the control decision in terms of both time complexity and decision accuracy. Although `UFrame` will decide to make a mixed control decision instead of using the model's prediction, since the correct class was given the largest softmax output, the mixed decision will be biased much to the correct decision.

4. Experiments

In order to validate the performance of `UFrame` with a real-world IoT application, we have carried out an experiment. Please note that the use of the proposed framework is not limited to IoT devices. It can also be used for general-purpose and programmable low-power sensor devices. The considered use case here is making control decisions for indoor self-driving toy cars. For the low computing power of single board computers (SBC) such as Raspberry Pi (RPi), we simplified the self-driving task to image classification. At the beginning, a series of manual driving tasks were carried out by a human, during which images through the USB camera and the human controller's input key strokes, i.e., left, forward and right, are collected. The acquired image dataset was then increased by flipping horizontally and shifting by a small amount of pixels. The entire dataset was divided into three, i.e., training (42,446 samples), validation (5305 samples) and testing (5306 samples), before training the CNN model (see Figure 6). To speed up the real-time control decision-making process, the trained model to run on RPi was converted to a TensorFlow Lite equivalent. The trained model was an image classifier that mapped the incoming camera image into one of the three different classes indicating steering wheel directions, i.e., left, forward or right.

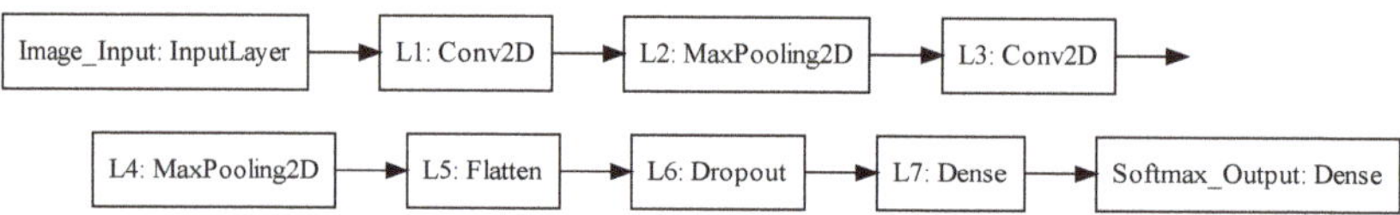

Figure 6. The displayed CNN model was used for the indoor self-driving task.

In fact, the self-driving task can be implemented without deep learning in many cases. For example, an IoT car can detect the lanes on both sides with a feature extraction technique, e.g., Hough transform [28]. Then, by comparing the centers of the lanes on both sides and the center of the car, an IoT car can drive autonomously. However, in this experiment, we considered a realistic situation where the drive or journey could be interrupted by other moving objects, such as humanoid robots, as in Figure 7c. Additionally, there are several intersections, and an IoT car can decide on which direction to go by the directional signs (see Figure 7d). For a self-driving car to successfully drive while complying with the simple rules of the road, i.e., following the directional signs and stopping when blocked by other objects, we chose to solve the self-driving task by CNN-based deep learning approach.

Each toy car shown in Figure 7 carried an RPi v4 as a controller and an L298N motor drive shield on its back. The RPi was powered by a battery, which is invisible in the figure, with 5.0 V and 2.0 A output. The IoT toy cars were connected via a built-in WiFi interface so that they could communicate with each other and with the road side unit (RSU). The RSU broadcast heavy-traffic and accident information, and a toy car receiving such information was to slow down. Toy cars could return to the normal speed only when another message indicating the clearance of the situation was received from RSU. If a car failed to receive any information from the RSU, the car which successfully received the information could forward it to other cars nearby. Please note that such reception failure can happen for many reasons, such as out of the transmission range of RSU, packet collision and packet drop, to name a few.

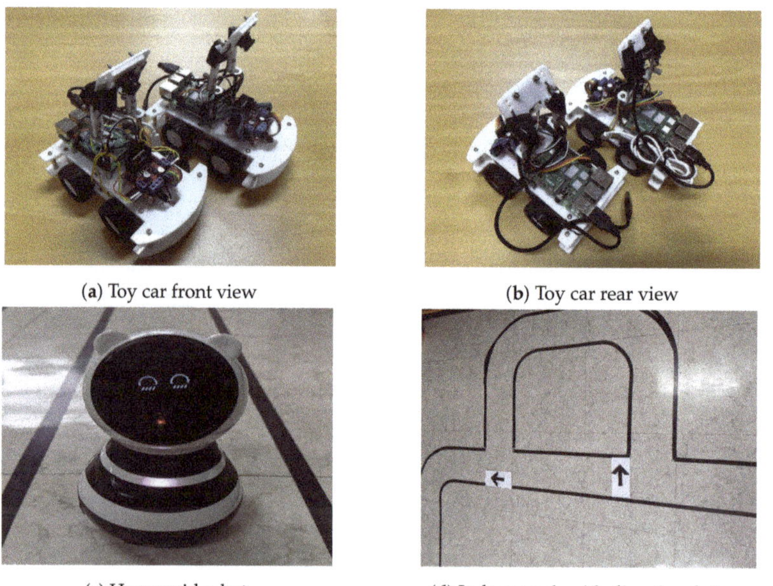

(**a**) Toy car front view

(**b**) Toy car rear view

(**c**) Humanoid robot

(**d**) Indoor track with directional signs

Figure 7. Toy cars with Raspberry Pi v4 and an L298N motor drive shield, a humanoid robot, were used for the experiments on an indoor track with directional signs.

Through a series of trainings with different configurations, five epochs with the mini-batch size of 128 were chosen to avoid over-fitting. The resulting model yielded an accuracy of about 95% on the test dataset. Again, we measured the max2 uncertainty score of the softmax output from the correctly classified samples in the validation dataset. The per-class means and standard deviations of the max2 uncertainty scores are reported in Table 3. We also evaluated the detection performance on the IoT car image dataset, and the results are shown in Figure 8. Please note that the same evaluation was carried out for the MNIST dataset as well (see Figure 5). It is clear from both figures that, although one dataset along with its application is very much different from the other in terms of the underlying neural network architecture, the number of classes and the contents in the images, the proposed framework can effectively identify misclassified samples in both applications (see Figures 5a and 8a), and the proportion of the incorrectly identified samples is insignificant overall (see Figures 5b and 8b). Although there is a difference in the patterns between Figures 5a and 8a—the orange bar exceeds the blue bar with a lesser value of α in the IoT car dataset than in the MNIST dataset—that was only because of the different number of classes and the samples in each neural network and dataset, respectively.

Table 3. Per-class means and standard deviations of the max2 uncertainty score acquired from feeding the validation dataset to the trained model.

Key Stroke or Direction (i.e., Class)	Left	Forward	Right
mean	0.530	0.547	0.520
standard deviation	0.085	0.097	0.066

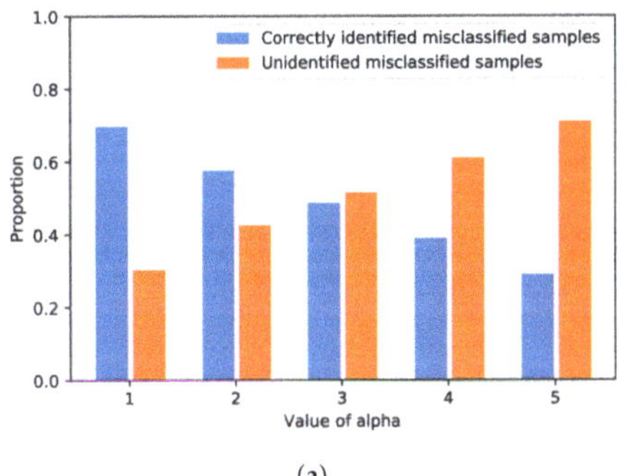
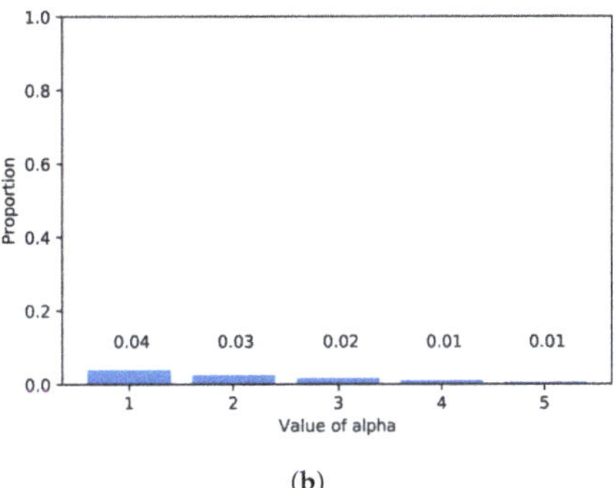

(a)   (b)

Figure 8. The detection performance of UFrame as to the IoT car image dataset is shown. (**a**) Among the misclassified samples by the CNN model, the proportions of the correctly identified samples (i.e., error flag was true) and those were are not (i.e., error flag was false) are shown in blue and orange, respectively. (**b**) Among the samples correctly identified by the CNN model, the proportion of the incorrectly identified samples (i.e., error flag was true) is shown with the corresponding value.

The misprediction of the model can be regarded as a steering wheel direction error by greater than or equal to 90 degrees. If the model misses a forward direction, for example, the wheel direction error is exactly 90 degrees no matter which direction the model mistakenly chooses. If the model misses a left direction, for example, the wheel direction error is either 90 or 180 degrees for mistakenly choosing forward or right direction, respectively.

For this steering wheel control task on an IoT car, when the model's prediction is uncertain, the proposed UFrame can make a mixed control decision in the following manner by levering the decision uncertainties, i.e., the softmax output. If the max2 uncertainty score of the softmax output is below the threshold of the class to which the model classified the current input, the model prediction is passed to the toy car an the control input as it is. On the other hand, if the uncertainty score exceeds the threshold, the control output of the car, i.e., the steering direction, becomes a probabilistic combination of the three directions as shown in Figure 9. Suppose the case described in the figure: for the given input image, the softmax output is [0.3, 0.6, 0.1] for left, forward and right directions. Additionally, suppose the max2 uncertainty score has exceeded the threshold. Then, instead of using the model prediction (i.e., moving straight since 0.6 is the largest among the softmax output) to steer the toy car, UFrame makes a mixed control decision as follows. Let [−1,0], [0,1] and [1,0] be the unit vectors representing left, forward and right directions, respectively. Additionally, let each softmax output be the probability of the given input image belonging to the corresponding direction. UFrame multiplies a unit vector and the corresponding probability for each direction, and then adds them together to produce a vector $[v_x, v_y]$. The v_x indicates the normalized velocity towards left or right depending on the sign (+ or −). Likewise, v_y indicates the normalized velocity towards the forward direction. The next step is to convert $[v_x, v_y]$ to the motor speed for the four wheels which will be passed to the corresponding motors via the motor driver shield.

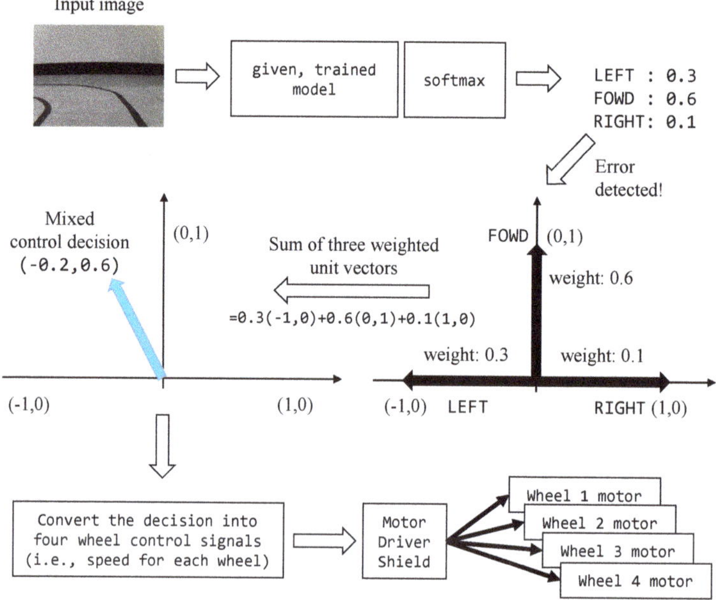

Figure 9. The way that UFrame makes a mixed control decision when the max2 uncertainty score exceeds the threshold.

To evaluate and compare the performance of UFrame, we have measured the errors in angle, i.e., the angle difference between the correct direction and the direction chosen by either the CNN model or UFrame (i.e., mixed control decision) for each sample in the test dataset. Please note that when the CNN model makes an incorrect decision on direction, the error in angle amounts to at least 90 degrees. The performance evaluation result with respect to the different values of α is shown in Table 4.

The CNN model which has nothing to do with α made correct decisions on 5058 test samples, and missed 248, resulting in about 95% accuracy on the test dataset. Those misses deviate from the correct steering angle by at least 90 degrees. On the other hand, for any values of α, UFrame made only eight misses of such large-degree mistakes. As aforementioned, smaller values of α make the error detector more strict. In the case of $\alpha = 1$, for example, UFrame did not set the error flag only for 4636 samples, which is the smallest. On the other hand, in the case of $\alpha = 5$, 5014 samples resulted in taking the model's decision as it was without making a mixed control decision due to the large threshold. As α decreases, UFrame makes more mixed control decisions, and results in the smallest the number of control mistakes with 50+ degrees of angle differences to the correct angle. That proves that making a mixed control even in the case wherein the model makes a correct prediction does not degrade the quality of the decision, since in such cases the softmax output is biased to the correct decision and so does the mixed decision. On the contrary, as α increases, UFrame makes less mixed control decisions, but suffers from having many control mistakes with the same amount of angle errors (i.e., >50 degrees). However, for any values of α, UFrame outperforms the CNN model.

Table 4. Performance evaluation results: for different values of α, the number of samples with respect to the angle difference between the correct direction and the chosen direction by the CNN model and UFrame is shown below. The number of samples in the test dataset was 5306.

Model	UFrame					CNN
Value of α	1	2	3	4	5	n/a
no difference	4636	4775	4860	4925	5014	5058
≤10 degrees	104	23	9	6	4	0
>10 degrees	566	508	437	375	288	0
>20 degrees	419	419	415	367	287	0
>30 degrees	337	337	337	337	273	0
>40 degrees	270	270	270	270	255	0
>50 degrees	228	228	228	228	236	0
>60 degrees	184	184	184	186	218	0
>70 degrees	143	143	145	173	213	0
>80 degrees	91	104	132	171	213	0
≥90 degrees	8	8	8	8	8	248

The CNN model produced 248 errors, or in other words, the model misclassified 248 input images. However, due to the high accuracy of the model (i.e., about 95%), even when the model made an incorrect prediction, the softmax output for the correct direction was still large, increasing the uncertainty score. The high uncertainty score lets the proposed framework make a mixed decision. By mixing the three unit vectors with the softmax output being the weight, the mixed control decision leans towards the correct decision. This is why the proposed framework makes a much smaller number of errors than the CNN model. On the other hand, regardless of α, the proposed framework produced eight cases with large angle errors, i.e., ≥90 degrees. This happens when the softmax output of the model is completely incorrect, giving almost zero probability to the correct direction. When the majority of the probability is given to one of the incorrect directions, the error flag becomes off, preventing the proposed framework from making a mixed decision. When the other two incorrect decisions receive a similar amount of probability and the correct direction is either left or right, the proposed framework makes a mixed decision, but it deviates much from the correct decision. Such cases happened for the input images that did not have any meaningful information for the CNN model to predict which direction to go—for example, images having no track/lane at all and blurry images (from camera shake).

5. Discussion and Notes

5.1. Possible Variation on the Proposed Framework

One possible variation of making a mixed control decision is using stochastic sampling which is well-known in generative deep learning [29]. To be specific, when the max2 uncertainty score exceeds the threshold, UFrame can randomly sample the final output, e.g., model decision or prediction, from the distribution specified by the softmax output. Although it is a viable strategy, it may result in making large errors in the case of the IoT car control, for example, due to the randomness. The proposed mixed control decision module also makes a decision in a probabilistic manner, but the resulting decision is biased according to the model's prediction. Therefore, as long as the model has learned features well enough and results in a high accuracy on the test dataset, the proposed mixed control decision model will outperform the stochastic-sampling-based approach.

5.2. Working with Advanced Softmax Losses

The proposed `UFrame` does not require any changes when used with different softmax loss models. The advantages of the advanced softmax losses only contribute to enhancing the performance of the proposed `UFrame`. Sophisticated softmax loss models tend to maximize the inter-class distances while minimizing the intra-class distances on a given metric space. Therefore, the number of incorrect decisions, i.e., misclassifications—our interest, is expected to be reduced. Additionally, due to the enhanced capability of discriminating input data belonging to one class from others, the uncertainty score distribution of the correctly classified input is expected to be better distinguishable from those of the misclassified ones.

5.3. Complexity of the Proposed Framework

The proposed `UFrame` incurs the following two different types of complexity. One is the time complexity caused by TLA. TLA feeds the entire validation dataset to the given, trained model, and thus, the time complexity for threshold learning depends on the complexity of the model and the number of validation samples. However, since TLA runs offline, it does not slow down the real-time performance of the model operating with `UFrame`. On the other hand, the other operations of `UFrame`, i.e., computing the max2 uncertainty, setting the error flag and making a mixed control, may delay the model to some degree. It takes $O(N_c)$ and $O(1)$ to compute the max2 uncertainty and to determine whether or not to set the error flag, respectively. In the case of making a mixed control decision, `UFrame` needs an extra $O(N_c)$ time. Overall, `UFrame` adds a time complexity of $O(N_c)$ to the entire decision-making process, which is negligible.

5.4. Performance of the Proposed Framework in General

From the evaluation studies presented in both Section 3 for the MNIST image and Section 4 for the IoT car image dataset, we have shown the effectiveness of the proposed framework in different applications. To be specific, for a small value of α the proposed framework can identify the majority of the misclassified samples from the deep learning model by using the threshold-based error detection algorithm. Considering the large difference in the two datasets and in the corresponding applications (see Section 4 for our discussion on how different one set is from the other), the results we have presented in the previous two sections suggest that the proposed framework can be applied to a wide-range of applications using softmax decision-making models, which is why we have considered two very different applications in this work.

However, the performance of the error detector depends on the accuracy of the underlying deep learning model. As aforementioned, the threshold learning algorithm uses the correctly classified samples from the model. A poor accuracy of a model will result in a relatively small number of correctly identified samples with a high uncertainty score. Then, the learned threshold can easily be biased, degrading the performance of the error detector for having an incorrect threshold. Therefore, the proposed framework should be used with a model which guarantees a high level of classification accuracy. The same applies to the control decision algorithm in the proposed framework.

6. Conclusions

In this paper, we have proposed an effective framework that enhances the performance of softmax decision-making deep learning models. Inspired by the idea that the uncertainty score of the softmax output indicates how uncertain the model is as to its decision, the proposed uncertainty-score-based framework effectively identifies the majority of the misclassified or misrecognized samples. In other words, the proposed framework makes use of the distribution of the probabilities or uncertainties in the softmax output to discover incorrect decisions made by the model. In addition, we showed by an empirical study how to effectively enhance the performance of the given, trained model by making a mixed control decision when the model's output was likely to be incorrect. Additionally, the proposed UFrame does not make any modification to or put any computationally heavy burden on the existing model.

Due to being low in complexity and compatible with general softmax-based deep learning models, the proposed framework can boost the field of deep learning with IoT. Additionally, the proposed algorithm that makes a mixed control decision can be applied to the fields wherein precise control decisions are required, especially for robots. The capability of the proposed framework is not limited to IoT or general-purpose sensors, and thus, it can be used for enhancing the performances of sophisticated deep learning models for classification/recognition by identifying incorrectly classified samples. We envision that the proposed framework will play an important role in mission-critical applications with or without IoT, where the tolerable error rate is strictly limited.

As for the future work, we plan to deploy the proposed framework on various IoT devices with different deep learning applications to evaluate the performance of the proposed framework with respect to the computing capacities of IoT devices, the complexity of the deep learning model and its classification/recognition performance. We also plan to carry out studies on performance comparisons between the proposed framework and other similar approaches in terms of time/space complexity and accuracy (or error rate) for classification/recognition or control decisions. We also plan to extend our work to solve difficult control tasks wherein the decision domain is continuous and wherein the control decisions are affected by external random factors.

Author Contributions: Conceptualization, C.C., W.C. and T.K.; methodology, C.C. and T.K.; software, C.C. and T.K.; validation, C.C. and T.K.; formal analysis, T.K.; investigation, T.K.; resources, C.C. and T.K.; data curation, C.C. and T.K.; writing—original draft preparation, C.C. and T.K.; writing—review and editing, W.C. and T.K.; visualization, C.C. and T.K.; supervision, T.K.; project administration, T.K.; funding acquisition, T.K. All authors have read and agreed to the published version of the manuscript.

Funding: This research was funded by Hallym University Research Fund, 2019 (HRF-201905-012).

Conflicts of Interest: The authors declare no conflict of interest.

Abbreviations

The following abbreviations are used in this manuscript:

ANN   Artificial Neural Network
CNN   Convolutional Neural Network
IoT   Internet of Things

References

1. Xu, L.D.; He, W.; Li S., Internet of Things in industries: A survey. *IEEE Trans. Ind. Inform.* **2014**, *10*, 2233–2243. [CrossRef]
2. Perera, C; Liu, C.H.; Jayawardena, S.; Chen, M. A survey on Internet of Things from industrial market perspective. *IEEE Access* **2014**, *2*, 1660–1679. [CrossRef]
3. Laplante P.A.; Laplante, N. The Internet of Things in healthcare: Potential applications and challenges. *IEEE IT Prof.* **2016**, *18*, 2–4. [CrossRef]
4. Baker, S.B.; Xiang, W.; Atkinson, I. Internet of Things for smart healthcare: Technologies, challenges, and opportunities. *IEEE Access* **2017**, *5*, 26521–26544. [CrossRef]
5. Liu, Y.; Yang, C.; Jiang, L.;Xie, S.; Zhang, Y. Intelligent edge computing for IoT-based energy management in smart cities. *IEEE Netw.* **2019**, *33*, 111–117. [CrossRef]
6. Morello, R.; Capua, C.D.; Fulco, G.; Mukhopadhyay, S.C. A smart power meter to monitor energy flow in smart grids: The role of advanced sensing and IoT in the electric grid of the future. *IEEE Sens. J.* **2017**, *17*, 7828–7837. [CrossRef]

7. Zantalis, F.; Koulouras, G.; Karabetsos, S.; Kandris, D. A review of machine learning and IoT in smart transportation. *Future Internet* **2019**, *11*, 94. [CrossRef]

8. Masek, P.; Masek, J.; Frantik, P.; Fujdiak, R.; Ometov, A.; Hosek, J.; Andreev, S.; Mlynek, P.; Misurec, J. A harmonized perspective on transportation management in smart cities: The novel IoT-driven environment for road traffic modeling. *Sensors* **2016**, *16*, 1872. [PubMed]

9. Kim, T.; Qiao, D.; Choi, W. Energy-efficient scheduling of Internet of Things devices for environment monitoring applications. In Proceedings of the IEEE International Conference on Communications (ICC), Kansas City, MO, USA, 20–24 May 2018; pp. 1–7

10. Ian, G.; Bengio, Y.; Courville, A. *Deep Learning*; MIT Press: Cambridge, MA, USA, 2016; ISBN 978-026-203-561-3.

11. LeCun, Y.; Bengio, Y.; Hinton, G. Deep learning. *Nature* **2015**, *521*, 436–444. doi:10.1038/nature14539. [CrossRef] [PubMed]

12. Tang, J.; Sun, W.; Liu, S.; Gaudiot, J.-L. Enabling deep learning on IoT devices. *IEEE Comput.* 2017, 50, 92–96. [CrossRef]

13. Yao, S.; Zhao, Y.; Zhang, A.; Hu, S.; Shao, H.; Zhang, C.; Su, L.; Abdelzaher, T. Deep learning for the Internet of Things. *IEEE Comput.* **2018**, *51*, 32–41. [CrossRef]

14. Krizhevsky, A.; Sutskever, I.; Hinton, G.E. Imagenet classification with deep convolutional neural networks. In Proceedings of the Annual Conference on Neural Information Processing Systems (NIPS), Lake Tahoe, NV, USA, 3–8 December 2012; pp. 1097–1105.

15. Simonyan, K.; Zisserman, A. Very deep convolutional networks for large-scale image recognition. In Proceedings of the International Conference on Learning Representations (ICLR), San Diego, CA, USA, 7–9 May 2015; pp. 1–14.

16. Szegedy, C.; Liu, W.; Jia, Y.; Sermanet, P.; Reed, S.; Anguelov, D., Erhan, D.; Vanhjoucke, V.; Rabinovich, A. Going deeper with convolutions. In Proceedings of the IEEE Conference on Computer Vision and Pattern Eecognition (CVPR), Boston, MA, USA, 7–12 June 2015; pp. 1–9.

17. Liu, W.; Wen, Y.; Yu, Z.; Li, M.; Raj, B.; Song, L. Sphereface: Deep hypersphere embedding for face recognition. In Proceedings of the IEEE Conference on Computer Vision and Pattern Eecognition (CVPR), Honolulu, HI, USA, 21–26 July 2017; pp. 212–220.

18. Liu, W.; Wen, Y.; Yu, Z.; Yang, M. Large-margin softmax loss for convolutional neural networks. In Proceedings of the International Conference on Machine Learning (ICML), New York City, NY, USA, 19–24 June 2016; pp. 1–10.

19. Wang, F.; Cheng, K.; Liu, W.; Liu, H. Additive margin softmax for face verification. *IEEE Signal Proc. Lett.* **2018**, *25*, 926–930. [CrossRef]

20. Huang, G.B.; Learned-Miller, E. *Labeled Faces in the Wild: Updates and New Reporting Procedures*; Technical Report; Department Computer Science, University Massachusetts Amherst: Amherst, MA, USA, 2014; pp. 1–5.

21. Kemelmacher-Shlizerman, I.; Seitz, S.M.; Miller, D.; Brossard, E. The megaface benchmark: 1 million faces for recognition at scale. In Proceedings of the IEEE Conference on Computer Vision and Pattern Recognition (CVPR), Las Vegas, NV, USA, 27–30 June 2016; pp. 4873–4882.

22. Wang, M.; Lu, S.; Zhu, D.; Lin, J.; Wang, Z. A high-speed and low-complexity architecture for softmax function in deep learning. In Proceedings of the IEEE Asia Pacific Conference on Circuits and Systems (APCCAS), Chengdu, China, 26–30 October 2018; pp. 223–226. [CrossRef]

23. Cho, J.; Park, K.; Karki, M. Improving sensitivity on identification and delineation of intracranial hemorrhage lesion using cascaded deep learning models. *J. Digit. Imaging* **2019**, *32*, 450–461. [CrossRef]

24. Jain, R.; Chiu, D.; Hawe, W. *A Quantitative Measure of Fairness and Discrimination for Resource Allocation in Shared Computer Systems*; DEC Research Report TR-301; Digital Equipment Corporation: Hudson, MA, USA, 1984; pp. 1–37.

25. Kim, T.; Chun, C.; Choi, W. Optimal User Association Strategy for Large-Scale IoT Sensor Networks with Mobility on Cloud RANs. *Sensors* **2019**, *19*, 4415. [PubMed]

26. Wen, Y.; Zhang, K.; Li, Z.; Qiao, Y. A discriminative feature learning approach for deep face recognition. In Proceedings of the European Conference on Computer Vision (ECCV), Amsterdam, The Netherlands, 8–16 October 2016; pp. 499–515. [CrossRef]

27. LeCun, Y.; Cortes, C.; Burges, C.J. MNIST Handwritten Digit Database. Available online: http://yann.lecun.com/exdb/mnist (accessed on 3 July 2020).
28. Duda, R.O.; Hart, P.E. Use of the Hough transformation to detect lines and curves in pictures. *Commun. ACM* **1972**, *15*, 11–15. [CrossRef]
29. Chollet, F. *Deep Learning with Python*; Manning Publications: Shelter Island, NY, USA, 2017; ISBN 978-161-729-443-3.

Article

Implementing Deep Learning Techniques in 5G IoT Networks for 3D Indoor Positioning: DELTA (DeEp Learning-Based Co-operaTive Architecture)

Brahim El Boudani [1,*], **Loizos Kanaris** [2], **Akis Kokkinis** [2], **Michalis Kyriacou** [3], **Christos Chrysoulas** [4], **Stavros Stavrou** [3] and **Tasos Dagiuklas** [1]

[1] Division of Computer Science and Informatics, London South Bank University, London SE1 0AA, UK; tdagiuklas@lsbu.ac.uk

[2] Department of Electrical Engineering, Eindhoven University of Technology, 5612 AE Eindhoven, The Netherlands; l.kanaris@sigintsolutions.com (L.K.); a.kokkinis@sigintsolutions.com (A.K.)

[3] Faculty of Pure and Applied Sciences, Open University of Cyprus, 2252 Nicosia, Cyprus; m.kyriacou@sigintsolutions.com (M.K.); stavros.stavrou@ouc.ac.cy (S.S.)

[4] School of Computing, Edinburgh Napier University, Edinburgh EH11 4DY, UK; C.Chrysoulas@napier.ac.uk

[*] Correspondence: elboudab@lsbu.ac.uk

Received: 1 July 2020; Accepted: 22 September 2020; Published: 25 September 2020

Abstract: In the near future, the fifth-generation wireless technology is expected to be rolled out, offering low latency, high bandwidth and multiple antennas deployed in a single access point. This ecosystem will help further enhance various location-based scenarios such as assets tracking in smart factories, precise smart management of hydroponic indoor vertical farms and indoor way-finding in smart hospitals. Such a system will also integrate existing technologies like the Internet of Things (IoT), WiFi and other network infrastructures. In this respect, 5G precise indoor localization using heterogeneous IoT technologies (Zigbee, Raspberry Pi, Arduino, BLE, etc.) is a challenging research area. In this work, an experimental 5G testbed has been designed integrating C-RAN and IoT networks. This testbed is used to improve both vertical and horizontal localization (3D Localization) in a 5G IoT environment. To achieve this, we propose the DEep Learning-based co-operaTive Architecture (DELTA) machine learning model implemented on a 3D multi-layered fingerprint radiomap. The DELTA begins by estimating the 2D location. Then, the output is recursively used to predict the 3D location of a mobile station. This approach is going to benefit use cases such as 3D indoor navigation in multi-floor smart factories or in large complex buildings. Finally, we have observed that the proposed model has outperformed traditional algorithms such as Support Vector Machine (SVM) and K-Nearest Neighbor (KNN).

Keywords: 5G IoT; indoor positioning; deep learning; tracking; localization; navigation; positioning accuracy; single access point positioning; Internet of Things

---

1. Introduction

In the era of 5G IoT [1], real-time positioning is becoming increasingly required by context-aware applications and location-based services. Typical scenarios include locating doctors and patients inside a hospital, advertising commercial products to mall visitors, monitoring gas and oil plants status, pinpointing dead crops in vertical farms, identifying victims' location in Public Protection and Disaster Recovery (PPDR), etc. Moreover, several advanced applications can further provide cellular phone fraud detection, location-sensitive billing, as well as navigation from and to almost everywhere, through the utilization of heterogeneous wireless technologies, fusion of sensor and IoT data [2–5]. A recent report published by IEEE has estimated 50 billion [6] mobile devices will be connected to the cloud by the end of 2020. These devices will need constant access to data anywhere. Cisco has predicted that 26 billion [7] of these devices will be IoT or Wireless Sensor Network (WSN) devices. In this respect, technologies like Cloud Radio Access Network (C-RAN), Millimeter Wave (mm-Wave) communication, ultra dense communication [8], device-to-device (D2D) communication and Vehicle-to-everything (V2X) [9,10] and protocols like IEEE 802.11be (Extremely high Throughput WLAN) [11], IEEE 802.11az (Next Generation Positioning) [12] are not only introduced to increase the bandwidth of communication but also to offer the possibility of co-operative and precise localization. Additionally, with 5G paving the path for a seamless collaboration among heterogeneous wireless systems (cellular, WiFi, WSN, IoT, etc.), a great opportunity has risen in the area of indoor localization in urban areas under the framework of smart cities. Such high dense networks could be utilized to solve multi-agent positioning and offer agility and scalability for accurate positioning as a service. In this direction, we propose a DEep Learning-based with Co-operative Architecture (DELTA) algorithm to enhanced 3D indoor localization. The contributions of this paper can be summarized as follows:

- A realistic 3D indoor localization scenario for 5G IoT networks has been designed using an emulated 5G C-RAN and Zolertia IoT nodes.
- We present a novel approach to Received Signal Strength (RSS)-based fingerprint using 3D multi-layered radiomap to enhance the learning of network signal behaviour.
- A deep learning cooperative algorithm is implemented on the constructed multi-layered radiomap for an improved 3D localization indoor localization. The proposed method targets improving vertical and horizontal localization for use case scenarios such as indoor navigation or people tracking in multi-floor smart or large complex buildings. Based on the results of the emulated realistic radio-planning, we have shown how the DELTA outperformed KNN and SVM.

The remaining of this paper is organized as follows: Section 2 covers related research to this paper. Section 3 describes the problem related to indoor positioning in a 3D environment. Section 4 gives a detailed description of the underlying architecture of the DELTA model. Section 5 consists of a discussion and analysis of the performance results produced by our proposed approach compared with other traditional models. Lastly, Section 6 summarises a conclusion and spots possible future work.

2. Related Work

Indoor positioning techniques can be divided into two main categories: fingerprint and multilateration. In the latter, given a known propagation speed, the distance between a receiver and a group of transmitters is measured using techniques such as Direction of Arrival (DoA), Time of Arrival (TOA)/Time of Flight (TOF), Angle of Arrival (AoA), Time Difference of Arrival (TDOA) and Return Time of Flight (RTOF). These techniques are commonly used in Global Navigation Satellite Systems (GNSS) [13], such as Global Positioning System (GPS) and Galileo, but surprisingly they are also found in IoT indoor navigation solutions [14]. However, multilateration relies mainly on the travelling time or the direction of the signal rays. This makes indoor localization a complex task especially with many issues rising such as synchronization errors and multi-path fading [15–17].

In the fingerprint-based technique, a set of RSS measurements are taken and linked to specific Reference Points (RP) (also known as fingerprints or signatures). Localization using this approach works in two phases: offline and online. During the offline phase, a site survey is conducted with the purpose of linking the measured signal strength values to predefined RPs. The outcome of this measurements campaign is then stored in a radiomap database. During the online phase, a user equipment receives real-time signals and tries to match them with existing records stored in the radiomap database using a matching algorithm. In the context of IoT localization, the RSS signal is collected from wireless technologies such as Zigbee, LoRA, Wifi, Raspberry Pi, BLE, RFID. Since it does not require any specialised equipment or time synchronization to obtain the RSS signal, this technique is usually preferred to multilateration. For instance, authors in [18] have studied how robust localization for robots and IoT can be achieved using RSS fingerprint. Additionally, another interesting approach has been introduced in [19] where the authors have focused on the use of IoT and Wifi-enabled devices to improve fingerprinting in an indoor environment. Recently, a new concept has been developed by Ali et al. [20] using raster maps instead of traditional offline scene analysis. Furthermore, a hybrid solution implemented on LoRa devices, which combines RSS fingerprinting with AoA methods is discussed in [14]. The proposed idea is very promising but it has inherited synchronization issues from multilateration. From these examples, it is undoubtedly clear that the RSS-based fingerprint method is widely used in the research community. This is due to improved localization and reduced computational complexity, as concluded by Amr et al. [19]. A detailed comparison of technologies and algorithms implementing the fingerprint technique for IoT indoor positioning has been carried out by [15,21–23].

In the fingerprint-based approach, deep learning techniques have been widely used to extract common patterns from a sparse radiomap database and to improve localization. In recent years, it has gained a huge popularity among the indoor localization researchers, in particular, due to its robustness and high accuracy [24]. Supervised and unsupervised deep learning algorithms have been recently implemented in 2D localization [25] and multi-floor localization [26]. Recently, Wafa et al. [27] studied the use of Convolutional Neural Networks (CNN) on IoT-Sensor System to determine the node location. In this simulation, the authors converted the 2D localization problem into a 3D image tensor identification problem. The 3D tensor has been constructed using a 2D matrix of RSS signals and 1D kurtosis. This concept has achieved 2 m average error accuracy but a similar system was also implemented in [28] and usually requires a large number of access points deployed in a small space to achieve this result. In [29], authors have implemented a Deep Belief Network (DBN) on an active RFID tag system for accurate location estimation. Their solutions consisted of a set of stacked Restricted Boltzmann Machine (RBM) layers called autoencoders trained using Contrastive Divergence with one-step iteration (CD-1). This algorithm has improved the 2D positioning. To achieve this, the authors have deployed a large number of RFID tags in a 12 m × 12 m indoor environment, which does not take into account the power consumption of the devices. Finally, Wang et al. [30] have suggested a hybrid deep learning solution combining a regression Deep Neural Network (DNN) with a Convolutional AutoEncode (CAE) using Visible Light Communication (VLC). To overcome the issue of fluctuated signal reading in the RSS-based fingerprint method, the authors have proposed an algorithm taking into account a set of consecutive signal readings and converting them to an RSS Temporal Image (RTI), instead of implementing the traditional RSS measurement processing technique. However, despite having been used in several works [31,32], VLC suffers from issues such as interference with other ambient lights, signal shadowing and usually requires the receiver to be in Line-Of-Sight (LOS), which can affect the accuracy of the location estimation. A detailed comparison of deep learning and other machine learning algorithms used in localization for IoT environment is covered in [33,34].

Until now, most of the existing IoT-based indoor localization solutions have mainly focused on either 2D localization or floor detection. However, in some special use cases scenarios such as indoor navigation for Unmanned Aerial Vehicle (UAV) or Automated Guided Vehicle (AGV) in a smart factory or big supermarket, precise 3D positioning is indispensable for daily operations. To address this issue, we suggest the DELTA to maximize the localization accuracy and minimize the distance error in a 3D indoor environment.

3. System Model and 3D Localization Problem

In this section, we introduce our proposed system model using Deep Neural Networks (DNN) and multi-layered radiomap to perform 3D Indoor Localization. To the best of our knowledge, this is a novel approach to implement deep learning on multi-layered radiomap for localization purposes. The main benefit of the proposed method is improved localization accuracy, and computational complexity minimization during online fingerprinting through the adoption of deep learning techniques, while at the same time utilizing the widely spreading WSN and/or IoT infrastructure making it an economical solution. To realize these steps, we considered N to be the number of transmitters in the environment and x, y and z, the corresponding coordinates of each fingerprint entry on the constructed radiomap. The 3D multi-layered fingerprint database has been constructed by linking the RSS values received from the transmitters to a 3D location on the radiomap [35]. This can be mathematically expressed as:

$$M = \{(L_1, S_1), (L_2, S_2), \ldots, (L_{n-1}, S_{n-1}), (L_n, S_n)\} \tag{1}$$

where M is the ratio-map database, $S \in \mathbb{R}^{NXM}$ is a vector of RSS signal values and L is a vector of three values: $L \equiv \{x, y, z\}$ and L_n represents the total number of the sample location of x_n, y_n and z_n associated with each signal vector sample S_n collected during the offline-phase.

In this respect, the estimation problem is defined by solving the 3D localization problem using a matrix of historical location points and their corresponding signal values. However, the challenge is to model the non-arbitrary relationships between N transmitters members of S signal matrix to predict accurately the 3D location L using a deep learning algorithm. To achieve this, the 3D localization has been segmented to two sets of problems:

Problem 1. *Given a matrix of S signal sent from N transmitters, predict the x and y coordinates of a 2D mobile station location. This can be written as:*

$$\lambda_1 = f(\bar{S}_{ij}) \tag{2}$$

where λ_1 represents the x_i and y_i 2D location, which we would like to estimate, and $f(S_{ij})$ represents the function that utilizes RSS values received by the transmitters to predict the location of the mobile station.

Problem 2. *Given a matrix of S signal sent from N transmitters to the mobile station and x_i, y_i, known from problem 1, estimate the z_i coordinate. This can be mathematically expressed as:*

$$\lambda_2 = f(\bar{S}_{ij}, \lambda_1) \tag{3}$$

where λ_2 is the z_i location, λ_1 is the output of problem 1 solution and S_{ij} represents a matrix of signal values S as previously stated in problem 1.

4. DELTA 3D Localization for 5G WSN Network

In this section, the DELTA System has been developed for 3D multi-layered indoor environment localization. Figure 1 depicts the steps undertaken to realize a co-operative system for accurate 3D prediction.

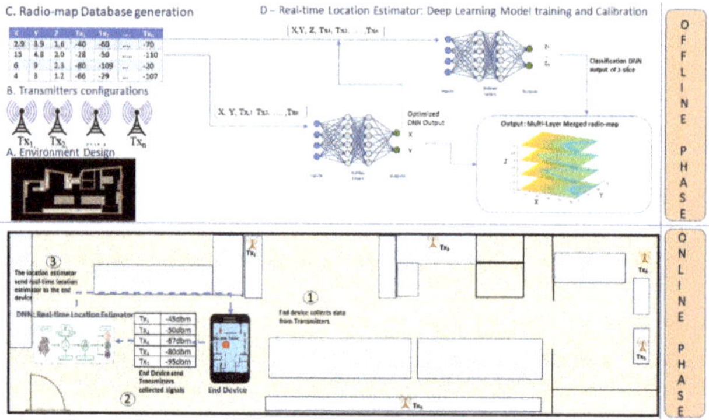

Figure 1. Detailed architecture of deep learning-based co-operative architecture (DELTA).

4.1. Test Environment Description

In this subsection, we describe the test environment. The area of interest is a typical laboratory, with open spaces as well as private rooms defined by the following dimensions: 8 m width × 16 m depth × 2.75 m height. The lab environment was dynamic during this experiment.

4.1.1. Step I: The Physical Network Setup

For the physical setup, an indoor test environment was deployed where a 5G network was emulated by a typical IoT network with Zolertia RE-Mote Revision B nodes connected to a LoWPAN Border Router, as illustrated in Figure 2.

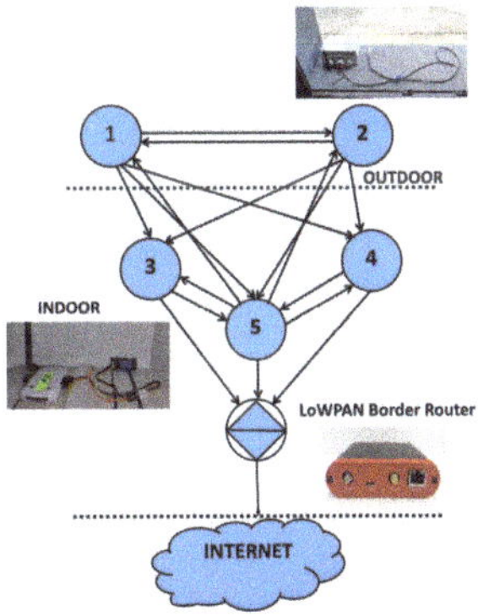

Figure 2. Network setup topology.

We randomly placed 5 Zolertia nodes, with their antennas at vertical polarization, as shown in Figure 2. The nodes and the ray tracing propagation mechanisms have been configured as per Tables 1 and 2.

Table 1. Wireless Sensor Network (WSN) and radio propagation parameters.

Parameter	Value
Rx sensitivity (dBm)	−120
Tx power (dBm)	3
antenna Type	omni
Max refractions	12
Max reflections	12
Max diffractions	1

4.1.2. Step II: Connecting IoT to 5G C-RAN

To simulate the 5G WSN environment, each Zolertia node was connected to an experimental 5G C-RAN. The setup was built using a GNS3 network simulator [36] and OpenDaylight Software Defined Controller [37]. These two can control the network setup behaviour at the network layer level. Figure 3 shows a setup built using a GNS3 network simulator and a Software Defined Controller OpenDaylight dashboard for the network topology. These two elements can control the network setup behaviour at the network layer level.

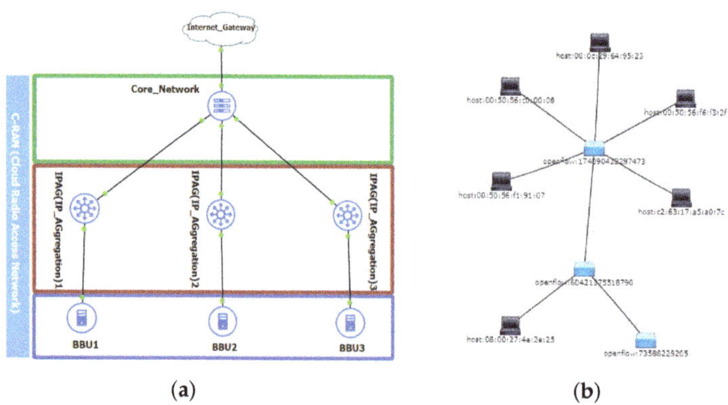

(a)   (b)

Figure 3. 5G C-RAN setup on GNS3 and WSN Network Connected to OpenDaylight SDN Controller. (**a**) 5G emulated C-RAN testbed on GNS3; (**b**) WSN and GNS3 emulated 5G C-RAN connected on OpenDaylight.

4.1.3. Step III: Simulating the Test Environment

Using a 3D deterministic simulator called TruNET Wireless [38], we constructed a multi-layered fingerprint radiomap dataset, in order to conduct the offline training phase, as illustrated in Figure 1. During this procedure, in addition to the network setup configuration, the constitutive parameters of all environment object materials were also configured as per Table 2, in order to retrieve realistic results [39]. The benefits of utilizing a deterministic simulation are to construct radiomaps instead of launching measurement campaigns as analysed in [40]. The summary of correlation results of this study is covered in Section 5. The simulation environment for this study is shown in Figure 4.

Table 2. Material constitutive parameters of the test environment.

Material	El. Per. (*F/m*)	L. Tangent
Concrete	3.9	0.23
Wood	2	0.025
Brick	5.5	0.03
Metal	1	1,000,000
Plasterboard	3	0.067
Glass	4.5	0.007

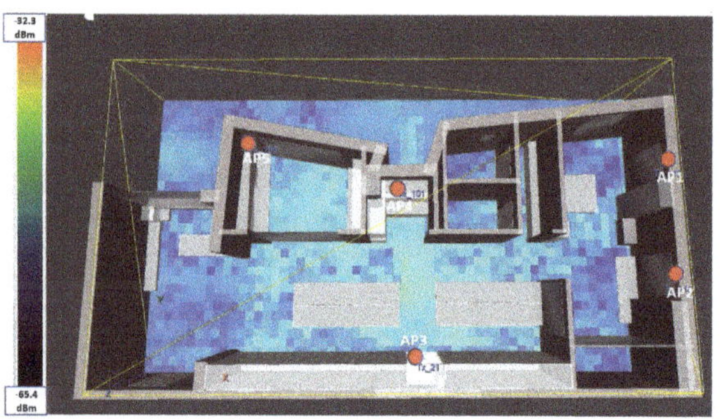

Figure 4. TruNET Wireless Simulator Radiomap for Access Point 3.

4.1.4. Physical Network Behaviour

The signal propagation can be affected by various factors leading to the degradation of the signal quality especially in low power radio networks such as Wireless Sensor Networks. For a successful simulation, it is always crucial to observe the physical network behaviour during the offline measurement campaign. The effects of the physical layer and the various factors contributing to changes in the environment have been extensively studied in [41]. Using Link Quality Estimation (LQE) metrics such as Packet Reception Ratio (PRR) and Signal-to-Noise-Ratio (SNR), Baccour et al. [41] have studied the factors affecting a transmitter chip similar to that used in this experiment. It is very crucial to note that the simulated environment can be affected by various changes happening at the physical network. For the nodes used in this simulation, Figure 5 shows how the change in the Received Signal Strength Indicator (RSSI) can affect the PRR.

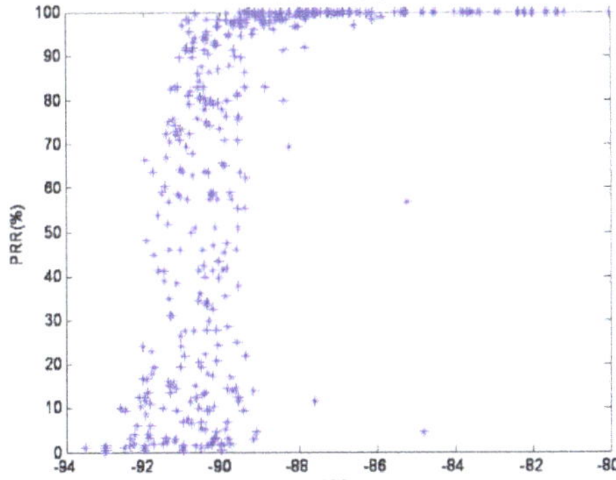

Figure 5. Packet Reception Ratio (PRR) vs. Received Signal Strength Indicator (RSSI) Signal CC2420 chip Baccour et al. [41].

Sometimes, measurement campaigns can be affected by various environmental noises, which may lead to unrealistic readings, either due to signal spikes or fluctuations. This noise can be either thermal noise or interference from other people's equipment operating at the same frequency. To ensure signal samples obtained from TruNET Wireless are realistic, Figure 6 depicts the RSS coverage correlation analysis experiment conducted in [42]. The samples have been collected from Zolertia nodes over a week period at different instances with an interval of 15 min for each sample. The produced measurements for each RP have been averaged using the mean value. During this experiment, the IoT nodes have always been fixed and the environment was dynamic with people moving around.

Finally, it is clearly indicated that the simulated RSS values from the TruNET wireless simulator highly approximates the measured ones reaching a correlation level of more than 73%.

4.2. DELTA Architecture

Deep learning is a fundamental building block of the proposed architecture. It allows computational models consisting of multiple processing layers to learn the representation of data within multiple abstract levels [43]. One of the most important elements of deep learning is deep neural networks. Bengio et al. [44] refer to this as either deep feed forward networks or Multiple Layers Perceptron (MLP) since they have more than two hidden layers.

Our proposed architecture, as illustrated in Figure 7, consists of two deep neural networks. The first is a regression model δ_1 used to predict the 2D location of a mobile device. The second is a classification model referred to as δ_2. Figure 7 illustrates the number of layers, neuron, input and output parameters used for both models. Based on numerous trials and hyper-parameters tuning, we observed that three hidden layers were the best fit model for both networks.

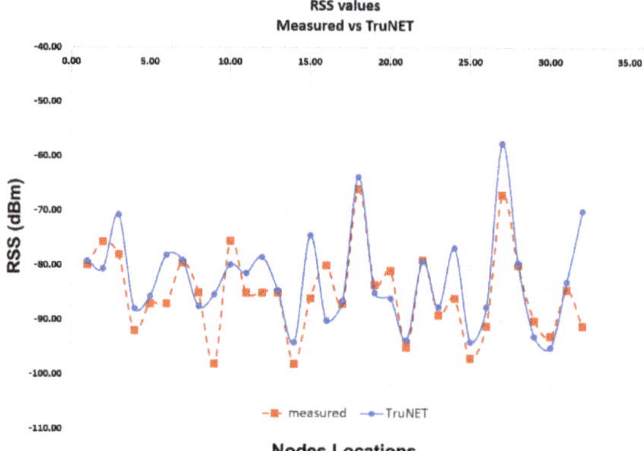

Figure 6. RSS Values Measured vs. TruNET Kanaris et al. [42].

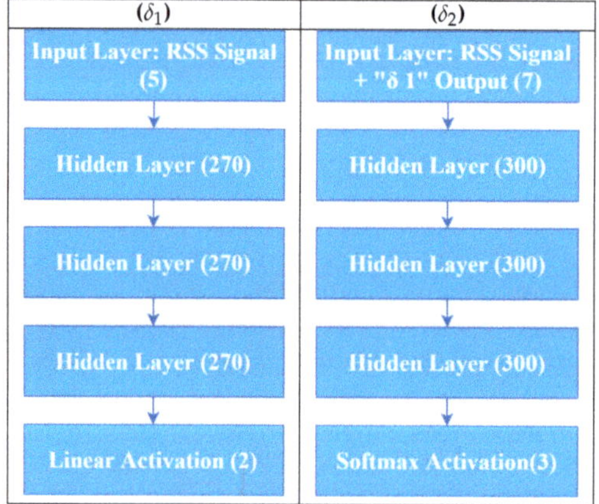

Figure 7. Layers of DELTA Architecture Network.

4.3. DELTA Layers

4.3.1. Input Layers

For δ_1, the input is a transposed vector of RSS signals that can be expressed as follows: $S = [s_1, s_2, \ldots, s_n]^T$. For δ_2, the input is slightly different to δ_1. It consists of RSS signal input S and the output of δ_1. Each observation has a set of signals and predicted locations. This can be written as:

$$\delta_{2\_input} = S \cup (L_x, L_y) \tag{4}$$

where S is the signal and L_x, L_y are the corresponding x and y locations. These two values are approximated using δ_1, as shown in Figure 7.

4.3.2. Hidden Layers

Each element of this input gets multiplied by its specific weight vector $\vec{w}$ and the product is added to a bias b. For the first hidden layer, this is expressed as follows:

$$h1 = \sum_{i=1}^{n} w_i^1 I_i + b_i^1 \tag{5}$$

where I_i is an element from the input vector. Each I_i represents an input from a transmitter in the constructed fingerprint database. A summation of all these inputs is then fed to an activation unit A. In this case, the type of activation function used is a called Rectified Linear Unit (ReLu).

$$A_1 = max(0, h1) \tag{6}$$

where A_1 is an activation unit for the first hidden layer. The output of this hidden layer is the number of hidden neurons specified in the first hidden layer. Similarly, Equation (7) for hidden layer 2 is expressed as follows:

$$h2 = \sum_{i=1}^{n} w_i^2 a_i^1 + b_i^2 \tag{7}$$

This result is then fed into a further activation unit A_2:

$$A_2 = max(0, h2) \tag{8}$$

The hidden layer three receives the output of Equation (8) and makes similar calculations to $h2$:

$$h3 = \sum_{i=1}^{n} w_i^3 a_i^2 + b_i^3 \tag{9}$$

Finally, the results returned in Equation (9) are fed into the activation unit A_3.

$$A_3 = max(0, h3) \tag{10}$$

4.3.3. Output Layers

For δ_1 model, since the desired output is a real-valued number, a linear function has been applied using the following equation:

$$g(y = j|a_i) = \sum_{i=1}^{n} w_i^4 a_i^3 + \epsilon_i \tag{11}$$

For δ_2 model, the output is multiple class labels, therefore the Softmax function equation below has been used:

$$\theta(a_i) = \frac{\exp(a_i^3)}{\sum_j \exp(a_j^3)} \tag{12}$$

To get the best final approximation, δ_1 supports δ_2. Algorithm 1 explains how both networks cooperate to make a final localization.

Algorithm 1: DELTA Algorithm for 3D Localization

Input :RSS ▷ Get Signal Vector
Output: 3D Location
Require: Signal UpperThreshold η;
Require: Signal LowerThreshold μ;
for RSS_i *in RSS* **do**
  $r \leftarrow \frac{RSS_i - \mu}{\mu - \eta}$ ▷ Normalize signal
  $2D \leftarrow \delta_1(r)$ ▷ Apply first model prediction
  $1D \leftarrow \delta_2(r, \delta_1)$ ▷ Apply second model prediction
  $3D \leftarrow 2D \cup 1D$ ▷ merge δ_1 and δ_2 results
endfor
return 3D Location

4.4. Prepossessing

4.4.1. Fingerprints Radiomap Database

As previously mentioned, we began by constructing the radiomap database using eight features. Table 3 gives a detailed explanation of each variable.

Table 3. The features used to construct the fingerprints database.

Variable	Min. Value	Max. Value	Type
X	0	8	coordinates
Y	0	16	coordinates
Z	0.25	1.75	coordinates
AP1	−120 dBm	−28 dBm	RSS value
AP2	−100 dBm	−30 dBm	RSS value
AP3	−100 dBm	−40 dBm	RSS value
AP4	−90 dBm	−50 dBm	RSS value
AP5	−100 dBm	−60 dBm	RSS value

The constructed radiomap consists of 2880 3D References Points (RPs) associated with RSS values from five different WSN Access Points(APs). Each AP is placed at least three meters away. The position of these APs is shown on the lab floor-plan illustrated in Figure 8.

To ensure that there is no redundancy in the information collected, a Pearson correlation test has been conducted between each AP and the result is shown in Figure 9. There is clearly no high negative or positive correlation between the APs used in this experiment.

In addition to this, Figure 10 shows each layer on the radiomap database constructed is significantly different from the other layer. The Figure shows the signal at 0.25, 0.75 and 1.75 m for Access Point 1.

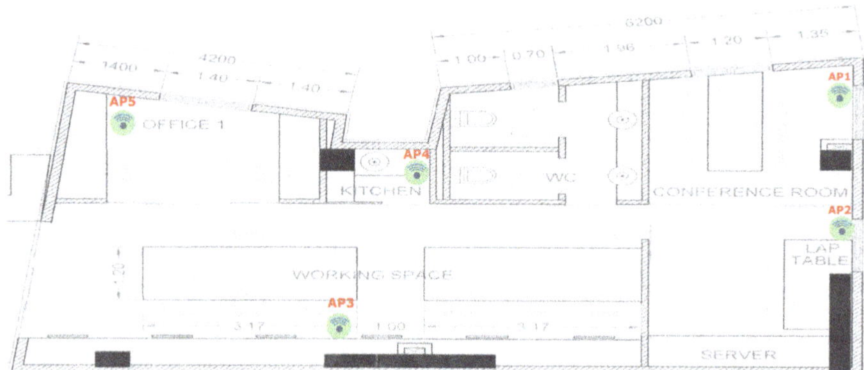

Figure 8. Access points position on the setup environment floor-plan.

4.4.2. One-hot Encoding

One-hot encoding is one of the most common techniques for converting a token into a vector [45]. The conversion is achieved by associating each unique integer with every unique value from the column z. This turns every unique value into a binary vector having the size of the unique values. As a result, every column will have zero except for where the unique value has occurred. In our case, we have used the steps followed in Algorithm 2 to one-hot encode our target variable:

Algorithm 2: One hot Encoding

Input   :Column Z   ▷ get Z columns
Output: Result Matrix of N binary vectors unique values from Z
Dictionary D=[];
Results R={[],[],[],[]};
for *i in Z.length* **do**
   if i $\notin$ D: ▷ If value not in dictionary add it
   key=D[i]
   D[i]=Z[i]
endfor
return D
Map D into results R columns as binary vector $\{[Z_1], [Z_2], \ldots, [Z_n]\}$

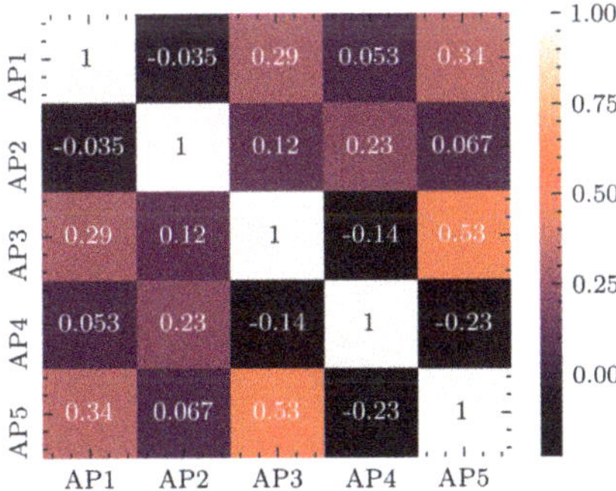

Figure 9. WSN access points correlation matrix.

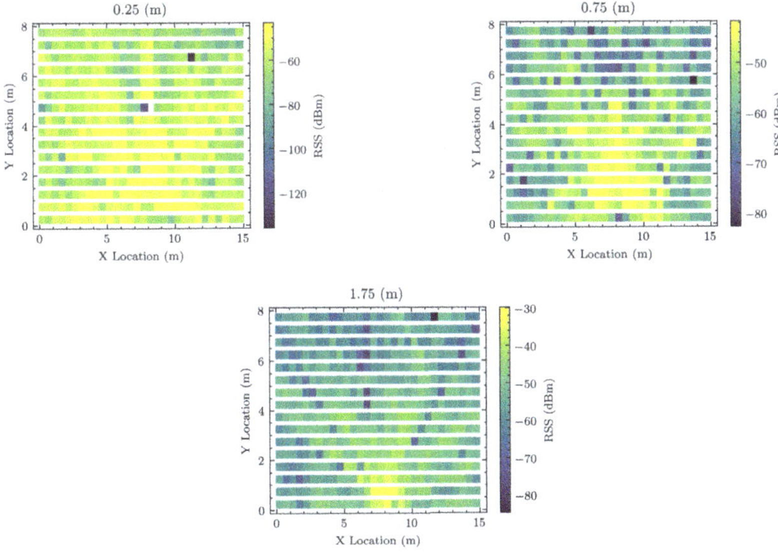

Figure 10. Signal strength map for WSN access point 1 for each Z layer.

4.4.3. Min–Max Normalization

Min–Max normalization has been implemented to make sure the learning of signal representation data is faster for DELTA architecture models to converge quickly. This concept works by fitting the original data into a new scale between 0 and 1. After this numeric transformation, the highest value becomes close to 1 and the lowest value is close to 0 as stated in [46]. The formula used to achieve this, is the following:

$$\frac{RSS_i - min(RSS)}{min(RSS) - max(RSS)} \tag{13}$$

where $min(RSS)$ represents the values minimum threshold signal specified during the training signal, i.e., -120 dBm and $max(RSS)$ represents the maximum value measured, i.e., -30 dBm. Each signal measurement we want to convert is denoted by RSS_i where i is the ith row in N Transmitter. For other scenarios, it is advisable to use the receiver sensitivity level as the minimum value and the strongest measured signal during the offline-phase as the maximum value.

4.5. Hyper-Parameters Fine-Tuning

4.5.1. Loss Functions

- Using Euclidean Distance as loss function for δ_1 model, the purpose is to train the model to minimize the Mean Euclidean Distance (MED) error between the actual and the predicted location.

$$D(\mathbf{L}^{act}, \mathbf{L}^{pred}) = \frac{1}{M} \sum_{n=1}^{m} \sqrt{(x_j^{act} - x_j^{pred})^2 + (y_j^{act} - y_j^{pred})^2} \tag{14}$$

L^{act} here denotes the actual location and L^{pred} denotes the predicted location.

- For the δ_2 model, Categorical Cross-entropy is implemented as a loss function. This can be written as:

$$H(\mathbf{L}^{act}, \mathbf{L}^{pred}) = -\sum_{j=0}^{M} \sum_{i=0}^{N} (z_{ij}^{Act} \cdot log(z_{ij}^{pred})) \tag{15}$$

where L^{act} denotes the actual location and L^{pred} denotes the predicted location. While z_{ij} denotes the ith observation in the jth z output class or level.

4.5.2. Hidden Layers and Neurons Size Determination

The number of hidden layers and neurons count used in the DELTA has been determined using the loss function specified the previous subsection. Figure 11 shows the performance of each network for each neuron count and layers number selected. As demonstrated in this figure, the categorical cross entropy loss is minimized after a third hidden layer has been added and the neurons count has been set to 300. Similarly, the average error was decreased in delta one after the parameters were changed to 300 neurons and three hidden layers.

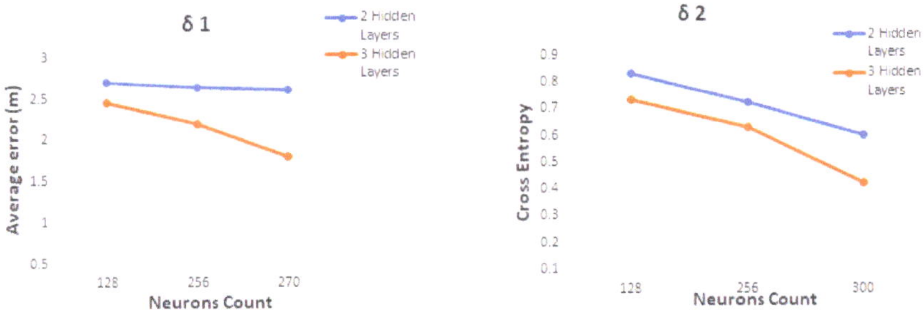

Figure 11. The number of hidden layers and neurons vs. each loss function.

4.5.3. Batch Normalization

A batch is the number of samples propagated through the neural network model before the parameters are updated. To train each neural network faster, we have supported each layer with a batch normalization. This sort of normalization is applied to input samples of the same batch size. This fine-tuning technique has been proven to speed up the training and learning process by 12 times faster than the normal architecture as described by authors in [47]. The formula for the batch normalization implemented on each Deep Neural Network of DELTA system is:

$$T_i = \frac{(T_i - \mu(T)}{\sqrt{\sigma^2(T) + \epsilon}} \tag{16}$$

where T is training batch, $\mu(T)$ is its mean, $\sigma^2(T)$ is its variance and ϵ is a small constant number added to support the variance. For this to work in Keras deep learning library [45], a layer of batch normalization with explicit parameters has to be added at the beginning of each hidden layer.

4.5.4. Regularization

To avoid overfitting, a regularization technique has been implemented to switch off certain neurons for some layers. This technique is called dropout. Details for this technique are provided by Nitish et al. in [48]. The dropout rate used in DELTA is 0.20 as suggested by [48]. After experimentation, we have concluded that for better results are achieved when implementing batch normalization before dropout.

4.6. Optimization

Optimization is the process of training a network using mini-batches and iterations to get the optimum configuration for its parameter. One of the widely used stochastic optimization algorithms in deep learning is ADAptive Momentum (ADAM). The algorithm can be viewed as a combination of RMSprop and Momentum [49]. It works by correcting the bias b and the weight w after each iteration. To get the best results from ADAM's parameters, we specified a learning rate $\alpha = 0.001$, $\beta_1 = 0.9$ for the momentum control, $\beta_2 = 0.99$ for squared weight in RMSprop section and $\epsilon = 10^{-8}$ as specified by the authors in [49]. To implement this in Keras, ADAM parameters have to be specified before the model is compiled.

4.7. Scoring

Using 900 hidden neurons and three hidden layers, we have constructed model δ_1 to predict x and y locations. This has yielded 279,302 parameters to be trained. Our cost function is the euclidean distance difference between each predicted observation and the original location. To minimize it, hyper-parameters have been fine-tuned such as the batch sizes and the number of times an algorithm will iterate through an entire training dataset. One iteration is referred to as epochs. The aforementioned methodology resulted in an average positioning error of 1.6 m average (less than 2 m error overall) in both training and validation phases. Figure 12 shows how the δ_1 model mean Euclidean distance error in meters decreases over the number of epochs chosen, in this case 3000 epochs. However, by the end of epoch 3000, the model has converged and stopped improving its accuracy.

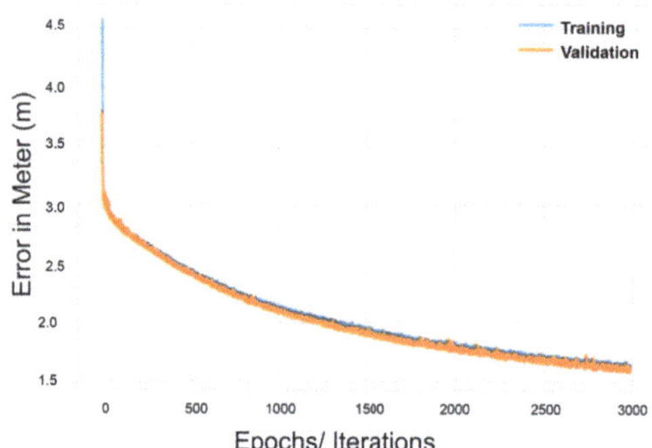

Figure 12. δ_1 Model mean Euclidean distance error in meter (m) vs. the number of Epochs.

Similarly, after an iterative tweaking of the architecture parameters, using 810 number of neurons and three hidden layers, we have constructed model delta 2 where z layer is the target variable. A total number of 235,592 parameters were trained in this model. The cost function is the multi-categorical cross entropy, which is used widely for classification scoring. Figure 13 shows how the categorical cross-entropy has been minimized after 2500 epochs.

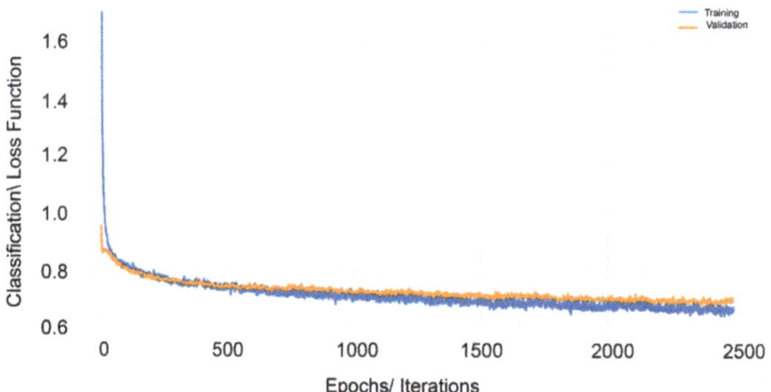

Figure 13. δ_2 Categorical cross-entropy vs. the number of Epochs.

5. Performance Evaluation Results

In this section, we explore, evaluate and critically analyse the simulation results against famous industry methods such as SVM and KNN. However, before going through the results analysis, it is worth mentioning that KNN and SVM modelling tasks have been carried out using Scikit-learn [50], a widely used Python library toolset for machine learning and statistics. More specifically, SVM models have developed using an SVM class from the Scikit-learn library and KNN models have been built using a classifier class called KNeighborsClassifier [51]. The DELTA models have been constructed using Keras API [52], a deep learning library also available in Python. During the evaluation phase, the three algorithms were implemented using python software on the same machine with Intel i7-4790@3.60GHz CPU and 16 GB of RAM. In terms of time complexity, KNN has finished after 230 ms while SVM has taken 450 ms. The proposed DNN has used 160 ms to execute, making it more efficient than KNN and SVM.

5.1. Results Analysis

5.1.1. δ_1s. KNN and SVM

Using 180 random samples [39], we have bench-marked and assessed DNN model δ_1 against KNN and Support Vector Regression (SVR) models. The samples have been obtained for each z layer making a total of 540 RPs. The SVR has been trained using a linear kernel, a degree of one and an epsilon value of one using 80% training and 20% validation data sets. Similarly, a KNN model has been trained with a K value set to three. The results in Figure 14 show the error distribution in meters for all three models. SVR has scored a rather worse error distribution where the peak of its distribution ranges between 4 and 6 m error. KNN has done slightly better compared to SVR. However, a large proportion of the distribution error falls between 3 and 5 m, which makes it the second worse performing after SVR. DNN δ_1 has performed better. The peak of its distribution error samples falls between zero and two meters with a mean error of 1.6 m. A detailed result is provided in Table 4.

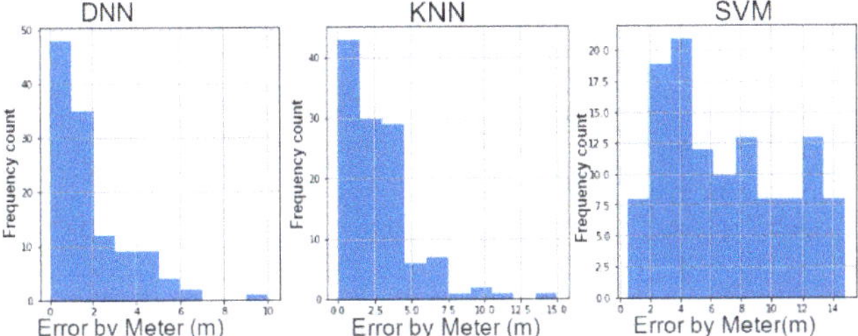

Figure 14. δ_1 vs. K-Nearest Neighbor (KNN) vs. Support Vector Regression (SVR).

Table 4. Frequency Count of Distance error (m) for each model.

	DNN	KNN	SVM
Less Than 2 m	79	51	9
Between 2 m and 7 m	39	64	60
More than 7 m	2	5	51

5.1.2. δ_2s. KNN and SVM

Using the aforementioned samples, the z layer (z coordinate) has been estimated. The results are depicted in Figure 15 illustrating a visual comparison of each classifier in a bar-chart using misclassification count as a measure. Each model has been given an equal number of three classes 0.25, 1.25 and 1.75 m. At first glance, Figure 15 shows that Support Vector Classifier (SVC) has performed very badly in terms of classification of observations. The model has failed to accurately classify during the online phase. More than 66%—circa 120 samples—have been wrongly classified. With a total of 40 misclassified samples, K-Nearest Neighbor (KNN) has performed better than SVC but still does not differentiate between certain classes properly. Our proposed δ_2 model of DNN, has made excellent classification compared to both later models. As an effect, 100% of the 0.25 m layer has been accurately classified while more than 95% of the other two classes, 1.25 and 1.75 m, have also been properly predicted. The total number of misclassified samples is 20 bringing the classification accuracy rate to 89%. This shows how the proposed 3-D multi-layered model has outperformed the traditional models.

Table 5 gives a detailed count of each model and its misclassification count. The worse performing model is highlighted in red and the best performing model is highlighted in blue.

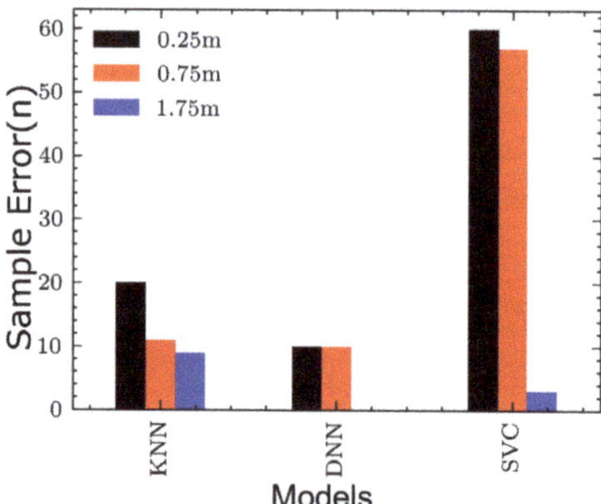

Figure 15. Model comparison: δ_2 vs. KNN vs. SVC.

Table 5. Misclassification count for each model.

Model	0.25 m	0.75 m	1.75 m
DNN	10	10	0
KNN	20	11	9
SVC	60	57	3

6. Conclusions

In this work, we have proposed a novel approach for 3D Indoor Localization using DNN cooperative network algorithms implemented on 3D multi-layer radiomaps. To emulate a 5G infrastructure IoT indoor scenario, an IoT network is interconnected to an experimental 5G C-RAN. Using only an offline fingerprint database, we have also demonstrated how the proposed model has outperformed traditional industry models such as KNN. We have accurately implemented this model to the indoor environment. If the steps shown in Figures 1 and 7 are properly followed, a reliable and fast 3D localization can be achieved. This concept can also be further developed to cover more complex indoor positioning scenarios, involving radio data from a heterogeneous network (HetNEt) such as 5G microinfrastructure (Microcells, femtocell, picocells, etc.). Finally, the proposed DELTA model works very well with RSS based IoT and Wireless Sensor Networks. Thus, our future work will be improving the model by including information fused from other networks such as WiFi and Bluetooth Low Energy (BLE) and experimenting with more vertical layers. Another research direction could be adding floor level detection for buildings with multiple floors.

Author Contributions: All authors contributed equally to this work. All authors have read and agreed to the published version of the manuscript.

Funding: This work is supported by the HORIZON2020 H2020-MSCA-RISE-2016 European Framework Program, the SONNET project Consortium, Sigint Solutions Ltd., and London South Bank University. Self-Organization toward reduced cost and energy per bit for future Emerging radio Technologies with contract number 734545 Program.

Conflicts of Interest: The authors declare no conflict of interest.

Abbreviations

The following abbreviations are used in this manuscript:

3/2D	3/2 Dimensions
5G	5th Generation
AoA	Angle of Arrival
ADAM	ADAptive Momentum
AP	Access Points
BLE	Bluetooth Low Energy
C-RAN	Cloud-Radio Access Network
CAE	Convolutional AutoEnconde
CD-1	Contrastive Divergence with one-step iteration
CNN	Convolutional Neural Networks
D2D	device-to-device
DBN	Deep Belief Network
DELTA	DEep Learning cooperaTive Architecture
DNN	Deep Neural Network
DoA	Direction of arrival
GNSS	Global Navigation Satellite System
GPS	Global Positioning System
HetNet	Heterogeneous Network
IoT	Internet of Things
KNN	K-Nearest Neighbor
LOS	Line-Of-Sight
MED	Mean Euclidean Distance
MDPI	Multidisciplinary Digital Publishing Institute
MLP	Multiple Layers Perceptron
mm-Wave	Millimeter Wave
PPDR	Public Protection and Disaster Recovery
PRR	Packet Reception Ratio
RP	Reference Point
RSS	Received Signal Strength
RSSI	Received Signal Strength Indicator
RTI	RSS Temporal Image
RTOF	Return Time of Flight
SNR	Signal-to-Noise-Ratio
SVC	Support Vector Classification
SVM	Support Vector Machine
SVR	Support Vector Regression
TDOA	Time Difference of Arrival
TOA	Time of Arrival
ToF	Time of Flight
V2X	Vehicle-to-Everything
VLC	Visible Light communication
WSN	Wireless Sensors Networks

References

1. Li, S.; Da Xu, L.; Zhao, S. 5G Internet of Things: A survey. *J. Ind. Inf. Integr.* **2018**, *10*, 1–9. [CrossRef]
2. Chen, M. Towards smart city: M2M communications with software agent intelligence. *Multimed. Tools Appl.* **2013**, *67*, 167–178. [CrossRef]
3. Zhang, D.; Yang, L.; Chen, M.; Zhao, S.; Guo, M.; Zhang, Y. Real-Time Locating Systems Using Active RFID for Internet of Things. *Syst. J. IEEE* **2014**, *10*, 1–10. [CrossRef]
4. Ji, H.; Xie, L.; Wang, C.; Yin, Y.; Lu, S. CrowdSensing: A crowd-sourcing based indoor navigation using RFID-based delay tolerant network. *J. Netw. Comput. Appl.* **2015**, *52*, 79–89. [CrossRef]

5. Macagnano, D.; Destino, G.; Abreu, G. Indoor positioning: A key enabling technology for IoT applications. In Proceedings of the 2014 IEEE World Forum on Internet of Things (WF-IoT), Seoul, Korea, 6–8 March 2014; pp. 117–118. [CrossRef]

6. Yang, C.; Li, J. *Interference Mitigation and Energy Management in 5G Heterogeneous Cellular Networks*; IGI Global: Hershey, PA, USA, 2016.

7. Cisco Visual Networking Index: Global Mobile Data Traffic Forecast Update, 2017–2022. Available online: https://davidellis.ca/wp-content/uploads/2019/12/cisco-vni-mobile-data-traffic-feb-2019.pdf (accessed on 28 July 2020).

8. Liu, Y.; Shi, X.; He, S.; Shi, Z. Prospective positioning architecture and technologies in 5G networks. *IEEE Netw.* **2017**, *31*, 115–121. [CrossRef]

9. Wymeersch, H.; Seco-Granados, G.; Destino, G.; Dardari, D.; Tufvesson, F. 5G mmWave positioning for vehicular networks. *IEEE Wirel. Commun.* **2017**, *24*, 80–86. [CrossRef]

10. Bartoletti, S.; Conti, A.; Dardari, D.; Giorgetti, A. 5G Localization and Context-Awareness. Available online: https://www.5gitaly.eu/2018/wp-content/uploads/2019/01/5G-Italy-White-eBook-5G-Localization.pdf (accessed on 28 July 2020).

11. López-Pérez, D.; Garcia-Rodriguez, A.; Galati-Giordano, L.; Kasslin, M.; Doppler, K. IEEE 802.11 be Extremely High Throughput: The Next Generation of Wi-Fi Technology Beyond 802.11 ax. *IEEE Commun. Mag.* **2019**, *57*, 113–119. [CrossRef]

12. Leonardo, L.; Yuhei, N.; Kurosaki, M.; Ochi, H. High Precision Localization Protocol with Diversity for 802.11 az. *IEICE Tech. Rep. IEICE Tech. Rep.* **2017**, *117*, 69–74.

13. Jaulin, L. 5-Instantaneous Localization. In *Mobile Robotics*; Jaulin, L., Ed.; Elsevier: Amsterdam, The Netherlands, 2015; pp. 171–196. [CrossRef]

14. Baik, K.J.; Lee, S.; Jang, B.J. Hybrid RSSI-AoA Positioning System with Single Time-Modulated Array Receiver for LoRa IoT. In Proceedings of the 2018 48th European Microwave Conference (EuMC), Madrid, Spain, 23–27 September 2018; pp. 1133–1136.

15. Zafari, F.; Gkelias, A.; Leung, K.K. A survey of indoor localization systems and technologies. *IEEE Commun. Surv. Tutor.* **2019**. [CrossRef]

16. McElroy, C.; Neirynck, D.; McLaughlin, M. Comparison of wireless clock synchronization algorithms for indoor location systems. In Proceedings of the 2014 IEEE International Conference on Communications Workshops (ICC), Sydney, Australia, 10–14 June 2014; pp. 157–162.

17. Boukerche, A.; Villas, L.A.; Guidoni, D.L.; Maia, G.; Cunha, F.D.; Ueyama, J.; Loureiro, A.A. A new solution for the time-space localization problem in wireless sensor network using uav. In Proceedings of the third ACM International Symposium on Design and Analysis of Intelligent Vehicular Networks and Applications, Barćelona, Spain, 3–8 November 2013; pp. 153–160.

18. Bae, Y. Robust Localization for Robot and IoT Using RSSI. *Energies* **2019**, *12*, 2212. [CrossRef]

19. Hilal, A.; Khalil, M.; Salman, A.; El-Tawab, S. Exploring the Use of IoT and WiFi-enabled Devices to Improve Fingerprinting in Indoor Localization. In Proceedings of the 2019 IEEE Global Conference on Internet of Things (GCIoT), Dubai, UAE, 4–7 December 2019; pp. 1–6.

20. Ali, M.U.; Hur, S.; Park, Y. Wi-Fi-based effortless indoor positioning system using IoT sensors. *Sensors* **2019**, *19*, 1496. [CrossRef]

21. Njima, W.; Ahriz, I.; Zayani, R.; Terre, M.; Bouallegue, R. Comparison of similarity approaches for indoor localization. In Proceedings of the 2017 IEEE 13th International Conference on Wireless and Mobile Computing, Networking and Communications (WiMob), Rome, Italy, 9–11 October 2017; pp. 349–354.

22. Ding, G.; Zhang, J.; Tan, Z. Overview of received signal strength based fingerprinting localization in indoor wireless LAN environments. In Proceedings of the 2013 5th IEEE International Symposium on Microwave, Antenna, Propagation and EMC Technologies for Wireless Communications, Chengdu, China, 29–31 October 2013; pp. 160–164.

23. Abualsaud, K.; Elfouly, T.M.; Khattab, T.; Yaacoub, E.; Ismail, L.S.; Ahmed, M.H.; Guizani, M. A survey on mobile crowd-sensing and its applications in the IoT era. *IEEE Access* **2018**, *7*, 3855–3881. [CrossRef]

24. Li, D.; Lei, Y.; Li, X.; Zhang, H. Deep Learning for Fingerprint Localization in Indoor and Outdoor Environments. *ISPRS Int. J. Geo-Inf.* **2020**, *9*, 267. [CrossRef]

25. Passafiume, M.; Maddio, S.; Collodi, G.; Cidronali, A. An enhanced algorithm for 2D indoor localization on single anchor RSSI-based positioning systems. In Proceedings of the 2017 European Radar Conference (EURAD), Nuremberg, Germany, 11–13 October 2017; pp. 287–290.
26. Kim, K.S.; Lee, S.; Huang, K. A scalable deep neural network architecture for multi-building and multi-floor indoor localization based on Wi-Fi fingerprinting. *Big Data Anal.* **2018**, *3*, 4. [CrossRef]
27. Njima, W.; Ahriz, I.; Zayani, R.; Terre, M.; Bouallegue, R. Deep CNN for Indoor Localization in IoT-Sensor Systems. *Sensors* **2019**, *19*, 3127. [CrossRef]
28. Shao, W.; Luo, H.; Zhao, F.; Ma, Y.; Zhao, Z.; Crivello, A. Indoor positioning based on fingerprint-image and deep learning. *IEEE Access* **2018**, *6*, 74699–74712. [CrossRef]
29. Jiang, H.; Peng, C.; Sun, J. Deep Belief Network for Fingerprinting-Based RFID Indoor Localization. In Proceedings of the ICC 2019-2019 IEEE International Conference on Communications (ICC), Shanghai, China, 20–24 May 2019; pp. 1–5.
30. Wang, Z.; Zhang, X.; Wang, W.; Shi, L.; Huang, C.; Wang, J.; Zhang, Y. Deep Convolutional Auto-Encoder based Indoor Light Positioning Using RSS Temporal Image. In Proceedings of the 2019 IEEE International Symposium on Broadband Multimedia Systems and Broadcasting (BMSB), Jeju, Korea, 5–7 June 2019; pp. 1–5.
31. Tran, H.; Ha, C. Fingerprint-Based Indoor Positioning System Using Visible Light Communication—A Novel Method for Multipath Reflections. *Electronics* **2019**, *8*, 63. [CrossRef]
32. Kanaris, L.; Kokkinis, A.; Liotta, A.; Stavrou, S. Combining smart lighting and radio fingerprinting for improved indoor localization. In Proceedings of the 2017 IEEE 14th International Conference on Networking, Sensing and Control (ICNSC), Calabria, Italy, 16–18 May 2017; pp. 447–452.
33. Janssen, T.; Berkvens, R.; Weyn, M. Comparing Machine Learning Algorithms for RSS-Based Localization in LPWAN. In *International Conference on P2P, Parallel, Grid, Cloud and Internet Computing*; Springer: Berlin/Heidelberg, Germany, 2019; pp. 726–735.
34. Sinha, R.S.; Hwang, S.H. Comparison of CNN applications for RSSI-based fingerprint indoor localization. *Electronics* **2019**, *8*, 989. [CrossRef]
35. Kaemarungsi, K. Efficient design of indoor positioning systems based on location fingerprinting. In Proceedings of the 2005 International Conference on Wireless Networks, Communications and Mobile Computing, Maui, HI, USA, 13–16 June 2005; Volume 1, pp. 181–186.
36. GNS3. Getting Started with GNS3 | GNS3 Documentation. Available online: https://docs.gns3.com/docs/ (accessed on 28 July 2020).
37. OpenDaylight. Documentation Guide—OpenDaylight Documentation Magnesium Documentation. 2020. Available online: https://docs.opendaylight.org/en/stable-magnesium/documentation.html (accessed on 28 July 2020).
38. Fractal Networx Limited. TruNET Wireless. 2017. Available online: www.fractalnetworx.com (accessed on 28 July 2020).
39. Kanaris, L.; Kokkinis, A.; Fortino, G.; Liotta, A.; Stavrou, S. Sample Size Determination Algorithm for fingerprint-based indoor localization systems. *Comput. Netw.* **2016**, *101*, 169–177. [CrossRef]
40. Kanaris, L.; Kokkinis, A.; Liotta, A.; Stavrou, S. Quality of Fingerprint Radiomaps for Positioning Systems. In Proceedings of the 2017 24th International Conference on Telecommunications (ICT) (ICT 2017), Limassol, Cyprus, 3–5 May 2017.
41. Baccour, N.; Koubâa, A.; Mottola, L.; Zúñiga, M.A.; Youssef, H.; Boano, C.A.; Alves, M. Radio link quality estimation in wireless sensor networks: A survey. *ACM Trans. Sens. Netw. (TOSN)* **2012**, *8*, 1–33. [CrossRef]
42. Kanaris, L.; Sergiou, C.; Kokkinis, A.; Pafitis, A.; Antoniou, N.; Stavrou, S. On the Realistic Radio and Network Planning of IoT Sensor Networks. *Sensors* **2019**, *19*, 3264. [CrossRef]
43. LeCun, Y.; Bengio, Y.; Hinton, G. Deep learning. *Nature* **2015**, *521*, 436. [CrossRef] [PubMed]
44. Goodfellow, I.; Bengio, Y.; Courville, A. *Deep Learning*; MIT Press: Cambridge, MA, USA; London, UK, 2016.
45. Chollet, F. *Deep Learning with Python*, 1st ed.; Manning Publications Co.: Greenwich, CT, USA, 2017.
46. Wye, K.F.P.; Kanagaraj, E.; Zakaria, S.M.M.S.; Kamarudin, L.M.; Zakaria, A.; Kamarudin, K.; Ahmad, N. RSSI-based Localization Zoning using K-Mean Clustering. In *IOP Conference Series: Materials Science and Engineering*; IOP Publishing: Bristol, UK, 2019.
47. Ioffe, S.; Szegedy, C. Batch normalization: Accelerating deep network training by reducing internal covariate shift. *arXiv* **2015**, arXiv:1502.03167.

48. Srivastava, N.; Hinton, G.; Krizhevsky, A.; Sutskever, I.; Salakhutdinov, R. Dropout: A simple way to prevent neural networks from overfitting. *J. Mach. Learn. Res.* **2014**, *15*, 1929–1958.

49. Kingma, D.P.; Ba, J. Adam: A method for stochastic optimization. *arXiv* **2014**, arXiv:1412.6980.

50. Scikit-learn. 1.4. Support Vector Machines—Scikit-Learn 0.23.2 Documentation. Available online: https://scikit-learn.org/stable/modules/svm.html (accessed on 8 December 2019).

51. Scikit-learn. Sklearn.Neighbors.KNeighborsClassifier—Scikit-Learn 0.23.2 Documentation. Available online: https://scikit-learn.org/stable/modules/generated/sklearn.neighbors.KNeighborsClassifier.html (accessed on 8 December 2019).

52. Keras. Keras API Reference. Available online: https://keras.io/api/ (accessed on 8 December 2019).

Article

A Novel Hybrid Algorithm Based on Grey Wolf Optimizer and Fireworks Algorithm

Zhihang Yue [1,2]**, Sen Zhang** [1,2,*] **and Wendong Xiao** [1,2]

[1]   School of Automation and Electrical Engineering, University of Science and Technology Beijing, Beijing 100083, China; zhyue0101@163.com (Z.Y.); wdxiao@ustb.edu.cn (W.X.)

[2]   Key Laboratory of Knowledge Automation for Industrial Processes of Ministry of Education, School of Automation and Electrical Engineering, University of Science and Technology Beijing, Beijing 100083, China

*   Correspondence: zhangsen@ustb.edu.cn

Received: 17 February 2020; Accepted: 5 April 2020; Published: 10 April 2020

Abstract: Grey wolf optimizer (GWO) is a meta-heuristic algorithm inspired by the hierarchy of grey wolves (Canis lupus). Fireworks algorithm (FWA) is a nature-inspired optimization method mimicking the explosion process of fireworks for optimization problems. Both of them have a strong optimal search capability. However, in some cases, GWO converges to the local optimum and FWA converges slowly. In this paper, a new hybrid algorithm (named as FWGWO) is proposed, which fuses the advantages of these two algorithms to achieve global optima effectively. The proposed algorithm combines the exploration ability of the fireworks algorithm with the exploitation ability of the grey wolf optimizer (GWO) by setting a balance coefficient. In order to test the competence of the proposed hybrid FWGWO, 16 well-known benchmark functions having a wide range of dimensions and varied complexities are used in this paper. The results of the proposed FWGWO are compared to nine other algorithms, including the standard FWA, the native GWO, enhanced grey wolf optimizer (EGWO), and augmented grey wolf optimizer (AGWO). The experimental results show that the FWGWO effectively improves the global optimal search capability and convergence speed of the GWO and FWA.

Keywords: Grey Wolf Optimizer; Fireworks Algorithm; hybrid algorithm; exploitation and exploration

---

1. Introduction

Finding an optimal solution in high-dimensional complex space is a common issue in many engineering fields [1]. When solving such problems, deterministic algorithms find it difficult to find the optimal value, the calculation cost is too high, and the calculation time is too long [2]. The meta-heuristic optimization algorithm has, due to its simplicity and strong searching ability, been widely used in solving complex problems. In recent years, scholars around the world have done a lot of research on these algorithms. Many natural-inspired meta-heuristic algorithms have been proposed, such as Particle Swarm Optimization (PSO) [3], Ant Colony Optimization (ACO) [4], Artificial bee colony algorithm (ABC) [5], Whale Optimization Algorithm (WOA) [6], Bird Swarm Algorithm (BSA) [7], Grey Wolf Optimizer (GWO) [8], Fireworks Algorithm (FWA) [9], Biogeography-based optimization (BBO) [10], and Moth Flame Optimization (MFO) [11].

Inspired by the leadership hierarchy and the mechanism of hunting of grey wolves, Gray Wolf Optimizer, a new meta-heuristic optimization algorithm, was proposed by Mirjalili in 2014 [8]. The GWO algorithm mimics the hunting mechanism to search the optima. In [8], several benchmark functions are used to evaluate the performance of GWO. The experimental results show that the GWO algorithm is feasible and superior to PSO, Gravitational Search Algorithm (GSA) [12], and Differential Evolution (DE) [13] in the accuracy of the solution and convergence speed. Due to the advantages of fewer adjustment parameters and a faster convergence, the GWO algorithm has been applied in a series of engineering problems such as the speed control of DC motors [14], parameter estimation in surface waves [15], and load frequency control of the Multi-microgrid System [16]. However, when facing problems with multi optimal solutions like multidimensional feature selection, GWO converges to local optima and fails to find the global optimal solution. Inspired by observing fireworks explosions, Tan and Zhu proposed the fireworks algorithm (FWA) in 2010. FWA mimics the explosion of fireworks that produce sparks and illuminate the surrounding area. The global optimal solution was found by controlling the amplitude of the explosion and the number of sparks generated by explosion. Several benchmark functions are employed to test the competence of FWA. The experimental results [9] show that FWA has a better solution accuracy and faster convergence speed than SPSO (Standard particle swarm optimization) [17] and CPSO (Clonal particle swarm optimization) [18].

When searching for the optimal value in a high-dimensional space, the single meta-heuristic algorithm always has some disadvantages, such as a low accuracy, poor generalization ability, and poor local optima avoidance ability. The hybrid algorithm utilizes the differences between the two optimization algorithms, combines their advantages, and makes up for shortcomings to improve the overall performance of solving complex optimization problems [19]. The solution searched for by the genetic algorithm (GA) [20] is not accurate enough, and the evolutionary strategy (ES) tends to fall into a local minimum when searching for the optimal value. Therefore, a hybrid algorithm in [21] combined GA and ES to make up for these shortcomings. This algorithm is used for electromagnetic optimization problems and achieved satisfactory results. In [22], two hybrid models were established by Mafarja to design feature selection techniques based on WOA. In the first model, the simulated annealing (SA) algorithm is inserted into the WOA algorithm. In the second model, SA and WOA are used separately. SA is used to exploit the search space found by the WOA algorithm. In [23], Alomoush proposed a hybrid algorithm based on Gray Wolf Optimizer (GWO) and Harmony Search (HS). GWO is used to update the pitch adjustment rate and bandwidth in the HS to improve the global optimization capabilities of the hybrid algorithm. In [24], the grey wolf optimizer and crow search algorithm are combined by Sankalap Arora. The hybrid algorithm is tested on 21 data sets as a feature selection technique. The results show that the algorithm has many advantages in solving complex optimization problems. In [25], biogeography-based optimization (BBO) and differential evolution (DE) are fused by sharing population information. In BBO/DE, the migration operators in BBO are changed as the increase of the iteration count. The results indicate that the hybrid algorithm is more effective when compared to traditional BBO and DE. In [26], a novel hybrid algorithm combining grey wolf optimizer (GWO) with particle swarm optimization (PSO) is used as a load balancing technique in the cloud computing environment. The conclusions indicate that the hybrid algorithm improves the convergence speed and simplicity when compared with other algorithms. In [27], ZHU proposed a novel hybrid algorithm based on the grey wolf optimizer (GWO) and differential evolution (DE). This algorithm is tested on 23 benchmark functions and a non-deterministic polynomial hard problem. The experimental results show that this algorithm has a good performance in exploration. A global optimization algorithm combining biogeography-based optimization (BBO) with fireworks algorithm (FWA) is proposed in [28]. The BBO migration operators have been inserted into the FWA to enhance the information exchange between the population and to improve the global optimization capability. In [29], WOA-SCA, composed of the whale optimization algorithm (WOA) and sine cosine algorithm (SCA) is proposed. SCA and WOA are used for exploration and exploitation, respectively. The WOA-SCA optimization algorithm is used to distribute DGs (distributed generators) and DSTATCOMs (distribution flexible alternating current transmission devices) to enhance the voltage profile by minimizing the total power losses in the system.

As mentioned above, GWO advances itself strongly on exploitation. However, in some cases it converges prematurely and gets trapped in a local optimum. FWA not only has a high exploration capability, but also has the disadvantage of a slow convergence speed compared to the recent algorithms. This paper proposes a new hybrid algorithm which combines the exploration ability of FWA with the exploitation ability of GWO to increase the convergence characteristics.

The major work of this paper is summarized as follows:

(1)   A novel hybrid algorithm based on GWO and FWA is proposed.
(2)   The proposed algorithm is tested on 16 benchmark functions with a wide range of dimensions and varied complexities.
(3)   The performance of the proposed approach is compared with standard GWO, FWA, Moth Flame Optimization (MFO), Crow Search Algorithm (CSA) [30], improved Particle Swarm Optimization(IPSO) [31], Biogeography-based optimization (BBO), Particle Swarm Optimization (PSO), Enhanced GWO (EGWO) [32], and Augmented GWO (AGWO) [33].

The rest of paper is arranged as follows: Section 2 introduces the GWO algorithm and FWA algorithm used in this paper. In Section 4, the FWGWO hybrid algorithm is proposed. Section 5 shows the experimental results and comparison of the algorithms used in the test function. Finally, the conclusions are given in Section 6.

2. Algorithms

2.1. Grey Wolf Optimizer

The Grey Wolf Optimizer (GWO) is a meta-heuristic algorithm proposed by Mirgalili et al. [8] in 2014. The GWO algorithm mimics the hunting mechanism and the leadership hierarchy of wolves to search the optima. Grey wolves are social animals, with an average of 5 to 12 wolves in each group. They also have a strict hierarchy. There are four levels in wolves' hierarchy, called alpha (α), beta (β), delta (δ), and omega (ω). Alpha is the first level. It is responsible for making decisions like hunting, finding a place to sleep, the waking time, and so on. The second level is alpha's candidate, the beta wolves, which help alpha in making decisions or in engaging in other activities. The third level is delta. The delta wolves are under the command of the first two levels, mainly responsible for reconnaissance, sentry, guard, and other tasks. The last level of a pack is omega. Omega wolves have to submit to the wolves in the first three levels. The omega wolves maintain the integrity of the hierarchical structure. The mathematical model of the GWO algorithm's hierarchy and hunting behavior is as follows:

2.1.1. Hierarchical Structure

The GWO algorithm is a mathematical model based on the social hierarchy of wolves. The fittest solution found is considered as the alpha (α). The second and third best solutions are beta (β) and delta (δ), respectively. The rest of the solutions are omega (ω). In the GWO algorithm, alpha, beta, and delta collectively command omega to search for the solution space.

2.1.2. Encircling Prey

The grey wolf encircles the prey when hunting. The mathematical equations for this behavior are shown in Equations (1) and (2):

$$\vec{D} = \left| \vec{C} \cdot \vec{X_P}(t) - \vec{X}(t) \right| \tag{1}$$

$$\vec{X}(t+1) = \vec{X_P}(t) - \vec{A} \cdot \vec{D} \tag{2}$$

where $\vec{D}$ is the distance from the wolf to the prey. $\vec{X}$ and $\vec{X_p}$ represent the position vector of the grey wolf and the position vector of the prey, respectively. t indicates the current iteration. $\vec{A}$ and $\vec{C}$ are coefficient vectors and are calculated as follows:

$$\vec{A} = 2\vec{a} \cdot \vec{r_1} - \vec{a} \tag{3}$$

$$\vec{C} = 2 \cdot \vec{r_2} \tag{4}$$

where $\vec{a}$ decreases linearly with the number of iterations from 2 to 0. $\vec{r_1}$ and $\vec{r_2}$ are the random vectors in [0,1].

2.1.3. Hunting

The grey wolves can easily encircle the prey with the ability to recognize its location. The whole hunting process is usually led by the alpha. However, in a complex search space, it is impossible to get the location of the prey at the beginning. Therefore, GWO consider that the first three best solutions, alpha, beta, and delta, have more information about the location of the prey. Then, the other wolves update their positions based on these three positions, as shown in Figure 1.

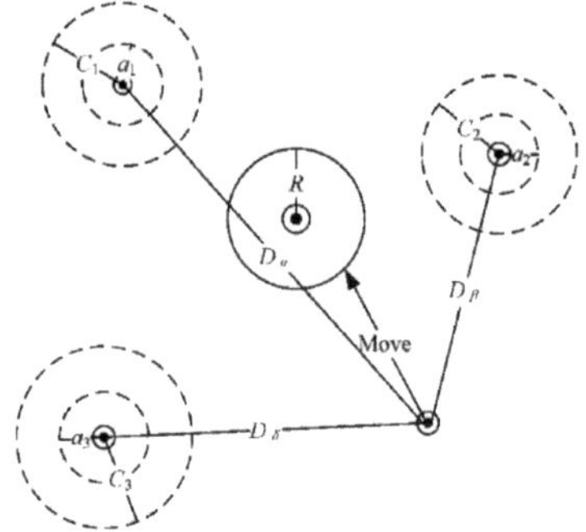

Figure 1. Position updating in GWO.

The mathematical equations of this phase are as follows:

$$\vec{D}_\alpha = \left| \vec{C_1}{\cdot}\vec{X}_\alpha - \vec{X} \right|, \vec{D}_\beta = \left| \vec{C_2}{\cdot}\vec{X}_\beta - \vec{X} \right|, \vec{D}_\delta = \left| \vec{C_3}{\cdot}\vec{X}_\delta - \vec{X} \right| \tag{5}$$

$$\vec{X}_1 = \vec{X}_\alpha - \vec{A}_1{\cdot}(\vec{D}_\alpha), \vec{X}_2 = \vec{X}_\beta - \vec{A}_2{\cdot}(\vec{D}_\beta), \vec{X}_3 = \vec{X}_\delta - \vec{A}_3{\cdot}(\vec{D}_\delta) \tag{6}$$

$$\vec{X}_{(t+1)} = \frac{\vec{X}_1 + \vec{X}_2 + \vec{X}_3}{3} \tag{7}$$

where, $\vec{X}_\alpha$, $\vec{X}_\beta$, and $\vec{X}_\delta$ are the position vectors of alpha, beta, and delta; $\vec{D}_\alpha$, $\vec{D}_\beta$, and $\vec{D}_\delta$ represent the distances from the search agent to alpha, beta, and delta respectively; $\vec{C}_1$, $\vec{C}_2$, and $\vec{C}_3$ are coefficient vectors; and $\vec{X}_1$, $\vec{X}_2$, and $\vec{X}_3$ are the step lengths in the direction toward alpha, beta, and delta. $\vec{X}$ and $\vec{X}_{(t+1)}$ indicate the position of the search agent before and after the update.

2.1.4. Search for Prey (Exploration) and Attacking Prey (Exploitation)

When the prey stops moving, the grey wolf completes the hunting process by attacking. In order to mimic the process of the grey wolf approaching the prey, the GWO algorithm causes $\vec{a}$ to linearly decrease from 2 to 0, as shown in Equation (8):

$$\vec{a} = 2 - (2 \times t/Max_{iter}) \tag{8}$$

According to Equation (3), $\vec{A}$ is a random value that lies in the range [−2a, 2a]. GWO uses $\vec{A}$ to force wolves to move closer or farther away from their prey. If $\vec{A} < 1$, the wolf will be forced to attack towards the prey. If $\vec{A} > 1$, the wolf will be forced to diverge from the prey (local minimum) to find a new fitter prey. $\vec{C}$ is a random value that lies in the range [0,2] which is employed to help GWO avoid being trapped in the local optima. The pseudo code of the GWO algorithm is presented in Algorithm 1.

Algorithm 1 Pseudo Code of GWO

1.Initialize the wolf population $X_i (i = 1, 2, \ldots, n)$
2.Initialize $a, A,$ and C
3. Calculate the fitness of each search agent
4. $X_\alpha =$ the best search agent
5. $X_\beta =$ the sec ond best search agent
6. $X_\delta =$ the third best search agent
7. **while** $(t <$ Max number of iterations)
8.   **for** each search agent
9.   Update the position of the current search agent by equation(7)
10.   **end for**
11.   Update$a, A,$ and$C$
12.   Calculate the fitness of all search agents
13.   Update$X_\alpha, X_\beta,$ and X_δ
14. $t = t + 1$
15. **end while**
16. **return** X_α

---

3. Fireworks Algorithm

Inspired by the behavior according to which fireworks burst into sparks in the night sky and illuminate the surrounding area, Tan and Zhu proposed the Fireworks Algorithm (FWA) in 2010. In FWA, a firework is considered as a viable solution in the search space of the optimization problem. Furthermore, the process of sparking fireworks is deemed as the process of searching the neighborhood of these fireworks. The specific steps of the fireworks algorithm are as follows:

(1)   N fireworks are randomly generated in the search space, and each firework represents a feasible solution.

(2)   Evaluate the quality of these fireworks. Consider that the optimal location may be close to the fireworks with a better fitness; these fireworks get a smaller search amplitude and more explosion sparks to search the surrounding area. On the contrary, those fireworks with a bad fitness will get fewer explosion sparks and a larger search amplitude. The number of explosion sparks and the explosion amplitude of fireworks are calculated as shown in Equations (9) and (10):

$$S_i = m \cdot \frac{y_{\max} - f(x_i) + \xi}{\sum_{i=1}^{n} (y_{\max} - f(x_i)) + \xi} \tag{9}$$

$$A_i = \hat{A} \cdot \frac{f(x_i) - y_{\min} + \xi}{\sum_{i=1}^{n} (f(x_i) - y_{\min}) + \xi} \tag{10}$$

where S_i is the number of explosion sparks. A_i represents the amplitude of the explosion. $y_{\min} = \min(f(x_i)), (i = 1, 2, \ldots, N)$ is the minimum fitness among the N fireworks. $y_{\max} = \max(f(x_i)), (i = 1, 2, \ldots, N)$ is the maximum fitness value among the N fireworks. m and $\hat{A}$ are parameters controlling the total number of sparks and the maximum explosion amplitude, respectively. ξ is the smallest constant in the computer, utilized to avoid a zero-division-error.

To avoid that the good fireworks produce far more explosive sparks than the fireworks with a poor fitness, Equation (11) is used to bound the number of sparks that are generated:

$$\hat{S}_i = \begin{cases} round(a{\cdot}m) & \text{if } S_i < am \\ round(b{\cdot}m) & \text{if } S_i > bm, a < b < 1 \\ round(S_i) & \text{if } S_i \text{ otherwise} \end{cases} \tag{11}$$

where a, b are constant parameters, and $round()$ represents the rounding function.

(3)   To guarantee the diversity of the fireworks, another method of generating sparks, Gaussian explosion, is designed in FWA. For the randomly selected fireworks, a number of dimensions are randomly selected and updated, as shown in Equation (12), to get the position of a Gaussian spark at the dimension k:

$$\hat{x}_{ik} = x_{ik} \times e \tag{12}$$

where $e \sim N(1,1)$, $N(1,1)$ is a Gaussian random value with mean 1 and standard deviation 1. The generated explosion sparks and Gaussian sparks may exceed the boundaries of the search space. Equation (13) maps the sparks beyond the boundary at dimension k to a new position:

$$\tilde{x}_k^j = x_k^{\min} + \left| \tilde{x}_k^j \right| \% (x_k^{\max} - x_k^{\min}) \tag{13}$$

where $x_k^{\max}$ represents the upper bound of the search space at dimension k, and $x_k^{\min}$ is the lower bound of the search space at dimension k.

(4) N locations should be selected as the next generation of fireworks from the explosion sparks, Gaussian sparks, and the current fireworks. In FWA, the location with the best fitness is always kept for the next iteration. Then, $N - 1$ locations are chosen, determined by their distance to other locations. The distance and the selection probability of x_i is defined as follows:

$$R(x_i) = \sum_{j \in k} d(x_i, x_j) = \sum_{j \in k} \|x_i - x_j\| \tag{14}$$

$$p(x_i) = \frac{R(x_i)}{\sum\limits_{j \in k} R(x_j)} \tag{15}$$

In this selection strategy, if there are many other locations around x_i, the selection probability will be reduced to keep the diversity of the next generation.

The execution flow of the FWA algorithm is shown in Algorithm 2.

4. Hybrid FWGWO Algorithm

4.1. Establishment of FWGWO

As mentioned in Section 2, GWO is strong at exploitation but weak at avoiding a premature convergence and local optimum. The FWA algorithm has a strong exploration capability, but it lacks in exploitation. In this section, the FWGWO hybrid algorithm is proposed to combine the GWO exploitation capability with the FWA exploration capability to obtain a better global optimization capability. FWGWO alternately uses the FWA algorithm for exploration in the search space and the GWO algorithm for exploitation to search the global optimum without changing the general operation of the GWO and FWA algorithms. In order to balance the exploration with the exploitation, an adaptive equilibrium coefficient is proposed in this paper. Updating the position X_α means that the best fitness has been changed. When the current position is closer to the optimal solution, the coefficient p will be updated to change the search strategy, as shown in Equation (16):

$$p = 0.9 \times (1 - \cos(\frac{\pi}{2} \cdot \frac{t}{Max_{iter}}))$$ (16)

where p represents the adaptive balance coefficient. t is the current number of iterations. Max_{iter} indicates the maximum number of iterations.

Algorithm 2 Pseudo Code of FWA

---

1. Randomly select n locations for fireworks
2. **while** stop criteria = flase **do**
3.   Set off n fireworks respectively at the n locations
4.   **for**each fireworkx_i**do**
5.   Calculate the number of sparks that the firework yields by equation (9)
6.   Obtain locations of each sparks of the fireworkx_i
7.   **end for**
8. **for** $k = 1$: number of Gaussian sparks **do**
9.   Randomly select a fireworkx_j
10.   Generate a Gaussian spark for the firework
11. **end for**
12. Select the best location and keep it for next explosion generation
13. Randomly select n − 1 locations from the two types of sparks and the current fireworks
14. according to the probability given in equation (15)
15. **end while**

A random value r in [0, 1] is set for a comparison with the adaptive balance coefficient p. If $r > p$, the next iteration will be executed using the FWA algorithm. Otherwise, the GWO algorithm is used for this iteration. The function curve of p is shown in Figure 2. The value of p is small and slowly increases in the early optimization stage to make sure that FWGWO explores a huge search space by multiple calls of FWA to avoid being trapped in the local minimum. In the later optimization phases, the algorithm exploits small regions to efficiently search the optimum with the rapidly increasing p. To avoid that only GWO is executed in the final stage of the FWGWO algorithm, the value of p is growing in [0, 0.9]. In some cases, after one iteration of the FWA algorithm, the FWGWO algorithm will proceed to the next iteration of the FWA algorithm without further exploitation so as to escape the current local optimal space and miss the global optimal solution. To avoid these cases, the FWGWO algorithm exploits the current region with at least T iterations of the GWO algorithm before proceeding to the next FWA algorithm. In this paper, T is set to 10. The variable k is defined to count how many GWO iterations have occurred since the last FWA iteration. k is initialized at the beginning of FWGWO. After each GWO iteration, k increases itself by 1. If $k > T$ and $r > p$, FWA will be used to execute the iteration. At the last of these iterations, k will be set to 0.

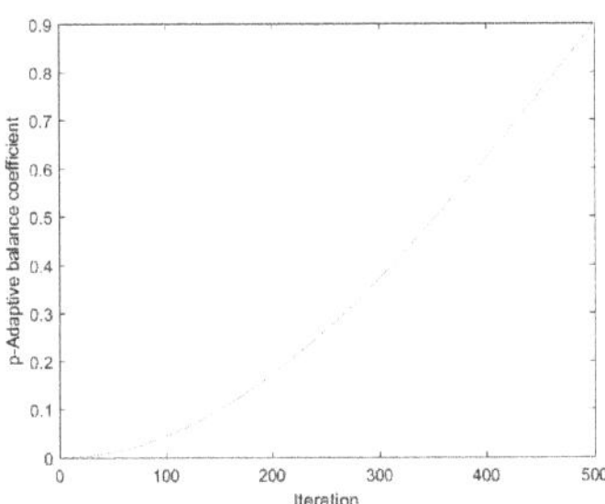

Figure 2. Adaptive balance coefficient.

The pseudo code of the hybrid FWGWO algorithm is shown in Algorithm 3.

Algorithm 3 Pseudo Code of FWGWO

---

1. **Initialize** the wolf population $X_i (i = 1, 2, \ldots, n)$
2. **Initialize** $a,\ A,\ k,\ t$ and C
3. Calculate the fitness of each search agent
4. X_α = the best search agent
5. X_β = the sec ond best search agent
6. X_δ = the third best search agent
7. **while** (t < Max number of iterations)
8.   **foreach** search agent
9.   Update the position of the current search agent by equation (7)
10.   **end for**
11.   Update $a, A,$ and C
12.   Calculate the fitness of all search agents
13.   Update $X_\alpha, X_\beta,$ and X_δ
14.   **if** X_α changed **then**
15.   Update the adaptive balance coefficient (p) by equation (16)
16.   **end if**
17.   **if** $k >= T$ and $rand() > p$ **then**
18.   **for** $each search agent\ x_i$ **do**
19.   Generate explosion sparks
20.   **end for**
21.   Generate m Gaussian spark for the search agents
22.   Selec new search agents by equation (15)
23. $t = t + 1$
24. $k = 0$
25.   **end if**
26. $k = k + 1$ $t = t + 1$
27. **end while**
28. **return** X_α

4.2. Time Complexity Analysis of FWGWO

The time complexity of FWGWO is mainly determined by the population size n, dimensions of the solution space d, and max number of iterations t. O (FWGWO) = O (population initialization) + O (Calculation of fitness for the entire population) + O (Update of population position). The time complexity of the population initialization is O $(n \times d)$, and the time complexity for calculating the fitness of the entire population is O $(t \times n \times d)$. As for the part with the updating position, the time complexity consists of two parts: O = O (position update with GWO) + O (position update with FWA). The time complexity of this part is O $(10/11 \times t \times n \times d + 1/11 \times t \times 3 \times n \times d) \approx$ O$(t \times n \times d)$. Therefore, the time complexity of the FWGWO algorithm is O $(n \times d + t \times n \times d + t \times n \times d) =$ O $((2t + 1) \times n \times d)$. For comparison, the time complexity of GWO is O $((2t + 1) \times n \times d)$, and the time complexity of FWA is O $((4t + 1) \times n \times d)$. The time complexity of FWGWO is the same as that of GWO and smaller than that of FWA.

5. Experimental Section and Results

In this section, 16 benchmark functions with different characteristics are used to evaluate the proposed FWGWO hybrid algorithm. The experiment results are compared with nine other algorithms to verify the superiority of FWGWO.

5.1. Compared Algorithms

A total of nine algorithms, including IPSO, PSO, BBO, CSA, MFO, FWA, GWO, AGWO, and EGWO, were selected for comparison with the FWGWO algorithm proposed in this paper. These nine algorithms contain both classical algorithms and new algorithms proposed in recent years. The parameter settings of these algorithms are shown in Table 1. These parameters are chosen based on the parameters of these algorithms in the original papers. For all algorithms in this experiment, the size of the population is 20, the dimension is 100 dimensions, and the maximum number of iterations is 500. This experiment has been carried out with different experimental parameters. The size of the population has been set to 10, 20, 30, and 50. The dimension has been set to 20, 50, 100, and 500. The results of these experiments are similar. Considering the length of this paper, the typical parameters mentioned above are taken as an example.

Table 1. Parameter settings.

	Parameter	Value(s)
GWO	a	Linearly decreased from 2 to 0
	Total number of sparks	50
	Maximum explosion amplitude	40
FWA	Number of Gaussian sparks	5
	a	0.04
	b	0.8
ISPO	Inertia w(wMin,wMax)	[0.4,0.9]
	Acceleration constants(c1,c2)	[2,2]
PSO	Inertia w(wMin,wMax)	[0.6,0.9]
	Acceleration constants(c1,c2)	[2,2]
AGWO	a	$a = 2 - (\cos(rand()) \times t / Max\_iter)$
EGWO	a	$a = rand()$
	Immigration probability	[0,1]
	Mutation probability	0.005
	Habitat modification probability	1.0
BBO	Step size	1.0
	Migration rate	1.0
	Maximum immigration	1.0
	Flight length	2
CSA	Awareness probability	0.1
MFO	a	$a = -1 + Iteration \times ((-1) / Max\_iteration)$

5.2. Benchmark Functions

Sixteen benchmark functions listed in Table A1 (see in Appendix A) are utilized to evaluate the optimized performance of FWGWO. These benchmark functions can be divided into two categories. In the first category, the unimodal functions f1-f8 with only one global optimum are used to test the exploitation capability. In the second category, the multimodal functions f9–f16 with more than two local optima are utilized to assess the ability of FWGWO to find global optima.

5.3. Performance Metrics

Three performance indicators, the mean fitness, fitness variance, and best fitness, are selected in this paper to evaluate the results of the experiments. The mathematical equations are defined as follows:

$$MeanFitness = \frac{1}{M}\sum_{i=1}^{M} G_i \tag{17}$$

$$std = \sqrt{\frac{\sum_{i=1}^{M} (G_i - MeanFitness)^2}{M}} \tag{18}$$

where M represents the total number of independently repeated experiments. M is 30 in this paper. G represents the function fitness of each experiment, and i is the count of repeated experiments.

In addition, a nonparametric statistical test, Wilcoxon's rank-sum test [34], is utilized to show that the proposed FWGWO algorithm provides a significant improvement over other algorithms. The test was carried out with the results of the FWGWO and other algorithms in each benchmark function at a 5% significance level. All experiments are carried out using MATLAB R2017a on a computer with Inter i5-5200U 2.2GHz and an 8 GB memory.

5.4. Comparison and Analysis of Simulation Results

Thirty independently repeated experiments were implemented using the 10 algorithms mentioned above on each benchmark function. Table 2 shows the test results of these algorithms on each benchmark function. The best, mean, and variance of the fitness obtained by each algorithm are listed in Table 2, and the best results are bolded. Compared with other algorithms, the average of the fitness obtained by the FWGWO algorithm after 500 iterations is better and closer to the global optimal value on most of the benchmark functions. In terms of the best fitness obtained in 30 repetitions, FWGWO has a better performance than other algorithms in all 16 functions. What is more, the best fitness obtained by FWGWO is 0 on functions f9, f11, and f16. This shows that the FWGWO algorithm has found the global optimal value on these functions. These cases verified the superiority of FWGWO over other algorithms.

Table 2. Results of the benchmark functions.

Function		GWO	FWA	IPSO	PSO	AGWO	EGWO	BBO	CSA	MFO	FWGWO
	Mean	1.99×10^{-10}	3.73×10^{2}	1.57×10^{4}	2.21×10^{2}	3.93×10^{-17}	1.45×10^{-13}	2.87×10^{2}	1.06×10^{3}	7.04×10^{4}	$\mathbf{5.16 \times 10^{-18}}$
F1	Std	1.19×10^{-10}	5.49×10^{3}	5.18×10^{3}	3.01×10^{1}	7.20×10^{-17}	6.23×10^{-14}	2.47×10^{1}	1.83×10^{2}	1.35×10^{4}	1.93×10^{-16}
	Best	2.70×10^{-11}	4.84	7.40×10^{3}	1.50×10^{2}	4.61×10^{-18}	6.15×10^{-16}	2.54×10^{2}	7.09×10^{2}	4.69×10^{4}	$\mathbf{6.68 \times 10^{-24}}$
	Mean	1.53×10^{3}	6.14×10^{4}	1.44×10^{5}	2.71×10^{4}	1.79×10^{4}	3.34×10^{4}	5.74×10^{4}	8.49×10^{3}	2.70×10^{5}	$\mathbf{1.74}$
F2	Std	1.35×10^{3}	7.11×10^{4}	3.11×10^{4}	6.72×10^{3}	1.81×10^{4}	1.41×10^{4}	8.78×10^{3}	1.52×10^{3}	5.48×10^{4}	6.01×10^{1}
	Best	1.20×10^{2}	1.32×10^{1}	8.33×10^{4}	1.55×10^{4}	2.54×10^{2}	6.75×10^{3}	4.38×10^{4}	5.54×10^{3}	1.64×10^{5}	$\mathbf{4.97 \times 10^{-7}}$
	Mean	3.03	3.36×10^{1}	6.00×10^{1}	1.54×10^{1}	4.99×10^{1}	7.28×10^{1}	2.99×10^{1}	1.57×10^{1}	9.41×10^{1}	$\mathbf{1.02 \times 10^{-3}}$
F3	Std	2.17	1.44×10^{1}	4.32	1.66	3.13×10^{1}	7.82	3.01	1.40	1.72	3.06×10^{-3}
	Best	2.24×10^{-1}	8.24×10^{-1}	4.84×10^{1}	1.26×10^{1}	6.56×10^{-1}	5.61×10^{1}	2.35×10^{1}	1.32×10^{1}	9.13×10^{1}	$\mathbf{5.61 \times 10^{-7}}$
	Mean	9.81×10^{1}	9.38×10^{3}	9.12×10^{6}	2.95×10^{5}	9.82×10^{1}	9.83×10^{1}	7.86×10^{3}	4.01×10^{4}	2.04×10^{8}	$\mathbf{9.00 \times 10^{1}}$
F4	Std	5.22×10^{-1}	2.50×10^{6}	3.99×10^{6}	6.54×10^{4}	5.50×10^{-1}	6.68×10^{-1}	1.39×10^{3}	1.80×10^{4}	8.37×10^{7}	1.74×10^{1}
	Best	9.65×10^{1}	1.00×10^{2}	4.79×10^{6}	1.87×10^{5}	9.71×10^{1}	9.63×10^{1}	5.16×10^{3}	2.26×10^{4}	6.98×10^{7}	$\mathbf{1.66 \times 10^{1}}$
	Mean	1.16×10^{1}	2.61×10^{3}	1.59×10^{4}	2.13×10^{2}	1.50×10^{1}	1.56×10^{1}	2.86×10^{2}	1.03×10^{3}	7.46×10^{4}	$\mathbf{3.02 \times 10^{-2}}$
F5	Std	9.88×10^{-1}	2.14×10^{3}	4.71×10^{3}	3.34×10^{1}	6.26×10^{-1}	9.66×10^{-1}	2.88×10^{1}	1.07×10^{2}	1.39×10^{4}	2.42×10^{-2}
	Best	9.54	2.44×10^{1}	8.20×10^{3}	1.56×10^{2}	1.40×10^{1}	1.40×10^{1}	2.26×10^{2}	7.95×10^{2}	5.23×10^{4}	$\mathbf{1.27 \times 10^{-2}}$
	Mean	1.69×10^{-29}	1.73	3.58	3.72×10^{1}	$\mathbf{2.42 \times 10^{-39}}$	3.44×10^{-23}	1.58×10^{-4}	2.79×10^{-2}	1.80×10^{1}	2.09×10^{-35}
F6	Std	2.50×10^{-29}	4.11×10^{-1}	2.52	1.11×10^{1}	1.03×10^{-35}	6.04×10^{-22}	2.03×10^{-3}	1.65×10^{-2}	5.18	1.21×10^{-34}
	Best	4.86×10^{-34}	8.30×10^{-8}	1.05×10^{-1}	1.18×10^{1}	5.32×10^{-46}	1.38×10^{-35}	6.48×10^{-7}	8.35×10^{-3}	2.00	$\mathbf{1.80 \times 10^{-46}}$
	Mean	2.47×10^{-7}	2.39×10^{7}	4.13×10^{8}	8.83×10^{6}	$\mathbf{7.99 \times 10^{-14}}$	1.13×10^{-10}	1.14×10^{7}	2.89×10^{7}	1.46×10^{9}	1.06×10^{-13}
F7	Std	1.62×10^{-7}	8.81×10^{7}	2.66×10^{8}	2.57×10^{6}	8.94×10^{-14}	1.44×10^{-10}	2.50×10^{6}	9.00×10^{6}	8.26×10^{8}	2.12×10^{-12}
	Best	4.80×10^{-8}	1.31×10^{1}	6.05×10^{7}	4.00×10^{6}	3.74×10^{-15}	1.15×10^{-12}	7.24×10^{6}	1.75×10^{7}	1.16×10^{8}	$\mathbf{1.00 \times 10^{-18}}$
	Mean	5.38×10^{-11}	3.72×10^{2}	7.33×10^{3}	1.02×10^{4}	$\mathbf{1.80 \times 10^{-17}}$	2.54×10^{-14}	1.68×10^{2}	4.68×10^{2}	3.30×10^{4}	4.94×10^{-16}
F8	Std	4.61×10^{-11}	7.11×10^{2}	2.62×10^{3}	1.30×10^{3}	2.04×10^{-17}	1.31×10^{-14}	2.58×10^{1}	7.19×10^{1}	7.03×10^{3}	5.01×10^{-16}
	Best	1.19×10^{-11}	4.81×10^{-2}	3.38×10^{3}	7.58×10^{3}	5.21×10^{-19}	2.86×10^{-16}	1.35×10^{2}	3.14×10^{2}	1.40×10^{4}	$\mathbf{1.42 \times 10^{-21}}$
	Mean	1.09×10^{1}	2.66×10^{2}	5.88×10^{2}	1.24×10^{3}	2.54×10^{-1}	8.72×10^{2}	3.97×10^{2}	3.72×10^{2}	8.99×10^{2}	$\mathbf{4.43 \times 10^{-2}}$
F9	Std	8.83	1.04×10^{2}	6.85×10^{1}	9.61×10^{1}	6.46×10^{-5}	1.80×10^{2}	5.40×10^{1}	4.50×10^{1}	7.18×10^{1}	8.31×10^{-2}
	Best	5.96×10^{-7}	3.06×10^{-2}	4.79×10^{2}	1.05×10^{3}	$\mathbf{0.00}$	4.45×10^{2}	3.17×10^{2}	3.04×10^{2}	6.92×10^{2}	$\mathbf{0.00}$
	Mean	1.27×10^{-6}	4.98	1.58×10^{1}	6.96	1.06×10^{-9}	2.37×10^{-8}	3.71	6.80	1.99×10^{1}	$\mathbf{4.21 \times 10^{-10}}$
F10	Std	3.85×10^{-7}	2.93	9.91×10^{-1}	3.72×10^{-1}	7.18×10^{-10}	2.69×10^{-8}	1.18×10^{-1}	5.39×10^{-1}	1.52×10^{-1}	5.61×10^{-10}
	Best	6.16×10^{-7}	1.05×10^{-2}	1.43×10^{1}	6.06	1.97×10^{-10}	1.55×10^{-9}	3.49	6.07	1.95×10^{1}	$\mathbf{2.93 \times 10^{-14}}$
	Mean	4.78×10^{-3}	3.71×10^{-2}	2.52	1.03	1.20×10^{-3}	1.00×10^{-2}	8.33×10^{-1}	1.09	7.73	$\mathbf{1.11 \times 10^{-17}}$
F11	Std	1.09×10^{-2}	4.83×10^{-1}	3.68×10^{-1}	2.44×10^{-2}	9.13×10^{-3}	1.22×10^{-2}	3.47×10^{-2}	1.60×10^{-2}	1.52	1.20×10^{-16}
	Best	1.52×10^{-13}	3.47×10^{-2}	1.70	9.72×10^{-1}	$\mathbf{0.00}$	$\mathbf{0.00}$	7.60×10^{-1}	1.07	4.44	$\mathbf{0.00}$
	Mean	3.83×10^{-1}	3.34	3.57×10^{6}	1.87×10^{1}	5.37×10^{-1}	1.38×10^{1}	7.95	1.10×10^{1}	4.31×10^{8}	$\mathbf{2.13 \times 10^{-3}}$
F12	Std	6.78×10^{-2}	5.95×10^{5}	1.56×10^{6}	4.44	4.11×10^{-2}	8.57	2.37	3.03	1.65×10^{8}	6.89×10^{-3}
	Best	2.41×10^{-1}	1.20	1.56×10^{5}	7.79	4.06×10^{-1}	6.15×10^{-1}	2.33	6.06	1.34×10^{8}	$\mathbf{6.35 \times 10^{-4}}$
	Mean	7.37	1.21×10^{1}	1.69×10^{7}	8.76×10^{2}	8.32	4.50×10^{1}	1.68×10^{1}	1.67×10^{2}	8.61×10^{8}	$\mathbf{1.29 \times 10^{-1}}$
F13	Std	4.10×10^{-1}	7.62×10^{6}	1.19×10^{7}	8.25×10^{2}	3.16×10^{-1}	3.71×10^{1}	2.36	3.12×10^{1}	3.28×10^{8}	7.60×10^{-2}
	Best	6.36	1.01×10^{1}	1.87×10^{6}	1.97×10^{2}	7.59	9.89	1.26×10^{1}	8.56×10^{1}	2.15×10^{8}	$\mathbf{3.67 \times 10^{-2}}$
	Mean	4.50×10^{-3}	1.64	5.10×10^{1}	1.05×10^{2}	7.70×10^{-5}	1.25×10^{2}	2.19×10^{1}	2.46×10^{1}	8.11×10^{1}	$\mathbf{2.46 \times 10^{-6}}$
F14	Std	2.87×10^{-3}	9.69	1.17×10^{1}	1.26×10^{1}	8.55×10^{-4}	2.23×10^{1}	3.88	8.16	1.41×10^{1}	6.63×10^{-6}
	Best	7.96×10^{-7}	6.83×10^{-3}	3.26×10^{1}	7.72×10^{1}	5.32×10^{-12}	6.85×10^{1}	1.28×10^{1}	1.50×10^{1}	5.54×10^{1}	$\mathbf{1.06 \times 10^{-14}}$
	Mean	1.41×10^{-2}	4.92×10^{-1}	5.00×10^{-1}	3.95×10^{-1}	4.32×10^{-3}	1.30×10^{-2}	4.99×10^{-1}	4.88×10^{-1}	5.00×10^{-1}	$\mathbf{3.13 \times 10^{-3}}$
F15	Std	2.33×10^{-3}	1.46×10^{-1}	4.43×10^{-5}	1.71×10^{-2}	1.45×10^{-3}	2.74×10^{-3}	6.33×10^{-4}	2.78×10^{-3}	2.05×10^{-6}	1.18×10^{-3}
	Best	9.29×10^{-3}	1.91×10^{-2}	5.00×10^{-1}	3.56×10^{-1}	3.13×10^{-3}	6.22×10^{-3}	4.97×10^{-1}	4.82×10^{-1}	5.00×10^{-1}	$\mathbf{2.58 \times 10^{-6}}$
	Mean	1.38×10^{-10}	8.50	1.07×10^{3}	7.13×10^{2}	$\mathbf{7.40 \times 10^{-18}}$	2.48×10^{-13}	7.18×10^{1}	1.58×10^{2}	4.92×10^{3}	1.87×10^{-15}
F16	Std	7.46×10^{-11}	7.48×10^{1}	3.89×10^{2}	9.40×10^{1}	7.97×10^{-17}	1.04×10^{1}	3.26	9.91	1.30×10^{3}	2.09×10^{-15}
	Best	2.63×10^{-11}	4.28×10^{-1}	5.92×10^{2}	4.96×10^{2}	$\mathbf{0.00}$	$\mathbf{0.00}$	6.25×10^{1}	1.23×10^{2}	2.89×10^{3}	$\mathbf{0.00}$

Furthermore, the variances of the results of these 30 independent experiments are also smaller. As can be seen in Table 2, the optimization results of FWGWO on the unimodal functions f1–f5, whether the mean or the variance, are superior to other algorithms. As mentioned above, the unimodal function is utilized to test the exploitation ability of the algorithm. From these results, it can be seen that, compared with FWA, GWO, and other algorithms, FWGWO can find a better solution in a limited number of iterations and has a better exploitation ability. It can also be seen from Table 2 that FWGWO has better results in seven of the eight functions on the multimodal functions f9–f16. The multimodal functions are used to test the exploration ability of the algorithm. It can be known from the experiment results that the FWGWO algorithm has a better global optimization ability than other algorithms used in this paper, including the original GWO, the standard FWA, and the enhanced GWO.

Figures 3 and 4 show the convergence process of FWGWO and other algorithms mentioned in this paper on 16 reference benchmark functions. The curves we provided in Figures 3 and 4 are the average fitness curves of 30 repetitions. As shown in Figure 3, at the first few iterations, FWGWO's convergence is slightly slower than that of other algorithms. In this process, the algorithm explores a lot to search for a better exploitation of the local areas. In the following convergence process, FWGWO converges quickly in this region, searching for a better solution than other algorithms after a total of 500 iterations. Figure 4 shows the comparison of the convergence of FWGWO with other algorithms on multimodal functions. As mentioned above, the multimodal functions are used to test the local optima avoidance ability of the algorithm.

In Figure 4, we can see that unlike other algorithms that have been trapped in the local optimum, FWGWO continues to search for a better fitness on most benchmark functions and keeps approaching global optima. This proves that the FWGWO algorithm has a better global optimization searching ability when compared with other algorithms. According to the above comprehensive description, the hybrid algorithm FWGWO is superior to other algorithms in both the accuracy of the solution and the convergence. It can be concluded that the FWGWO algorithm has better overall performances on most optimization problems than other algorithms.

The mean running time of 30 repetitions has been shown in Table 3. The results show that the running time of the FWGWO algorithm is a little bigger than that of GWO and much smaller than that of FWA. As mentioned in Section 4, FWA has a larger time complexity, while GWO has a smaller time complexity. The location update strategy of FWA has been used in FWGWO a few times. Therefore, FWGWO has a bigger running time than GWO. Considering the improvement of the convergence performance, it is acceptable to sacrifice a little running time.

Wilcoxon's nonparametric statistical test is conducted at the 5% significance level in order to determine whether the FWGWO provides a significant improvement compared to other algorithms or not. The results of different algorithms on the benchmark function were employed to test the Wilcoxon rank sum, and p and h values were obtained as significant level indicators. If the p value is less than 0.05, the null hypothesis is rejected. At this time, the h value is 1, and the two algorithms tested are considered significantly different. Conversely, when the p value is greater than 0.05, the h value is 0, and the two algorithms tested are considered to not be significantly different. In this paper, the Wilcoxon rank sum is tested with the results of 30 repeated experiments on 16 benchmark functions by the FWGWO algorithm and other algorithms. The test results are shown in Table 4. In most cases, the h values of the test results are 1, except that the results' p values for AGWO and FWGWO on f8, f11, f15, and f16 are greater than 0.05 and the h values are 0. This means that the optimization efficiency of FWGWO and AGWO is similar in f8, f11, f15, and f16. The results show that in most cases the performance of the FWGWO algorithm is significantly improved when compared with other algorithms.

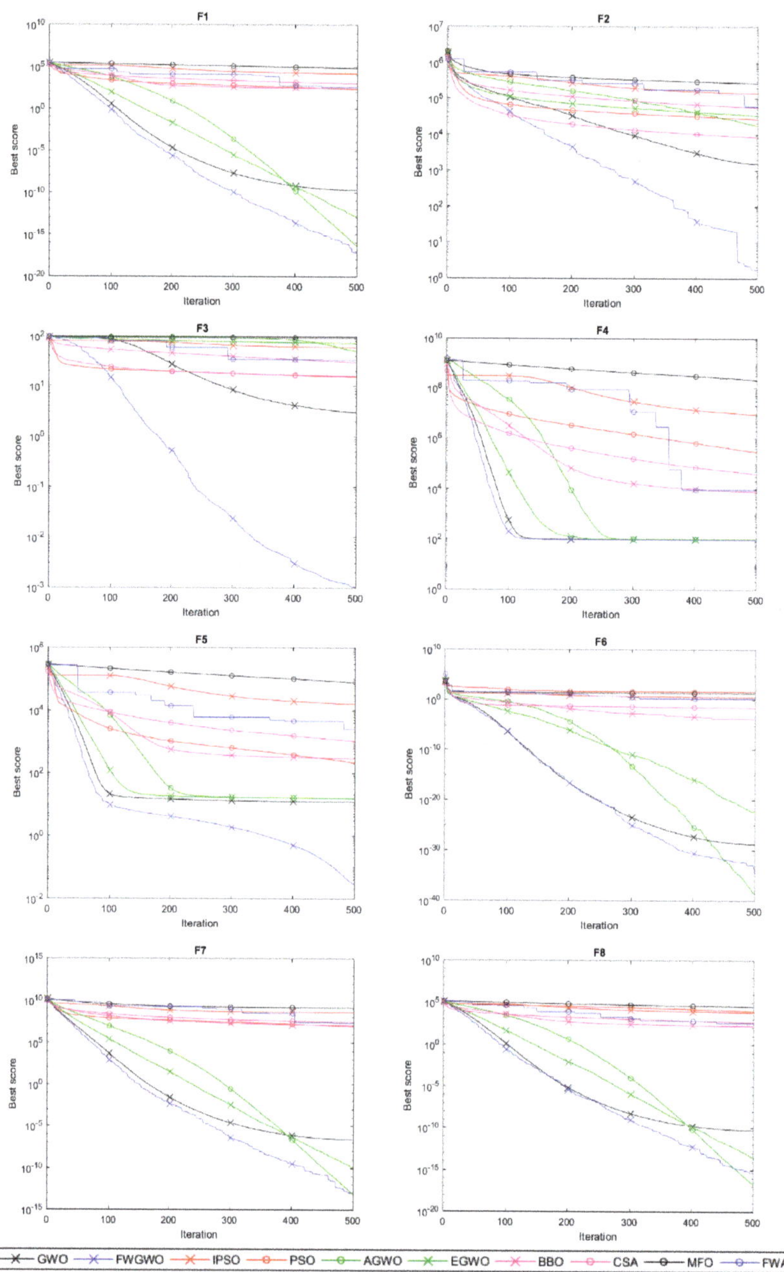

Figure 3. Convergence curves of the unimodal functions.

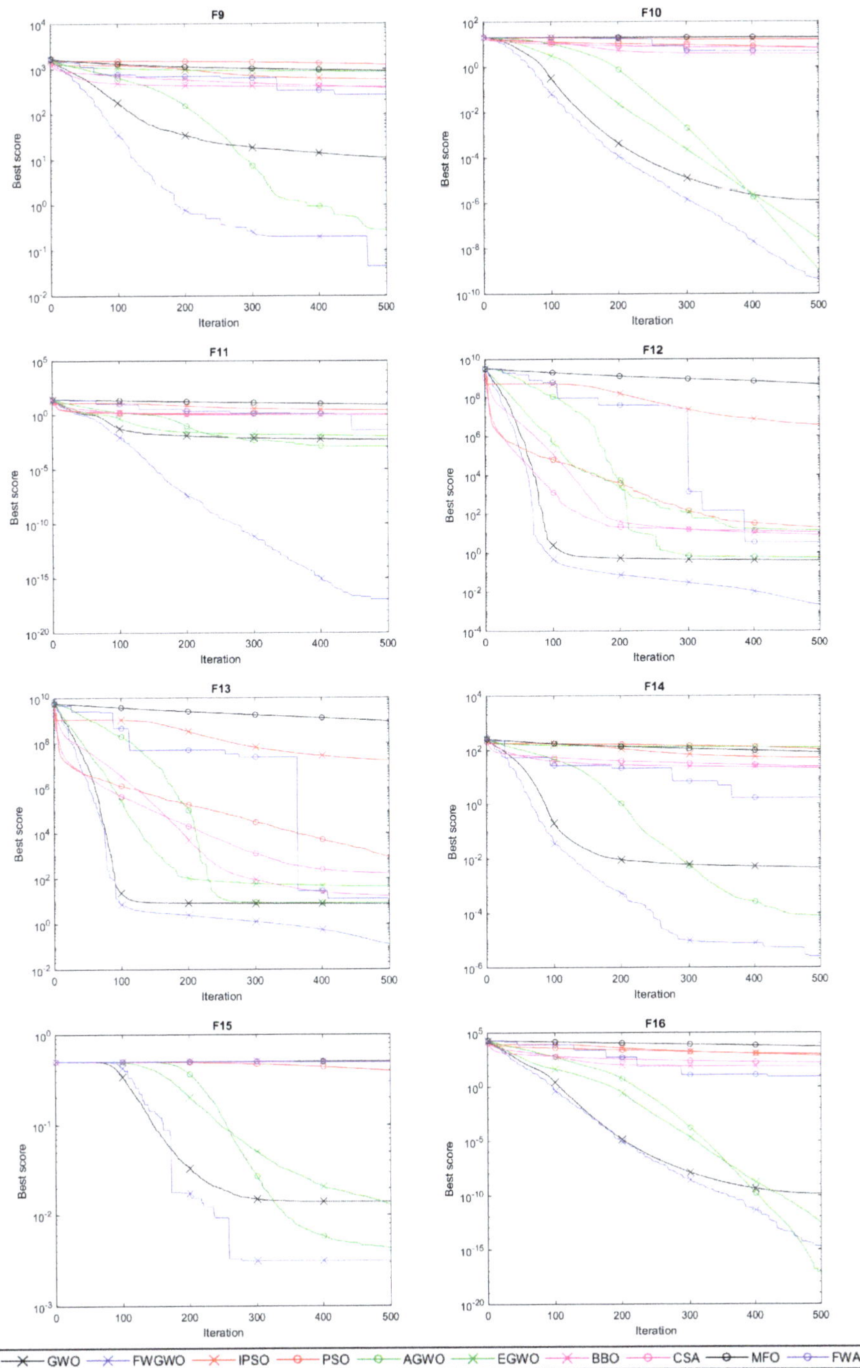

Figure 4. Convergence curves of the multimodal functions.

Table 3. Running time (seconds) of different algorithms.

Function	GWO	FWA	IPSO	PSO	AGWO	EGWO	BBO	CSA	MFO	FWGWO
F1	0.65625	6.1875	0.828125	0.5625	0.59375	0.53125	10.65625	1.296875	0.953125	1.390625
F2	2.0625	11.5	1.8125	1.828125	2.09375	1.734375	13.89063	4.890625	2.34375	2.46875
F3	0.375	6.75	0.234375	0.15625	0.25	0.171875	9.46875	0.890625	0.453125	0.8125
F4	0.453125	5.5625	0.25	0.1875	0.328125	0.203125	8.671875	0.890625	0.453125	0.671875
F5	0.40625	6.21875	0.21875	0.15625	0.25	0.140625	10.57813	0.75	0.421875	0.671875
F6	1.03125	8.53125	0.8125	0.796875	0.84375	0.71875	9.09375	2.640625	1.0625	2.5
F7	0.609375	6.640625	0.4375	0.375	0.515625	0.375	9.671875	1.296875	0.625	0.90625
F8	0.359375	5.375	0.21875	0.15625	0.234375	0.140625	9.4375	0.71875	0.453125	0.671875
F9	0.421875	6.328125	0.3125	0.25	0.265625	0.1875	8.28125	1.046875	0.453125	0.734375
F10	0.421875	7.171875	0.3125	0.25	0.28125	0.1875	8.3125	0.96875	0.421875	0.703125
F11	0.453125	6.265625	0.328125	0.25	0.34375	0.1875	8.609375	0.90625	0.484375	1.484375
F12	0.953125	7.5625	0.828125	0.71875	1	0.65625	9.09375	2.3125	1.046875	1.28125
F13	1.015625	8.046875	0.796875	0.734375	0.890625	0.703125	9.078125	2.359375	1.046875	1.34375
F14	0.515625	5.875	0.234375	0.21875	0.21875	0.15625	8.5625	0.953125	0.4375	0.625
F15	0.53125	7.25	0.265625	0.1875	0.328125	0.15625	8.125	0.921875	0.421875	0.65625
F16	0.453125	6.109375	0.296875	0.234375	0.28125	0.1875	8.3125	1.015625	0.484375	0.78125

Table 4. Wilcoxon's rank test of FWGWO and other algorithms on 16 benchmark functions.

Function		GWO	FWA	IPSO	PSO	AGWO	EGWO	BBO	CSA	MFO
F1	p-value	6.51×10^{-12}	6.51×10^{-12}	6.51×10^{-12}	6.51×10^{-12}	1.82×10^{-7}	7.86×10^{-12}	6.51×10^{-12}	6.51×10^{-12}	6.51×10^{-12}
	h-value	1	1	1	1	1	1	1	1	1
F2	p-value	7.86×10^{-12}	1.65×10^{-11}	6.51×10^{-12}	6.51×10^{-12}	7.15×10^{-12}	6.51×10^{-12}	6.51×10^{-12}	6.51×10^{-12}	6.51×10^{-12}
	h-value	1	1	1	1	1	1	1	1	1
F3	p-value	6.51×10^{-12}	6.51×10^{-12}	6.51×10^{-12}	6.51×10^{-12}	6.51×10^{-12}	6.51×10^{-12}	6.51×10^{-12}	6.51×10^{-12}	6.51×10^{-12}
	h-value	1	1	1	1	1	1	1	1	1
F4	p-value	2.05×10^{-10}	6.51×10^{-12}	6.51×10^{-12}	6.51×10^{-12}	1.72×10^{-10}	6.28×10^{-10}	6.51×10^{-12}	6.51×10^{-12}	6.51×10^{-12}
	h-value	1	1	1	1	1	1	1	1	1
F5	p-value	6.51×10^{-12}	6.51×10^{-12}	6.51×10^{-12}	6.51×10^{-12}	6.51×10^{-12}	6.51×10^{-12}	6.51×10^{-12}	6.51×10^{-12}	6.51×10^{-12}
	h-value	1	1	1	1	1	1	1	1	1
F6	p-value	7.15×10^{-12}	6.51×10^{-12}	6.51×10^{-12}	6.51×10^{-12}	6.29×10^{-3}	1.14×10^{-11}	6.51×10^{-12}	6.51×10^{-12}	6.51×10^{-12}
	h-value	1	1	1	1	1	1	1	1	1
F7	p-value	6.51×10^{-12}	6.51×10^{-12}	6.51×10^{-12}	6.51×10^{-12}	2.05×10^{-2}	7.08×10^{-11}	6.51×10^{-12}	6.51×10^{-12}	6.51×10^{-12}
	h-value	1	1	1	1	1	1	1	1	1
F8	p-value	6.51×10^{-12}	6.51×10^{-12}	6.51×10^{-12}	6.51×10^{-12}	8.45×10^{-2}	4.13×10^{-11}	6.51×10^{-12}	6.51×10^{-12}	6.51×10^{-12}
	h-value	1	1	1	1	0	1	1	1	1
F9	p-value	3.14×10^{-12}	1.73×10^{-12}	1.16×10^{-12}	1.16×10^{-12}	1.19×10^{-3}	1.16×10^{-12}	1.16×10^{-12}	1.16×10^{-12}	1.16×10^{-12}
	h-value	1	1	1	1	1	1	1	1	1
F10	p-value	6.51×10^{-12}	6.51×10^{-12}	6.51×10^{-12}	6.51×10^{-12}	1.27×10^{-7}	7.86×10^{-12}	6.51×10^{-12}	6.51×10^{-12}	6.51×10^{-12}
	h-value	1	1	1	1	1	1	1	1	1
F11	p-value	5.54×10^{-13}	5.54×10^{-13}	5.54×10^{-13}	5.54×10^{-13}	3.62×10^{-1}	6.72×10^{-7}	5.54×10^{-13}	5.54×10^{-13}	5.54×10^{-13}
	h-value	1	1	1	1	0	1	1	1	1
F12	p-value	6.51×10^{-12}	6.51×10^{-12}	6.51×10^{-12}	6.51×10^{-12}	6.51×10^{-12}	6.51×10^{-12}	6.51×10^{-12}	6.51×10^{-12}	6.51×10^{-12}
	h-value	1	1	1	1	1	1	1	1	1
F13	p-value	6.51×10^{-12}	6.51×10^{-12}	6.51×10^{-12}	6.51×10^{-12}	6.51×10^{-12}	6.51×10^{-12}	6.51×10^{-12}	6.51×10^{-12}	6.51×10^{-12}
	h-value	1	1	1	1	1	1	1	1	1
F14	p-value	9.47×10^{-12}	6.51×10^{-12}	6.51×10^{-12}	6.51×10^{-12}	2.36×10^{-3}	6.51×10^{-12}	6.51×10^{-12}	6.51×10^{-12}	6.51×10^{-12}
	h-value	1	1	1	1	1	1	1	1	1
F15	p-value	2.05×10^{-10}	6.51×10^{-12}	1.11×10^{-10}	6.51×10^{-12}	1.12×10^{-1}	4.81×10^{-8}	1.01×10^{-10}	8.47×10^{-11}	1.21×10^{-10}
	h-value	1	1	1	1	0	1	1	1	1
F16	p-value	1.71×10^{-12}	1.71×10^{-12}	1.71×10^{-12}	1.71×10^{-12}	6.75×10^{-2}	1.99×10^{-10}	1.71×10^{-12}	1.71×10^{-12}	1.71×10^{-12}
	h-value	1	1	1	1	0	1	1	1	1

6. Conclusions

In this paper, a hybrid algorithm FWGWO based on Grey Wolf Optimizer (GWO) and Fireworks Algorithm (FWA) is proposed for the optimization of multidimensional complex space. The FWGWO algorithm combines the good exploitation ability of GWO with the strong exploration ability of the FWA algorithm. In order to balance the exploitation with the exploration, an adaptive balance coefficient is employed in this algorithm. The probability of exploitation or exploration is controlled by the balance coefficient. By changing the balance coefficient, the FWGWO algorithm can avoid the local optimal value as much as possible and has a fast convergence speed. In order to verify the performance superiority of the FWGWO algorithm, the FWGWO algorithm was tested 30 times with IPSO, PSO, BBO, CSA, MFO, FWA, GWO, AGWO, and EGWO algorithms on 16 benchmark functions, and their test results were compared with each other. The compared results show that the FWGWO algorithm has a better global optimization ability and faster convergence speed compared with other algorithms. In addition, the Wilcoxon rank sum test was used to test the optimization results. The test results show that the FWGWO algorithm has a significant improvement compared to other algorithms.

This paper presents a new method for combining two algorithms. In the future, this method could be used in combination with other enhanced algorithms. Furthermore, the FWGWO can be applied to solve single- and multi-objective optimization problems like the feature selection problem.

Author Contributions: Data curation, Z.Y.; Formal analysis, Z.Y.; Funding acquisition, S.Z. and W.X.; Methodology, Z.Y.; Resources, Z.Y.; Software, Z.Y.; Supervision, S.Z.; Validation, Z.Y.; Writing – original draft, Z.Y.; Writing – review & editing, Z.Y. and S.Z. All authors have read and agreed to the published version of the manuscript.

Funding: This work is supported by the National Natural Science Foundation of China under grant No. 61673056 and 61673055, the Beijing Natural Science Foundation under grant No.4182039, and the National Key Research and Development Program of China under Grant 2017YFB1401203.

Acknowledgments: This work is supported by the National Natural Science Foundation of China under grant No. 61673056 and 61673055, the Beijing Natural Science Foundation under grant No.4182039, and the National Key Research and Development Program of China under Grant 2017YFB1401203. The authors would like to thank the anonymous reviewers for their insightful suggestions that will help us to improve the quality of this paper.

Conflicts of Interest: The authors declare no conflict interest.

Appendix A

Table A1. Benchmark functions.

Function	Range	$f_{\min}$
$F1 = \sum_{i=1}^{D} x_i^2$	$[-100, 100]$	0
$F2 = \sum_{i=1}^{D} \left(\sum_{j=1}^{i} x_j \right)$	$[-100, 100]$	0
$F3 = \max\{|x_i|, 1 \le i \le D\}$	$[-100, 100]$	0
$F4 = \sum_{i=1}^{D-1} \left[100\left(x_{i+1} - x_i^2\right)^2 + (x_i - 1)^2 \right]$	$[-30, 30]$	0
$F5 = \sum_{i=1}^{D} [x_i + 0.5]^2$	$[-100, 100]$	0
$F6 = 10^6 \cdot x_1^2 + \sum_{i=2}^{D} x_i^6$	$[-1, 1]$	0
$F7 = \sum_{i=2}^{D} \left(10^6\right)^{(i-1)/(n-1)} \cdot x_i^2$	$[-100, 100]$	0
$F8 = \sum_{i=2}^{D} i x_i^2$	$[-10, 10]$	0
$F9 = \sum_{i=1}^{D} \left[x_i^2 - 10\cos(2\pi x_i) + 10 \right]$	$[-5.12, 5.12]$	-418.982
$F10 = -20\exp\left(-0.2 \sqrt{\frac{1}{D} \sum_{i=1}^{D} x_i^2}\right) - \exp\left(\frac{1}{D} \sum_{i=1}^{D} \cos(2\pi x_i)\right) + 20 + e$	$[-32, 32]$	0
$F11 = \frac{1}{4000} \sum_{i=1}^{D} x_i^2 - \prod_{i=1}^{D} \cos\left(\frac{x_i}{\sqrt{i}}\right) + 1$	$[-60, 60]$	0
$F12 = \frac{\pi}{n}\left\{ 10\sin(\pi y_1) + \sum_{i=1}^{D-1} (y_i - 1)^2 \left[1 + 10\sin^2(\pi y_{i+1}) \right] + (y_n - 1)^2 \right\} + \sum_{i=1}^{D} u(x_i, 5, 100, 4)$	$[-50, 50]$	0
$F13 = 0.1\left\{ \sin(3\pi x_1) + \sum_{i=1}^{D-1} (x_i - 1)^2 \left[1 + 10\sin^2(3\pi x_{i+1}) \right] + (x_n - 1)^2 \right\} + \sum_{i=1}^{D} u(x_i, 5, 100, 4)$	$[-50, 50]$	0
$F14 = \sum_{i=1}^{D} |x_i \sin(x_i) + 0.1 x_i|$	$[-10, 10]$	0
$F15 = 0.5 + \left(\left(\sin\left(\sum_{i=1}^{D} x_i^2 \right) \right)^2 - 0.5 \right) \cdot \left(1 + 0.001\left(\sum_{i=1}^{D} x_i^2 \right) \right)^{-2}$	$[-100, 100]$	0
$F16 = \sum_{i=1}^{D-1} \left[x_i^2 + 2x_{i+1}^2 - 0.3\cos(3\pi x_i) - 0.4\cos(4\pi x_{i+1}) + 0.7 \right]$	$[-15, 15]$	0

References

1. Şenel, F.A.; Gökçe, F.; Yüksel, A.S.; Yiğit, T. A novel hybrid PSO–GWO algorithm for optimization problems. *Eng. Comput.* **2018**, *35*, 1359–1373. [CrossRef]
2. Brownlee, J. *Clever Algorithms: Nature-Inspired Programming Recipes*; Lulu Press: Morrisville, NC, USA, 2011.
3. Poli, R.; Kennedy, J.; Blackwell, T. Particle swarm optimization. *Swarm Intell.* **2007**, *1*, 33–57. [CrossRef]
4. Colorni, A.; Dorigo, M.; Maniezzo, V. Distributed optimization by ant colonies. In Proceedings of the European Conference on Artificial Life, Paris, France, 11–13 December 1991.
5. Karaboga, D. *An Idea Based on Honey Bee Swarm for Numerical Optimization*; Technical Report-TR06; Erciyes University: Kayseri, Turkey, 2005.
6. Mirjalili, S.; Lewis, A. The whale optimization algorithm. *Adv. Eng. Softw.* **2016**, *95*, 51–67. [CrossRef]
7. Meng, X.; Gao, X.Z.; Lu, L.; Liu, Y.; Zhang, H. A new bio-inspired optimization algorithm: Bird swarm algorithm. *J. Exp. Theor. Artif. Intell.* **2016**, *28*, 673–687. [CrossRef]
8. Mirjalili, S.; Mirjalili, S.M.; Lewis, A. Grey wolf optimizer. *Adv. Eng. Softw.* **2014**, *69*, 46–61. [CrossRef]
9. Tan, Y.; Zhu, Y. Fireworks algorithm for optimization. *Comput. Vis.* **2010**, *6145*, 355–364.
10. Simon, D. Biogeography-based optimization. *IEEE Trans. Evol. Comput.* **2008**, *12*, 702–713. [CrossRef]
11. Mirjalili, S. Moth-flame optimization algorithm: A novel nature-inspired heuristic paradigm. *Knowl.-Based Syst.* **2015**, *89*, 228–249. [CrossRef]
12. Rashedi, E.; Nezamabadi-Pour, H.; Saryazdi, S. GSA: A gravitational search algorithm. *Inf. Sci.* **2009**, *179*, 2232–2248. [CrossRef]
13. Storn, R.; Price, K. Differential evolution—A simple and efficient heuristic for global optimization over continuous spaces. *J. Glob. Optim.* **1997**, *11*, 341–359. [CrossRef]
14. Agarwal, J.; Parmar, G.; Gupta, R.; Sikander, A. Analysis of grey wolf optimizer based fractional order PID controller in speed control of DC motor. *Microsyst. Technol.* **2018**, *24*, 4997–5006. [CrossRef]
15. Song, X.; Tang, L.; Zhao, S.; Zhang, X.; Li, L.; Huang, J.; Cai, W. Grey wolf optimizer for parameter estimation in surface waves. *Soil Dyn. Earthq. Eng.* **2015**, *75*, 147–157. [CrossRef]
16. Yammani, C.; Maheswarapu, S. Load frequency control of multi-microgrid system considering renewable energy sources using grey wolf optimization. *Smart Sci.* **2019**, *7*, 198–217.
17. Bratton, D.; Kennedy, J. Defining a standard for particle swarm optimization. In Proceedings of the IEEE Swarm Intelligence Symposium, Honolulu, HI, USA, 1–5 April 2007; pp. 120–127.
18. Tan, Y.; Xiao, Z.M. Clonal particle swarm optimization and its applications. In Proceedings of the IEEE Congress on Evolutionary Computation, Singapore, 25–28 September 2007; pp. 2303–2309.
19. Mirjalili, S.; Wang, G.-G.; Coelho, L.D.S. Binary optimization using hybrid particle swarm optimization and gravitational search algorithm. *Neural Comput. Appl.* **2014**, *25*, 1423–1435. [CrossRef]
20. Goldberg, D.E. *Genetic Algorithms in Search Optimization and Machine Learning*; Addison-Wesley: Boston, MA, USA, 1989.
21. Choi, K.; Jang, D.-H.; Kang, S.-I.; Lee, J.-H.; Chung, T.-K.; Kim, H.-S.; Jung, T.-K. Hybrid algorithm combing genetic algorithm with evolution strategy for antenna design. *IEEE Trans. Magn.* **2015**, *52*, 1–4. [CrossRef]
22. Mafarja, M.; Mirjalili, S. Hybrid whale optimization algorithm with simulated annealing for feature selection. *Neurocomputing* **2017**, *260*, 302–312. [CrossRef]
23. Alomoush, A.A.; Alsewari, A.A.; Alamri, H.S.; Alamri, H.S.; Zamli, K.Z. Hybrid Harmony search algorithm with grey wolf optimizer and modified opposition-based learning. *IEEE Access* **2019**, *7*, 68764–68785. [CrossRef]
24. Arora, S.; Singh, H.; Sharma, M.; Sharma, S.; Anand, P. A new hybrid algorithm based on grey wolf optimization and crow search algorithm for unconstrained function optimization and feature selection. *IEEE Access* **2019**, *7*, 26343–26361. [CrossRef]
25. Kumar, S.; Pant, M.; Dixit, A.; Bansal, R. BBO-DE: Hybrid algorithm based on BBO and DE. In Proceedings of the International Conference on Computing, Communication and Automation (ICCCA), Greater Noida, India, 5–6 May 2017; pp. 379–383.
26. Gohil, B.N.; Patel, D.R. A hybrid GWO-PSO algorithm for load balancing in cloud computing environment. In Proceedings of the Second International Conference on Green Computing and Internet of Things (ICGCIoT), Bangalore, India, 16–18 August 2018; pp. 185–191.

27. Zhu, A.; Xu, C.; Li, Z.; Wu, J.; Liu, Z. Hybridizing grey wolf optimization with differential evolution for global optimization and test scheduling for 3D stacked SoC. *J. Syst. Eng. Electron.* **2015**, *26*, 317–328. [CrossRef]

28. Zhang, B.; Zhang, M.-X.; Zheng, Y. A hybrid biogeography-based optimization and fireworks algorithm. In Proceedings of the IEEE Congress on Evolutionary Computation (CEC), Beijing, China, 6–11 July 2014; pp. 3200–3206.

29. Selim, A.; Kamel, S.; Jurado, F. Voltage profile improvement in active distribution networks using hybrid WOA-SCA optimization algorithm. In Proceedings of the Twentieth International Middle East Power Systems Conference (MEPCON), Cairo, Egypt, 18–20 December 2018; pp. 1064–1068.

30. Alireza, A. A novel metaheuristic method for solving con- strained engineering optimization problems: Crow search algorithm. *Comput. Struct.* **2016**, *169*, 1–12.

31. Jiang, Y.; Hu, T.; Huang, C.; Wu, X. An improved particle swarm optimization algorithm. *Appl. Math. Comput.* **2007**, *193*, 231–239. [CrossRef]

32. Sharma, S.; Salgotra, R.; Singh, U. An enhanced grey wolf optimizer for numerical optimization. In Proceedings of the International Conference on Innovations in Information, Embedded and Communication Systems (ICIIECS), Coimbatore, India, 17–18 March 2017; pp. 1–6.

33. Hamour, H.; Kamel, S.; Nasrat, L.; Yu, J. Distribution network reconfiguration using augmented grey wolf optimization algorithm for power loss minimization. In Proceedings of the International Conference on Innovative Trends in Computer Engineering (ITCE), Aswan, Egypt, 2–4 February 2019; pp. 450–454.

34. Derrac, J.; García, S.; Molina, D.; Herrera, F.; Cabrera, D.M. A practical tutorial on the use of nonparametric statistical tests as a methodology for comparing evolutionary and swarm intelligence algorithms. *Swarm Evol. Comput.* **2011**, *1*, 3–18. [CrossRef]

Article

Passenger Flow Forecasting in Metro Transfer Station Based on the Combination of Singular Spectrum Analysis and AdaBoost-Weighted Extreme Learning Machine

Wei Zhou [1,2,3], **Wei Wang** [1,2,3,*] **and De Zhao** [1,2,3]

[1]   School of Transportation, Southeast University, Nanjing 211189, China; veager@seu.edu.cn (W.Z.); zhaode@seu.edu.cn (D.Z.)
[2]   Jiangsu Key Laboratory of Urban ITS, Nanjing 211189, China
[3]   Jiangsu Province Collaborative Innovation Centre of Modern Urban Traffic Technologies, Nanjing 211189, China
*   Correspondence: wangwei@seu.edu.cn

Received: 19 May 2020; Accepted: 19 June 2020; Published: 23 June 2020

Abstract: The metro system plays an important role in urban public transit, and the passenger flow forecasting is fundamental to assisting operators establishing an intelligent transport system (ITS). The forecasting results can provide necessary information for travelling decision of travelers and metro operations of managers. In order to investigate the inner characteristics of passenger flow and make a more accurate prediction with less training time, a novel model (i.e., SSA-AWELM), a combination of singular spectrum analysis (SSA) and AdaBoost-weighted extreme learning machine (AWELM), is proposed in this paper. SSA is developed to decompose the original data into three components of trend, periodicity, and residue. AWELM is developed to forecast each component desperately. The three predicted results are summed as the final outcomes. In the experiments, the dataset is collected from the automatic fare collection (AFC) system of Hangzhou metro in China. We extracted three weeks of passenger flow to carry out multistep prediction tests and a comparison analysis. The results indicate that the proposed SSA-AWELM model can reduce both predicted errors and training time. In particular, compared with the prevalent deep-learning model long short-term memory (LSTM) neural network, SSA-AWELM has reduced the testing errors by 22% and saved time by 84%, on average. It demonstrates that SSA-AWELM is a promising approach for passenger flow forecasting.

Keywords: automatic fare collection system; passenger flow forecasting; time series decomposition; singular spectrum analysis; ensemble learning; extreme learning machine

---

1. Introduction

As an import part in urban public transit, metro transit has developed rapidly and attracted a quantity of passengers in recent years. It is a great challenge for operators and design-makers to optimize the metro schedules and organize the passengers in the stations effectively. Accurate and timely short-term passenger flow forecasting is the fundament of intelligent transport systems (ITS) [1]. The prediction results not only offer evidence for passenger guidance to prevent congestion and trampling [2] but, also, provide necessary information for the metro schedule coordination scheme to match the metro capacity with the passenger flow demand.

As the connections of different metro lines, transfer stations are crucial in metro networks. Some researchers utilized the complex network theory to investigate the characteristics of the metro networks such as Beijing [3], Shanghai [4], Guangzhou [5], and some other cities [6]. The findings of their studies indicated that transfer stations played the most significant role in the networks. Some of them [3,4] suggested that the transfer stations should be paid more attention to. In addition, the passenger flow in the transfer station is usually much larger than that in a regular station, and the passenger flow increases more rapidly at the rush hours in the morning and evening. This is because transfer stations are usually located in areas with large travel demands—for instance, a transportation hub, business district, and so forth. Therefore, in order to avoid pedestrian congestion or early warnings of burst passenger flows for operators, it is vital to forecast the passenger flow accurately and timely in a transfer station.

The passenger flow is defined as the number of boarding or alighting pedestrians at the target station during a constant interval in the prediction tasks [7,8]. In previous studies, the collection of passenger flows mainly includes two ways, as follows:

1. **Videos.** The passenger flow videos are generally used to extract the passenger trajectories through image-processing techniques. The extracted data can help researchers to investigate and analyze passenger behaviors [9].

2. **Automatic Fare Collection (AFC) systems.** Based on AFC systems, the passenger boarding and alighting information is recorded by the sensors in turnstiles automatically, and the recorded data is easy to access. The AFC systems are initially designed and employed to charge the passengers automatically. Since the AFC systems can also record some extra information of the passengers (i.e., personal identification, boarding/alighting time, boarding/alighting station, etc.), the AFC data has been used in the researches of transportation engineering. These studies are mainly focused on four fields: prediction of passenger flow [2,7,10–12], analysis of passenger flow patterns [13], investigation of passenger behaviors [14,15], and evaluation of metro networks [3,6].

The task of passenger flow prediction is quite similar to traffic flow prediction [7,8,12,16], which is only different in the input data of the models. Therefore, many practical models of traffic flow prediction could be referred to as well. In the studies to date, the passenger/traffic flow prediction approaches are roughly classified into four categories, as listed below:

1. **Parametric models.** Due to a low computation complexity, parametric models are widely used in early studies—for instance, autoregressive integrated moving average (ARIMA) [17,18], Kalman filter (KF) [11], exponential smoothing (ES) [19], and so on. However, these models are sensitive to passenger flow patterns, since they are established based on the assumption of linearity.

2. **Nonparametric models.** In order to capture the nonlinearity of passenger flow, the nonparametric models are introduced in subsequent researches, such as K-nearest neighbor (KNN) [20,21], support vector regression (SVR) [7,10], artificial neural network (ANN) [1,22], etc. The empirical results from these studies have suggested that the nonparametric models usually performed better than parametric models when the data size was large. It is owing to the ability of nonlinearity modeling.

3. **Hybrid models.** The hybrid models are the combination of two or more individual methods. Due to both the linearity and nonlinearity of passenger flow, the hybrid models [2,23–26] are proposed to capture these two natures to increase the prediction accuracy. Both theoretical and empirical findings have demonstrated that the integration of different models can take full advantage of these models. Thus, this is an effective way to improve the predictive performance.

4. **Deep-learning models.** Besides the aforementioned three kinds of models, according to the latest researches, the deep-learning methods have been introduced and developed in the passenger flow forecasting problem, including long short-term memory (LSTM) [12,16,27], deep belief network (DBN) [28], stacked autoencoders (SAE) [29], convolutional neural network (CNN) [12,30], etc. Due to the universal approximation capability of complex neural networks, the deep-learning models can approximate any nonlinear function in theory [24,31]. From the findings of these studies, deep-learning models usually show a superiority of high forecasting accuracy to parametric and nonparametric models. However, because of high computation complexity,

the deep-learning models will require significant resources and training time [32]. In addition, these models are usually regraded as a "black box" [23] and lack interpretability of the results [32].

In recent studies, the combination of time series decomposition approaches is a novel research interest of the hybrid models to make a better predictive performance. The principle of this kind of model is that a complicated time series can be simplified through disaggregating the sequence into multiple frequency components. The decomposed components are forecasted separately, and then, these predicted results are summed as the final outcomes. The widely used time series decomposition methods include: wavelet decomposition (WD) [25,33], empirical mode decomposition (EMD) [2,26,34], Seasonal and Trend Decomposition Using Loess (STL) [35,36], singular spectrum analysis (SSA) [37–39], and so on. Sun et al. [25] and Liu et al. [33] employed the WD approach to decompose the original passenger flow into several high-frequency and low-frequency sequences, and then, these sequences were forecasted based on least squares SVR by Sun et al. [25] and extreme learning machine (ELM) by Liu et al. [33], respectively. Chen et al. [2], Wei and Chen [26], and Chen and Wei [34] all proposed that the passenger flow could be regarded as a nonlinear and nonstationary signal, and they utilized EMD to decompose the original passenger flow into nine intrinsic mode functions (IMF) components and one residue. Wei and Chen [26] predicted the disaggregated components through ANN, while Chen et al. [2] predicted them through LSTM. Qin et al. [35] utilized STL to disaggregate the monthly air passenger flow into three subseries: seasonal, trend, and residual series. Then, they developed the Echo State Network (ESN) to forecast each decomposed series. Chen et al. [36] also employed STL to decompose the daily metro ridership, and LSTM was used in the prediction stage. As for the SSA method, to the best of our knowledge, this method has never been introduced to an analysis passenger flow to date, although this method was devolved for traffic flow prediction. Mao et al. [37], Shang et al. [38], and Guo et al. [39] all have developed this method to analyze the traffic flow time series and obtained several components with different amplitudes and frequencies. Then, they reconstructed these components into a smoothing part and residue. In this way, the SSA could be regarded as a filter to remove noise from the original sequence. During the stage of forecasting, the denoise data was predicted by ELM [38] and a grey system model [39], respectively. Overall, these studies have clearly indicated that the hybridization of time series decomposition approaches can make an obvious improvement on the predictive accuracy. However, all the aforementioned literatures have failed to investigate the potential characteristic of passenger flow from the decomposed results.

In this study, a novel hybrid model (i.e., SSA-AWELM), SSA combined with an AdaBoost-weighted extreme learning machine (AWELM), is proposed to achieve more accurate predicted results for the metro passenger flow. The experimental data, recorded by the sensors in turnstiles, is collected from an AFC system. The main works of this paper are briefly described as follows:

1. The SSA approach is developed to decompose the original passenger into three components: trend, periodicity, and residue. Investigation of the three components can discover the inner characteristics of the original data.
2. The ELM improved by AdaBoost (i.e., AWELM) is developed to forecast the three components. ELM, a neural network famous for fast computer speeds, is implemented, and the prediction performance is enhanced through AdaBoost ensemble learning. Thus, the hybrid model SSA-AWELM has the advantage of both accuracy and speediness for passenger flow forecasting.
3. Multistep-ahead prediction of the passenger flow is established, which can offer more information of the future. A dataset collected from a metro AFC system is utilized to carry out the prediction tests and comparative analysis.

The rest of this paper is organized as follows: In Section 2, the problem is defined, and the proposed method is formulated. In Section 3, the procedures of data collection, data preprocessing, and design of the experiment are elaborated. The results and findings are analyzed and discussed in Section 4. At last, the conclusions are drawn in Section 5.

2. Materials and Methods

In this section, the AFC system is briefly introduced, and the passenger flow forecasting problem is explained in detail. In particular, the model SSA-AWELM is formulated to improve upon the performance of predictions.

2.1. Automatic Fare Collection Systems

The automatic fare collection (AFC) systems are established on the Internet of Things (IoT) and wireless sensor networks (WSN). As displayed in Figure 1, a typical AFC system consists of five hierarchical levels: cleaning center (CC), line centers (LC), station computers (SC), station equipment, and smart tickets and cards, from top to bottom [40]. A passenger touches a smart ticket or card, which has an integrated circuit (IC) clip (a type of microsensor) inside, to a turnstile when boarding or alighting; meanwhile, the sensor in the turnstile will respond and record some necessary information. Then, the information will be transmitted to the SC, LC, and, finally, to the CC. In addition, there are a few differences between boarding and alighting. When the passenger alights and passes a turnstile, the sensor will compute the traveling mileage and charge the fare automatically, and this transaction could be completed in milliseconds.

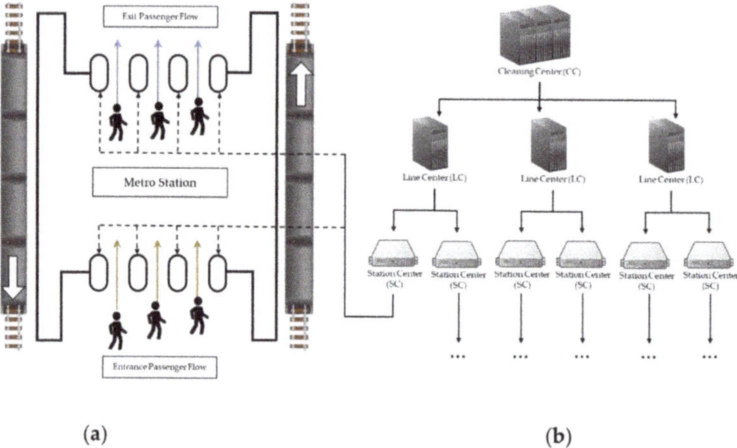

Figure 1. Brief structure of an automatic fare collection (AFC) system: (**a**) metro station and (**b**) computer cluster.

The AFC system can not only be employed by operators to collect the fares from passengers conveniently. For researchers, what is the most important is that the data mining results from the recorded information could assist with analyzing the operational quality, since the records include the personal identification, boarding/alighting station, boarding/alighting time, and some other useful information. Based on AFC systems, the passenger boarding and alighting information could be recorded by the sensors in turnstiles automatically, and the recorded data could be accessed easily. This makes it possible to realize real-time predictions of the metro passenger flow.

2.2. Passenger Flow Forecasting Problem

As mentioned in Section 1, passenger flow is the sum of boarding or alighting pedestrians during a constant interval (i.e., 5 min, 10 min, etc.) in the target station. Suppose x_t denotes the entrance or exit passenger flow at the time t, then it is obvious that x_t varies with the time. The passenger flow forecasting problem can be treated as a time series forecasting task, and the passenger flow time series takes the instinct of temporal dependence. In other words, the passenger flow is highly related to the historical data. Therefore, the research problem addressed in this paper is to forecast x_t by the historical passenger flow data $\{x_{t-1}, x_{t-2}, x_{t-3}, \ldots, x_{t-n}\}$, which is formulated as follows:

$$\hat{x}_t = E(x_{t-1}, x_{t-2}, \ldots, x_{t-n}) \tag{1}$$

where $\hat{x}_t$ represents the predictive value at time t, $E(\cdot)$ represents an established prediction model, and n represents the order of time lagging.

Although the single-step passenger flow forecasting has been widely studied, in order to provide travelers and managers with further information about passenger flow, multistep forecasting is necessary. In our study, the iterated strategy, which is widely used in time series predictions [41,42], is adopted for multistep passenger flow forecasting. As Equation (2) expresses, based on the established model with single-step prediction, the iterated strategy inputs a prediction value into the same model to forecast the value at the next time. It continues in this manner until reaching the maximum prediction horizon. The iterated strategy has two outstanding advantages. One is that the model just requires being trained once, and the other is that the prediction steps are unlimited.

$$\begin{aligned}
\hat{x}_{t+1} &= E(\hat{x}_t, x_{t-1}, \ldots, x_{t-n+1}) \\
\hat{x}_{t+2} &= E(\hat{x}_{t+1}, \hat{x}_t, \ldots, x_{t-n+2})
\end{aligned} \tag{2}$$

$$\cdots$$

2.3. The Proposed Hybrid Model

2.3.1. Singular Spectrum Analysis

Singular spectrum analysis (SSA) is a time series analysis approach without any statistical assumptions [43]. It can decompose the original data into several components. This method has been widely used to decompose the time series including traffic flow [37–39]. In this study, this approach is implemented to analyze the passenger flow. Suppose $Y(t)$ ($t = 1, 2, \ldots, N$) denotes the original passenger flow sequence with length N. The processes of the SSA approach contains four steps, as follows:

Step 1: Embedding

The original sequence $Y(t)$ is transformed into the trajectory matrix $F \in \mathbb{R}^{L \times K}$, which is calculated as the following equation:

$$F = \begin{bmatrix} f_1 & f_2 & \cdots & f_K \\ f_2 & f_3 & \cdots & f_{K+1} \\ \vdots & \vdots & \ddots & \vdots \\ f_L & f_{L+1} & \cdots & f_N \end{bmatrix} \tag{3}$$

where L is window length, $K = N - L + 1$, and f_i is the ith ($1 \leq i \leq N$) value of the original sequence.

Step 2: Singular Value Decomposition (SVD)

The SVD algorithm is conducted to decompose the trajectory matrix F, computed as follows:

$$F = U \cdot \Sigma \cdot V^{\mathrm{T}} = \sum_{i=1}^{d} \sqrt{\lambda_i} U_i V_i^{\mathrm{T}} \tag{4}$$

where Σ is diagonal matrix, and the diagonal elements ($\sqrt{\lambda_1} \geq \sqrt{\lambda_2} \geq \ldots \geq \sqrt{\lambda_d} \geq 0$) are the singular values of F. Vectors U_i and V_i, which are the ith column of matrix U and V, represent the left and right singular vectors, respectively. d represents the number of singular values, and it is also the rank of trajectory matrix F. The collection $\{U_i, \sqrt{\lambda_i}, V_i\}$ is denoted as the ith eigen triple of SVD.

Every eigen triple can reconstruct an elementary matrix F_i of trajectory matrix F:

$$F_i = \sqrt{\lambda_i} U_i V_i^{\mathrm{T}} \tag{5}$$

Thus, the sum of all elementary matrixes F_i is identical to the trajectory matrix F. The contribution of elementary matrix F_i is measured by the corresponding eigen value (equal to the square of the singular value) as the following equation:

$$\eta_i = \frac{\lambda_i}{\sum_{i=1}^{d} \lambda_i} \tag{6}$$

Step 3: Grouping

Indices set $D = \{1, 2, \dots, d\}$ is divided into M disjointed subsets $I_1, I_2, \dots, I_M$. Every indices subset I_m ($m = 1, 2, \dots, M$) is regarded as one group, and the elementary matrixes F_i ($i \in I_m$) in each group are summed. In previous papers, the w-correlation method [43] is prevalent to split the results set. However, this method is conducted from the perspective of signal analysis, which lacks the interpretability for passenger flow. In this study, the elementary matrixes F_i are grouped into three parts of trend F_T, periodicity F_P, and residue F_R, expressed as Equation (7), and this process is detailed in Section 4.1.

$$F = F_T + F_P + F_R \tag{7}$$

Step 4: Diagonal averaging

The grouped matrixes F_i ($F_i \in \{F_T, F_P, F_R\}$) are transformed into the one-dimensional time series format by diagonal averaging. Assume f_{ij} ($1 \leq i \leq L$, $1 \leq j \leq K$) is the element of matrix F_i, $L^* = \min(L, K)$, $K^* = \max(L, K)$, and $f_{ij}^* = f_{ij}$, if $K > L$; otherwise, $f_{ij}^* = f_{ji}$. Then, every element y_i of the time series $Y_i(t)$ is computed as the following equation:

$$y_t = \begin{cases} \frac{1}{t} \sum_{m=1}^{t} f_{m,t-m+1}^* & 1 \leq t < L^* \\ \frac{1}{L^*} \sum_{m=1}^{L^*} f_{m,t-m+1}^* & L^* \leq t \leq K^* \\ \frac{1}{N-t+1} \sum_{m=t-K^*+1}^{N-K^*+1} f_{m,t-m+1}^* & K^* < t \leq N \end{cases} \tag{8}$$

As such, the original passenger flow $Y(t)$ is disaggregated into three components of trend $T(t)$, periodicity $P(t)$, and residue $R(t)$.

2.3.2. AdaBoost Ensemble Learning

As a strategy of ensemble learning, AdaBoost was originally proposed by Freund and Schapire [44] for classification problems. Drucker [45] developed this algorithm in the application of a regression problem, and it was improved upon by Solomatine and Shrestha [46,47]. With the integration of a few homogenous models (called base learners), this method can improve the performance of base learners. In this study, the AdaBoost algorithm is utilized to assist the ELM to predict the passenger flow more accurately.

Supposing a dataset $\{(x_i, y_i)\}_{i=1}^{N}$ with N samples, T is the maximum iteration number. The specific steps of AdaBoost is presented as the following:

Step 1: Initialize the distribution of sample weights:

$$\Gamma_1 = [\gamma_{1,1}, \gamma_{1,2}, \dots, \gamma_{1,N}]^T, \text{where } \gamma_{1,n} = \frac{1}{N}, n = 1, 2, \dots, N \tag{9}$$

Step 2: For the training process of each iteration, $t = 1, 2, \ldots, T$.

Step 2.1: Use the dataset with a distribution of Γ_t to train the WELM and obtain the base learner $E_t(x)$.

Step 2.2: Calculate the absolute relative error of each sample and the error rate of $E_t(x)$:

$$\varepsilon_t = \sum_{n=1}^{N} \gamma_{t,n} n : \left| \frac{E_t(x_n) - y_n}{y_n} \right| > \varphi \tag{10}$$

where $\left| (E_t(x_n) - y_n) / y_n \right|$ represents the absolute relative error of each sample; ε_t is the error rate of $E_t(x)$; and $n = 1, 2, \ldots, N$ is the index of the sample. $n : \left| (E_t(x_n) - y_n) / y_n \right| > \varphi$ represents that only the error for any particular sample is greater than the preset error, the so-called threshold φ; the corresponding sample will be considered. φ is a preset parameter and will be discussed at the end of the present subsection. More details are described in [47].

Step 2.3: Calculate the coefficient for updating the sample weights:

$$\beta_t = \varepsilon_t^k \tag{11}$$

where k is the power coefficient of error rate ε_t requiring to be preset. According to the study of Solomatine and Shrestha [47], k is selected from 1 (linear law), 2 (square law), and 3 (cubic law). A high value of k may cause the algorithm to become unstable. Thus, k is set as 1 in our study.

Step 2.4: Update the distribution of sample weights:

$$\gamma_{t+1,n} = \frac{\gamma_{t,n}}{Z_t} \times \begin{cases} \beta_t, & if \left| \frac{E_t(x_n) - y_n}{y_n} \right| \leq \varphi \\ 1, & otherwise \end{cases} , n = 1, 2, \ldots, N \tag{12}$$

where Z_t is a normalization factor, such that $\sum_{n=1}^{N} \gamma_{t+1,n} = 1$.

Step 3: Update $t = t + 1$ and loop **Step 2.1** to **2.4** until reaching the maximum iteration number T. Finally, the output is computed as:

$$g(x) = \frac{1}{\sum\limits_{t=1}^{T} \ln \frac{1}{\beta_t}} \left[\sum_{t=1}^{T} \left(\ln \frac{1}{\beta_t} \right) E_t(x) \right] \tag{13}$$

The AdaBoost algorithm is sensitive to the threshold φ. If the φ is too low, the model will be underfitting. On the other hand, too high a value of φ will raise overfitting problems. In our study, the threshold φ is set adaptively according to the median of absolute relative errors ε_t during each iteration, expressed as the following equation:

$$\varphi = \text{median}\{\varepsilon_1, \varepsilon_2, \ldots, \varepsilon_N\} \tag{14}$$

As presented in the above steps, AdaBoost is an iteration process. The base learner will be trained, and the distribution of the sample weights will be updated during each iteration. Thus, if the base learner is complex and spends lots of computing time, the consuming time of AdaBoost will increase linearly. In this study, ELM is adopted as the base learner, which is famous for its fast training speed. This model is elaborated in the next subsection.

2.3.3. Weighted Extreme Learning Machine

Extreme learning machine (ELM) is a kind of single hidden layer feed-forward network (SLFN), which is proposed by Huang at el. [48]. Compared with traditional ANN models, ELM does not need to tune the input weights and hidden layer biases during training. After the initialization of the ELM, the input weights and hidden biases are fixed, and only the output weights are optimized. Therefore, the training process of ELM is faster than the traditional ANN model. Since weighted samples are used to train the base learners of AdaBoost, the weighted extreme learning machine (WELM) is developed in this study.

Assuming a weighted dataset $\left\{ \left(x_i, y_i, \gamma_i \right) \right\}_{i=1}^{N}$ with N samples, and $x_i = [x_{i,1}, x_{i,2}, \ldots, x_{i,P}]^T \in \mathbb{R}^{P \times 1}$ and $y_i = [y_{i,1}, y_{i,2}, \ldots, y_{i,Q}]^T \in \mathbb{R}^{Q \times 1}$, γ_i represent the input vector, output vector, and sample weights, respectively. The output of ELM with h hidden neurons is expressed as:

$$f(x_i) = \sum_{h=1}^{H} \beta_h g(w_h x_i + b_h), i = 1, 2, \ldots N \tag{15}$$

where $w_h = [w_{h,1}, w_{h,2}, \ldots, w_{h,3}]^T$ represents the connection weights from the input layer to the hth hidden neuron; b_h represents the bias in the hth hidden neuron; $\beta_h = [\beta_{h,1}, \beta_{h,2}, \ldots, \beta_{h,Q}]^T$ represents the connection weights from the hth hidden neuron to the output layer; and $g(\cdot)$ is the activation function, and the *sigmoid* function is adopted in this study, which is formulated as $g(\cdot) = 1/(1 + e^{-x})$. Since w_h and b_h are assigned initially, Equation (15) can be simplified as:

$$H\beta = Y \tag{16}$$

where $\beta = [\beta_1, \beta_2, \ldots, \beta_H]^T$; $Y = [y_1, y_2, \ldots, y_N]^T$; and H is the output matrix of the hidden layer, expressed as:

$$H = \begin{bmatrix} g(w_1 x_1 + b_1) & \cdots & g(w_H x_1 + b_H) \\ \vdots & \ddots & \vdots \\ g(w_1 x_N + b_1) & \cdots & g(w_H x_N + b_H) \end{bmatrix} \tag{17}$$

The purpose of ELM is to optimize β with the object of the minimum mean square error cost function, which is expressed as $\min_{\beta} \|H\beta - Y^2\|$. Furthermore, when the samples are weighted with Γ, the loss function of every sample requires multiplying with the corresponding sample weight, formulated as:

$$\min_{\beta} \mathrm{diag}(\Gamma) H\beta - Y^2 \tag{18}$$

where $\mathrm{diag}(\Gamma)$ is the diagonal matrix with the diagonal of Γ, and the solution of Equation (18) is:

$$\beta = \left(H^T \mathrm{diag}(\Gamma) H \right)^{-1} H^T \mathrm{diag}(\Gamma) Y \tag{19}$$

Overall, the output weights β of the WELM can be computed according to Equation (19) directly. It is different to the training process of traditional ANN, which is an iteration process to update connecting weights and neuron biases. This is the reason why ELM costs much less computing time than the traditional ANN.

2.3.4. The Hybrid Model

The model combination of a singular spectrum analysis and AdaBoost-weighted extreme learning machine is proposed to forecast the passenger flow in this paper, symbolized as SSA-AWELM. The flow chart of this hybrid model is displayed in Figure 2, and the special process of it is described as follows:

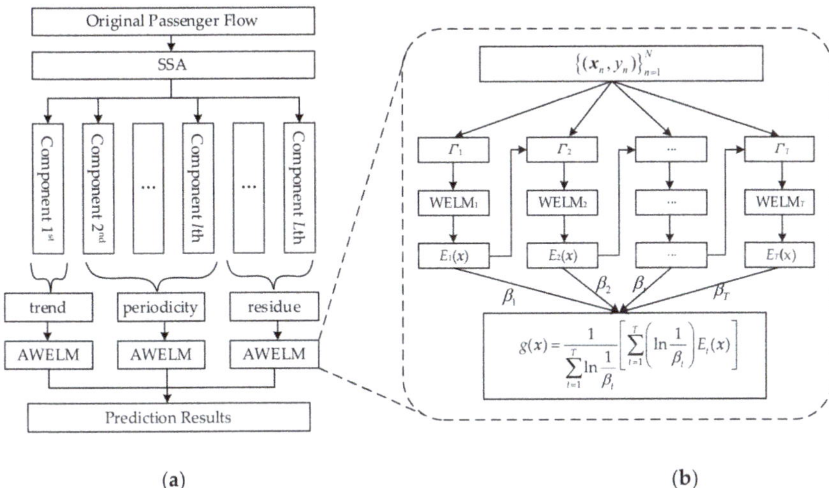

(a)   **(b)**

Figure 2. The flow chart of the hybrid singular spectrum analysis-AdaBoost-weighted extreme learning machine (SSA-AWELM) model: (**a**) SSA-AWELM and (**b**) AWELM.

Step 1: SSA for decomposition. The original passenger flow is decomposed into several components by SSA approach, and these components are grouped into three parts of trend, periodicity, and residue.

Step 2: AWELM for components forecasting. The WELM improved by AdaBoost (AWELM) is implemented to model and predict the three components, separately.

Step 3: Integration for final forecasting results. The final outcomes of forecasting the passenger flow are calculated by summing the predicted results of the three components.

3. Empirical Study

3.1. Data Collection

In this paper, the passengers' alighting and boarding dataset is collected from the AFC system of Hangzhou metro in China. The dataset is online and provided by Ali Tianchi [49]. This dataset recorded detailed information when the passengers passed the turnstiles. The duration of the data was from the 1st to 26th in January 2019. The dataset includes seven fields, and they are listed in Table 1. In addition, some samples of the dataset are provided in Table 2.

Table 1. Data fields collected from the automatic fare collection (AFC) system of Hangzhou metro.

	Field	Description
1	Time	Passenger boarding or alighting time
2	Line ID	Number assigned to every metro line
3	Station ID	Number assigned to every metro station
4	Device ID	Number assigned to every turnstile
5	Status	Boarding or alighting: 0 represents alighting, and 1 represents boarding
6	User ID	Personal identification information
7	Pay Type	Ticket type

Table 2. Some samples of the collected data.

	Time	Line ID	Station ID	Device ID	Status	User ID	Pay Type
1	2019-01-01 06:00:00	B	15	759	1	Baecf***	1
2	2019-01-01 06:00:00	B	32	1558	1	Da226***	3
3	2019-01-01 06:00:01	B	8	402	1	Bb8e6***	1
4	2019-01-01 06:00:02	B	32	1562	1	C03b9***	2
5	2019-01-01 06:00:02	B	9	446	0	Be9c9***	1

3.2. Data Preprocessing

The preprocessing is to obtain a passenger flow time series data from the raw AFC dataset. In this study, the passenger flow data of the Qianjiang Road Station (Q.R. Sta.) and JinJiang Station (J. Sta.) are selected to conduct experiments. As displayed in Figure 3, the Q.R. Sta. is a transfer station between Line 2 and Line 4, and it is located in the Qianjiang New Town Central Business District (CBD). The Jinjiang Station is a transfer station between Line 1 and Line 4, which is located in Wangjiang New Town.

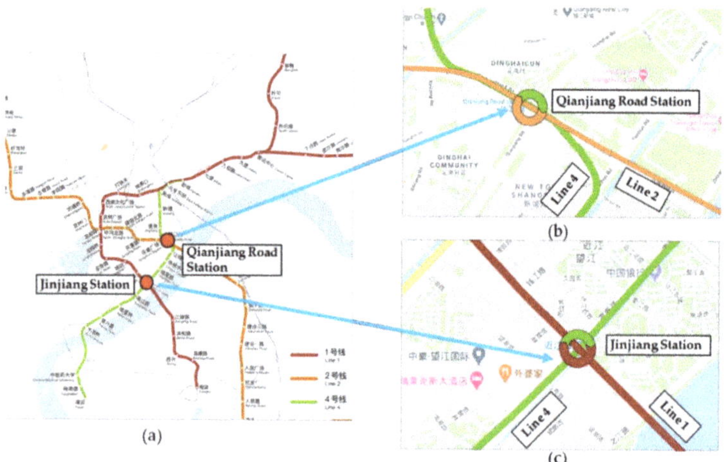

Figure 3. The location of the study metro transfer stations: (**a**) the Hangzhou metro network, (**b**) the Qiangjiang Road Station (Q.R. Sta.), and (**c**) the Jinjiang Road Station (J. Sta.).

According to previous studies [11,50], the raw recorded data are usually aggregated into 5-min intervals to obtain the passenger flow sequence. In order to keep the complete cycle periods of the sequence data, three continuous weeks, which were from the 6th to 26th of January, were selected from the AFC dataset. The time range selected was from 6:00 to 23:00 according to the operation time of the Hangzhou metro system, though a few records in the AFC dataset were out of this range. At last, there were 204 samples on average in one day and 4284 samples in total. Furthermore, the exit and entrance passenger flow sequences were computed separately. Hence, four experimental datasets were established, and they were used to test the proposed model, respectively.

The extracted passenger flow sequences are presented in Figure 4. Both the exit and entrance passenger flows on weekdays have distinct peaks in the morning (about from 8:00 to 9:00) and evening (about from 18:00 to 19:00) rush hours, while these patterns disappear on the weekends. Moreover, the peak patterns of exit and entrance passenger flows on weekdays are different. Taking the Q.R. Sta. as an example, the exit passenger flow in the morning rush hour (about 500 pedestrians per 5 min) is approximately 2.5 times larger than that in the evening rush hour (about 200 pedestrians per 5 min). On the contrary, the entrance passenger flow in the evening rush hour (about 300 pedestrians per

5 min) is approximately 1.5 times larger than that in the morning rush hour (about 200 pedestrians per 5 min). These results indicate that most passengers in this station are commuters. This finding agrees with the location of this station, i.e., it is in the Qianjiang New Town CBD and surrounded by numerous office buildings.

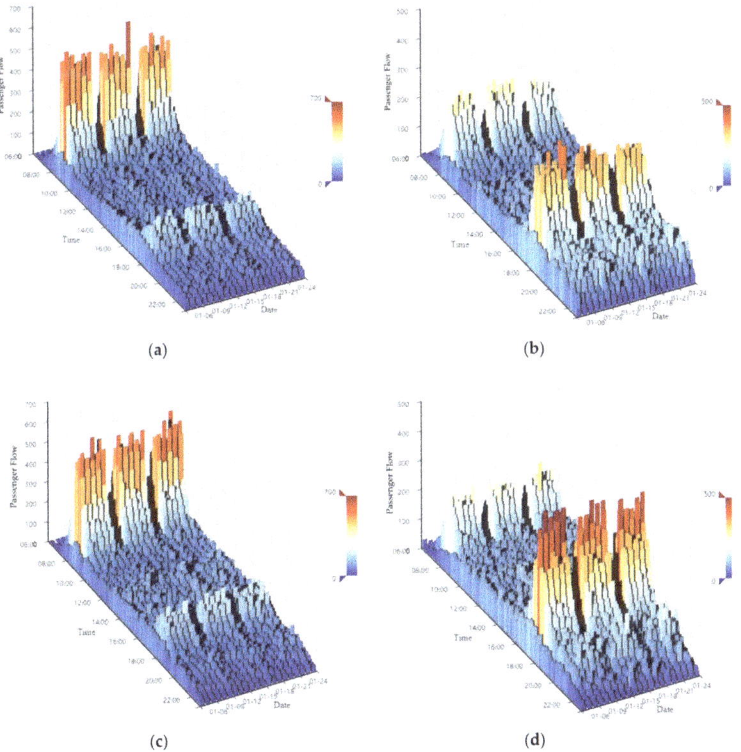

Figure 4. Passenger flow of the study metro transfer stations: (**a**) exit passenger flow of the Q.R. Sta., (**b**) entrance passenger flow of the Q.R. Sta., (**c**) exit passenger flow of the J. Sta, and (**d**) entrance passenger flow of the J. Sta.

The four datasets are all split into training datasets (i.e., the 6th to 19th of January) and testing datasets (i.e., the 20th to 26th of January). The grid search and 5-fold cross-validation methods are used to evaluate the training performance and determine the hyper-parameters of the models. Then, the models with determined hyper-parameters are evaluated by testing the datasets.

3.3. Comparison Models and Evaluation Measures

In order to demonstrate the contributions of the proposed SSA-AWELM model, the classical time series model ARIMA and four extra models based on the neural network, including ANN, LSTM, ELM, and AWELM, are tested as benchmarks. They are listed as follows:

- **ARIMA:** ARIMA is a classical statistical model for time series forecasting. It is widely used to predict traffic flow and passenger flow in early studies [17]. The performance of ARIMA is affected by three parameters: autoregressive order p, difference order d, and moving average order q. Generally, d is set based on the stationarity test, and the p and q are selected from the range of [0,12] based on the Bayesian information criterion (BIC) [51].

- **ANN:** Due to the ability of nonlinearity, the ANN model is widely used in time series modeling, including passenger flow forecasting. A typical ANN model consists of three parts: one input layer, one hidden layer, and one output layer and optimized through a back-propagation algorithm (thus, it also aliased as BPNN). In this study, the ANN model is optimized by a stochastic Adam algorithm with a mean square error (MSE) loss of function. The learning rate is set as 0.001, the batch-size is 256, and the epochs is 1000.
- **LSTM:** As a prevalent deep-learning model for time series modeling, the well-designed LSTM units replace traditional neurons in a hidden layer, which can assist the LSTM model to capture the temporal characteristics. This model has also been developed to predict the passenger flow. The parameters are set identically to the ANN model.
- **ELM:** The ELM model has been elaborated in Section 2.3.3.; the individual ELM model is utilized to forecast the passenger flow as a comparison.
- **AWELM:** The AWELM is combined by AdaBoost and WELM, which has been represented in Section 2.3.3. and Section 2.3.4.

To make sure that every model can achieve the best performance, the well-established grid search and 5-fold cross-validation methods are adopted to determine the hyper-parameters. The neuron number of the hidden layers in four neural network models are all selected from 2 to 50 with step 2, and the base learner of AWELM is selected from 1 to 20 with step 1. The determined hyper-parameters of each model are displayed in the Appendix A (see Table A1). In addition, the input and the output of the models are respectively set as 12 and 1 during training, and the horizon of the multistep-ahead prediction is set as 6. In other words, the passenger flow data at the last hour is used to forecast the next half-hour.

In order to accelerate learning and convergence during the training model, the min-max normalization approach (expressed as Equation (21)) is employed to scale the input data into the range of [0,1] before feeding it into the models. In addition, to obtain the final prediction results, the outputs of the models are rescaled by the reversed min-max normalization approach (expressed as Equation (21)).

$$x' = \frac{x - \min(x)}{\max(x) - \min(x)} \tag{20}$$

$$x = x' \times (\max(x) - \min(x)) + \min(x) \tag{21}$$

In order to evaluate the performances among models, two common measures are introduced in this study. They are the mean absolute error (MAE) and root mean square error (RMSE), computed as follows:

$$\text{MAE} = \frac{1}{N} \sum_{n=1}^{N} |y_n - \hat{y}_n| \tag{22}$$

$$\text{RMSE} = \sqrt{\frac{1}{N} \sum_{n=1}^{N} (y_n - \hat{y}_n)^2} \tag{23}$$

where y_n and $\hat{y}_n$ are the true value and predicted value, respectively, and N is the number of samples.

Besides the aforementioned two measures, the Diebold–Mariano (DM) test [52] is implemented to test the statistical significance between the proposed model and the benchmark models. The null hypothesis is that the prediction accuracy of the tested model $E_T(x)$ is equal to the reference model $E_R(x)$. In this study, the square error is adopted to measure the model loss, expressed as $e_i = (\hat{y}_i - y_i)^2$. Then, the DM statistic is defined as follows:

$$\text{DM} = \frac{g}{\sqrt{\hat{V}_g / N}} \tag{24}$$

where $g = \sum_{n=1}^{N} g_n / N$, $g_n = \left(\hat{y}_n^T - y_n\right)^2 - \left(\hat{y}_n^R - y_n\right)^2$, $\hat{V}_g = \gamma_0 + 2\sum_{k=1}^{P-1} \gamma_k$, and γ_k is the autocovariance at lag k, expressed as $\gamma_k = (1/N)\sum_{i=k+1}^{N}(g_i - g)(g_{i-k} - g)$. $\hat{y}_n^T$ and $\hat{y}_n^R$ respectively represent the predicted values of model $E_T(x)$ and $E_R(x)$, P is the prediction horizon, and N is the scale of the testing data.

4. Results Analysis

4.1. Analysis of SSA Decomposition

As mentioned in Section 2.3.1., window length L is the only parameter that requires to be determined before decomposition. From previous studies [37–39], if the time series data shows obvious periodicity, the window length L could be set as one period length. Thus, $L = 204$, because the passenger flow cycles daily (see Figure 4), and 204 samples on average are collected in one day (has been illuminated in Section 3.2.). Then, the original passenger flow can be disaggregated into 204 components. These components are grouped into the three parts of trend, periodicity, and residue, and this is inspired by the study [53] about using SSA to analyze the variants of electricity prices. To facilitate the analysis, taking the dataset of the Q.R. Sta. as an example, the eigen values of each component are plotted in Figure 5.

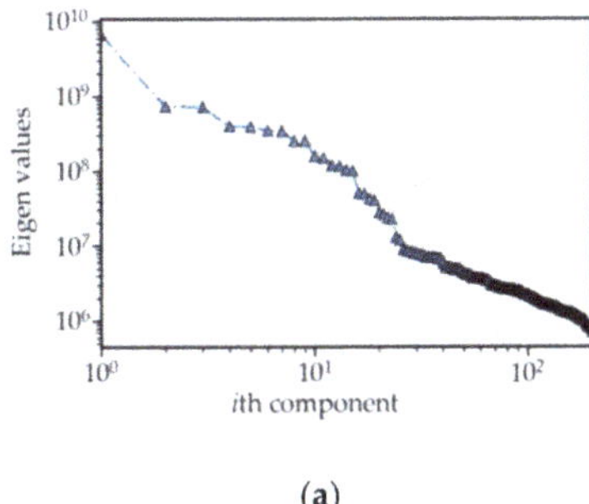
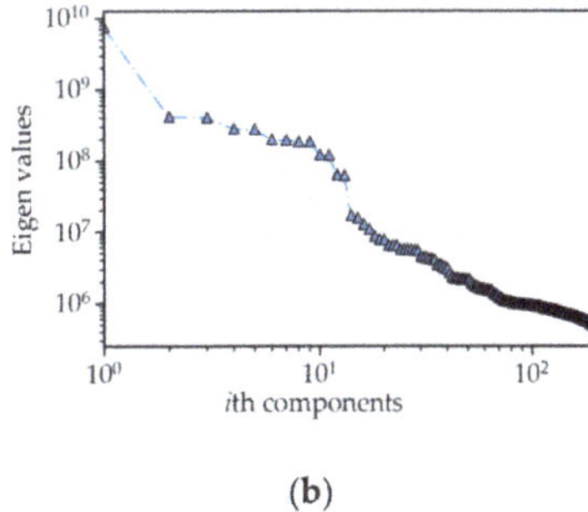

(a)   (b)

Figure 5. Eigen values of decomposed components (Q.R. Sta.): (**a**) exit passenger flow and (**b**) entrance passenger flow.

Taking Figure 5a as an example, it is clear that the first eigen value is significantly larger than the others, and the corresponding component is extracted separately as trend parts. Moreover, the eigen values curve declines slowly after the 23rd component, and the 23rd is regarded as the "break point". Then, the components from the 2nd to 23rd are reconstructed into periodic parts, and the remainder components from the 24th to 204th are reconstructed into residual parts. This is the same way as the entrance passenger flow in Figure 5b, and the "break point" is 13. Then, the components from the 2nd to 13th are reconstructed into periodic parts, and the remainder components from the 14th to 204th are reconstructed into residual parts. Finally, the obtained trend, periodicity, and residues of the original passenger flow are displayed in Figure 6.

As shown in Figure 6, every component can reveal different patterns of the original passenger flow. The trend represents the overall tendency, and the periodicity represents the variants within a day. Furthermore, it could be found in the trend that the passenger flow on weekdays is larger than that on weekends. In the periodical component, the passenger flow shows distinct peaks in the morning and evening rush hour on weekdays, but this is not obvious on the weekends. The peak patterns are different between the exit and entrance passenger flows: the exit passenger flow in the morning rush hour is much larger than that in the evening rush hour, and the entrance passenger flow is contrary to that. As for the residue, it fluctuates irregularly and can be treated as noise.

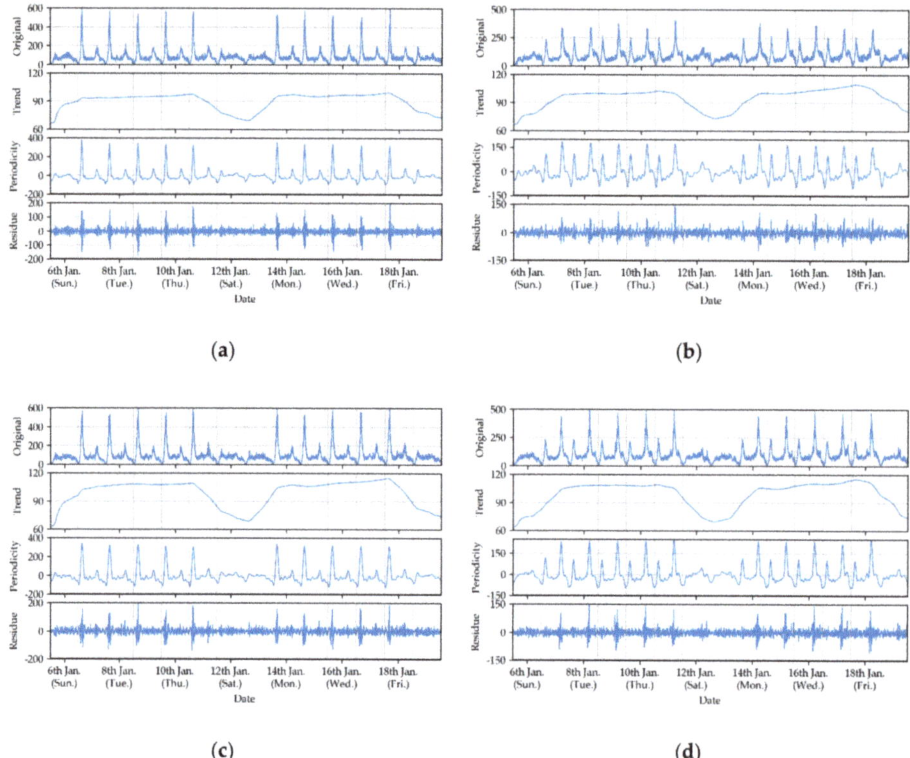

Figure 6. Decomposition results: (**a**) exit passenger flow of the Qianjiang Road Station, (**b**) entrance passenger flow of the Q.R. Sta., (**c**) exit passenger flow of the J. Sta., and (**d**) entrance passenger flow of the J. Sta.

4.2. Analysis of Hyper-Parameters

The performance of SSA-AWELM is highly dependent on the forecasting model AWELM of each component, and the AWELM has two hyper-parameters: the number of base learners (i.e., WELM) T and the hidden neurons of WELM H. The well-established grid search and five-fold cross-validation methods are adopted to determine the T and H. The H is selected from 2 to 50 with step 2, and T is selected from 1 to 20 with step 1. Taking the dataset of the Q.R. Sta. as an example, the process of hyper-parameter selection is displayed in Figure 7, and the log transformation is applied to the MSE to distinguish different values clearly. It could be found that AWELM is sensitive to the hidden neurons H but insensitive to base learner number T. Finally, the determined hyper-parameters H and T of AWELM are provided in Table A1 (see Appendix A).

4.3. Analysis of Forecasting Results

For the sake of a comparison analysis, the average evaluation measures of the forecasting results across all the six prediction horizons are presented in Table 3, and the scatter points of the true and predicted values are displayed in Figure 8. From Table 3, it is worth noting that the proposed SSA-AWELM performs best among all the models, followed by LSTM, ANN, AWELM, ELM, and ARIMA. Compared to LSTM, the RMSE and MAE of SSA-AWELM respectively reduced by 22.5% and 21.3% on average in the case of the Q.R. Sta. and reduced by 23.6% and 20.0% on average in the case of the J. Sta. AWELM performs a little better than ELM, which indicates the AdaBoost algorithm can reduce the prediction errors but with limitations. As expected, ARIMA is always inferior to other models, because it is a linear model. In addition, it can be seen in Figure 8 that the scatter points in SSA-AWELM are closest to the expectation line, and the corresponding coefficient of determination R^2 is largest. All the above findings can prove that the proposed SSA-AWELM is an effective approach to improve the accuracy of passenger flow forecasting. Furthermore, to compare the consuming time of different models, the training time is provided in Table A2 (See Appendix A).

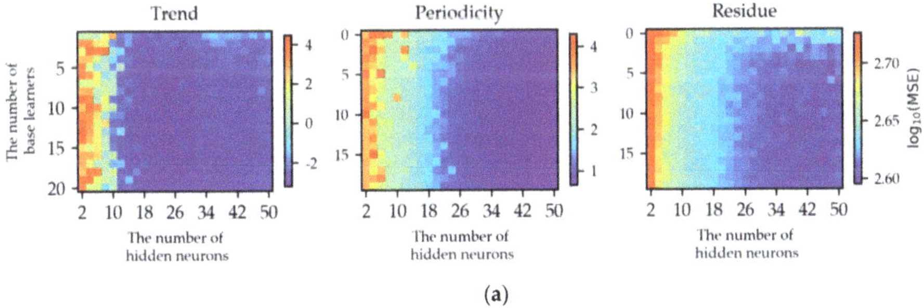

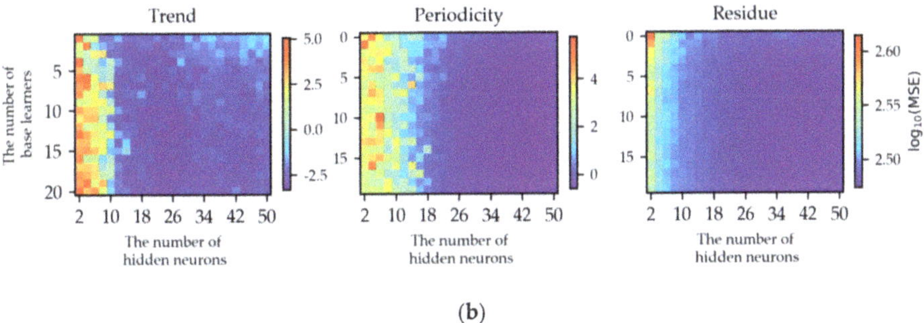

Figure 7. The process of hyper-parameter selection for SSA-AWELM (Q.R. Sta.): (**a**) exit passenger flow and (**b**) entrance passenger flow.

Table 3. Average evaluation measures across all six prediction horizons. RMSE: root mean square error and MAE: mean absolute error. Q. R. Sta.: Qiangjiang Road Station and J. Sta.: Jinjiang Road Station. ARIMA: Auto Regressive Integrated Moving Average. ANN: Artificial Neural Network. LSTM: Long Short-Term Memory neural network. ELM: Extreme Learning Machine. AWELM: AdaBoost-Weighted Extreme Learning Machine. SSA-AWELM: the proposed model combining Singular Spectrum Analysis with AdaBoost-Weighted Extreme Learning Machine.

Model	Exit Passenger Flow of Q.R. Sta.		Entrance Passenger Flow of Q.R. Sta.		Exit Passenger Flow of J. Sta.		Entrance Passenger Flow of J. Sta.	
	RMSE	MAE	RMSE	MAE	RMSE	MAE	RMSE	MAE
ARIMA	40.01	22.60	26.90	19.19	38.55	24.23	31.37	20.44
ANN	25.88	17.18	24.21	16.86	25.67	17.92	28.87	18.56
LSTM	23.34	15.78	22.24	15.69	24.01	16.76	27.71	17.74
ELM	29.14	19.28	24.49	17.35	28.32	19.67	29.52	18.98
AWELM	28.22	18.28	24.35	17.15	27.00	18.71	29.37	18.55
SSA-AWELM	19.53	12.93	15.77	11.84	22.24	15.10	17.25	12.49

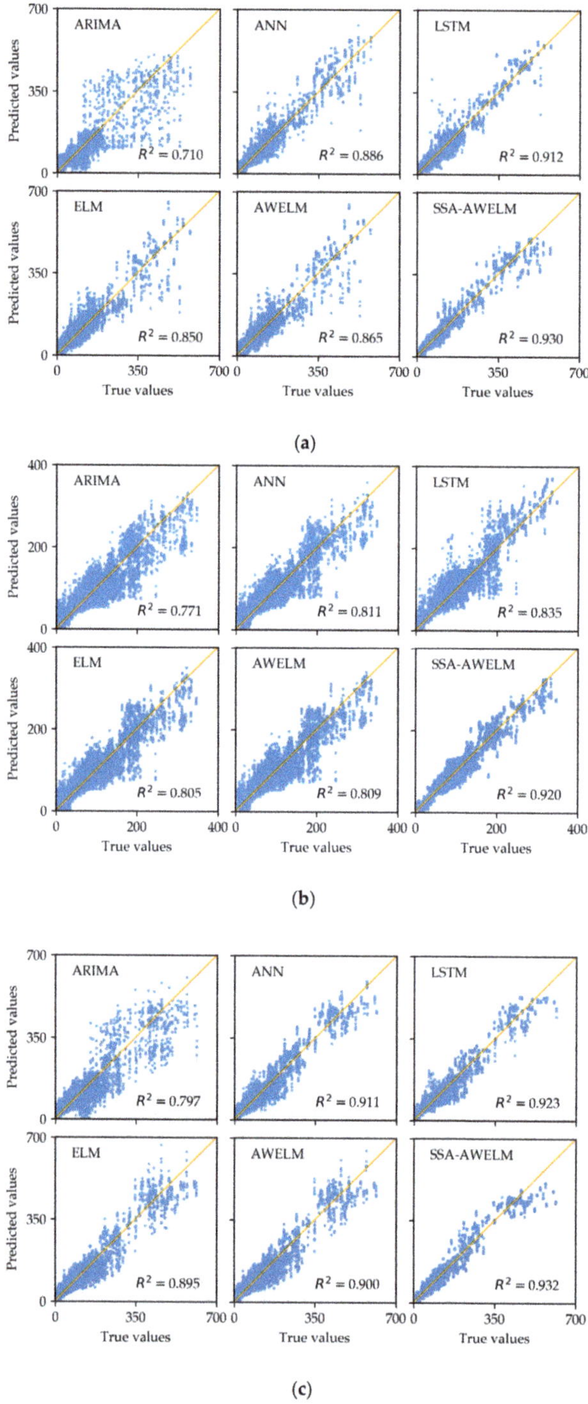

(a)

(b)

(c)

Figure 8. *Cont.*

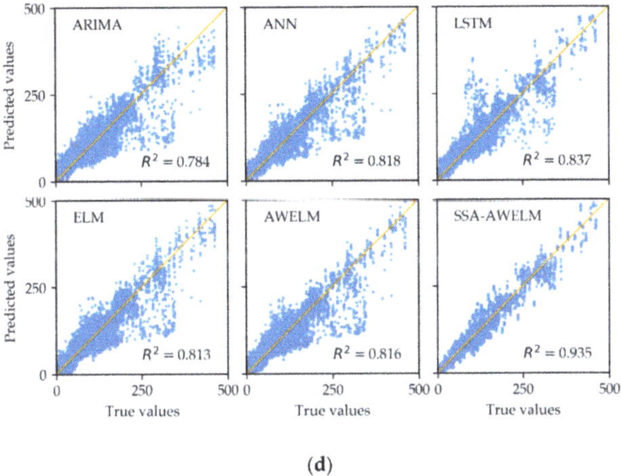

(d)

Figure 8. Prediction results: (**a**) exit passenger flow of the Q. R. Sta., (**b**) entrance passenger flow of the Q.R. Sta., (**c**) exit passenger flow of the J. Sta., and (**d**) entrance passenger flow of the J. Sta. ARIMA: Auto Regressive Integrated Moving Average. ANN: Artificial Neural Network. LSTM: Long Short-Term Memory neural network. ELM: Extreme Learning Machine. AWELM: AdaBoost-Weighted Extreme Learning Machine. SSA-AWELM: the proposed model combining Singular Spectrum Analysis and AdaBoost-Weighted Extreme Learning Machine.

4.4. Analysis of Multistep-Ahead Forecasting

In order to analyze the multistep forecasting errors, the evaluation measures of each prediction horizon are displayed in Figure 9. The DM test results of the comparison between the proposed SSA-AWELM and benchmarks are presented in Table 4. From Figure 9, it can be seen that the prediction errors of every model increase along the prediction horizons. This is caused by the cumulative errors, which stems from feeding prediction values into the models for multistep-ahead forecasting, and the cumulative errors are inevitable. What stands out in Figure 9 is that the proposed SSA-AWELM always performs best in every prediction horizon, and the errors increase slowest in comparison with the other models. This indicates that the SSA-AWELM can improve the robustness and restrict the propagation of the cumulative error during multistep-ahead forecasting. A reasonable explanation for this finding is that the SSA can decompose the original passenger flow into the three components of trend, periodicity, and residue. Each component holds individual characteristics that can be modeled more easily than the original complex data. Furthermore, compared with ELM, AWELM preforms slightly better. It suggests that AdaBoost can improve the accuracy of ELM but with limitations. Only combining with AdaBoost cannot promote the forecasting accuracy significantly. From Table 4, generally speaking, the proposed SSA-AWELM almost always outperforms the other models at a highly significant level. There are some exceptions when compared with LSTM for the exit passenger flow. In these situations, SSA-AWELM still performs better than LSTM but not always with a highly significant level. This might because LSTM has the advantage of capturing more temporal characteristics in terms of the exit passenger flow. Overall, these findings suggest that the proposed SSA-AWELM is outstanding during multistep-ahead predictions. These results prove that the SSA-AWELM is a robust approach for passenger flow forecasting.

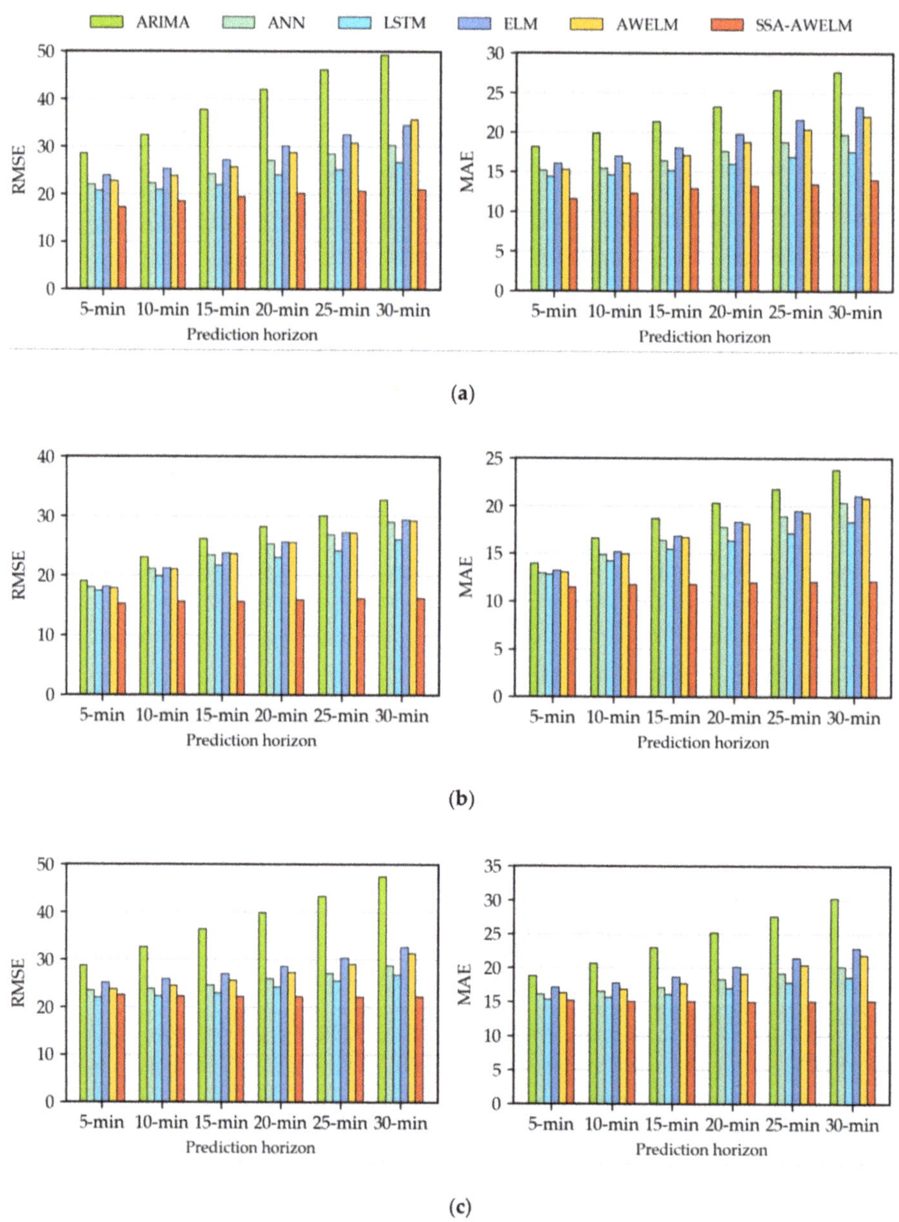

Figure 9. *Cont.*

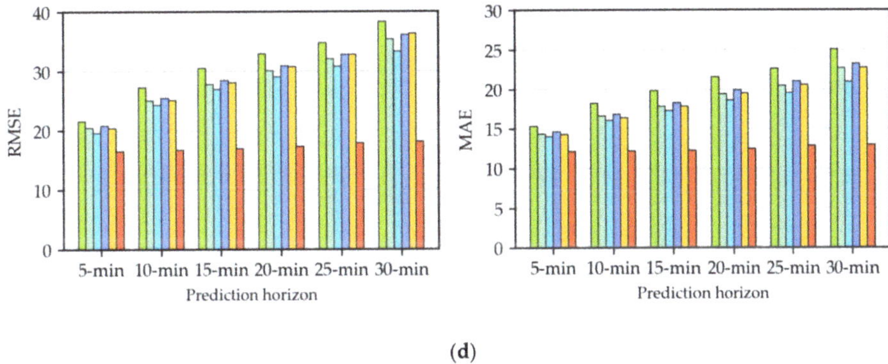

(d)

Figure 9. Evaluation of the multistep predictions: (**a**) exit passenger flow of the Q.R. Sta., (**b**) entrance passenger flow of the Q.R. Sta., (**c**) exit passenger flow of the J. Sta., and (**d**) entrance passenger flow of the J. Sta.

Table 4. Diebold–Mariano (DM) test results of the comparison between the proposed SSA-AWELM and benchmarks.

Prediction Horizon	Case: Exit Passenger Flow of Q.R. Sta.				
	ARIMA	ANN	LSTM	ELM	AWELM
5-min	−6.90 ***	−5.32 ***	−2.51 **	−5.54 ***	−5.07 ***
10-min	−5.58 ***	−3.75 ***	−1.88 *	−4.95 ***	−4.77 ***
15-min	−4.72 ***	−5.00 ***	−1.24	−4.98 ***	−4.91 ***
20-min	−4.13 ***	−3.29 ***	−1.77 *	−4.45 ***	−3.84 ***
25-min	−3.95 ***	−3.36 ***	−1.45	−5.06 ***	−4.39 ***
30-min	−3.78 ***	−3.00 ***	−1.42	−4.98 ***	−4.33 ***
	Case: Entrance Passenger Flow of Q.R. Sta.				
	ARIMA	ANN	LSTM	ELM	AWELM
5-min	−7.60 ***	−5.90 ***	−6.25 ***	−6.22 ***	−5.95 ***
10-min	−6.62 ***	−5.34 ***	−5.00 ***	−5.59 ***	−5.49 ***
15-min	−6.90 ***	−5.71 ***	−5.14 ***	−6.07 ***	−6.06 ***
20-min	−7.23 ***	−6.20 ***	−5.42 ***	−6.75 ***	−6.68 ***
25-min	−7.42 ***	−6.52 ***	−5.75 ***	−7.23 ***	−7.07 ***
30-min	−7.15 ***	−6.37 ***	−6.06 ***	−6.96 ***	−6.78 ***
	Case: Exit Passenger Flow of J. Sta.				
	ARIMA	ANN	LSTM	ELM	AWELM
5-min	−6.07 ***	−1.18	0.99	−2.29 **	−1.57
10-min	−5.80 ***	−2.31 **	0.15	−3.45 ***	−2.71 ***
15-min	−5.82 ***	−3.24 ***	−0.62	−4.45 ***	−3.95 ***
20-min	−5.52 ***	−4.13 ***	−1.59	−4.94 ***	−4.80 ***
25-min	−5.13 ***	−4.24 ***	−2.51 **	−4.69 ***	−5.18 ***
30-min	−4.68 ***	−4.61 ***	−2.72 ***	−4.89 ***	−5.11 ***
	Case: Entrance passenger flow of J. Sta.				
	ARIMA	ANN	LSTM	ELM	AWELM
5-min	−6.16 ***	−5.02 ***	−4.66 ***	−5.63 ***	−5.15 ***
10-min	−4.87 ***	−3.96 ***	−3.75 ***	−4.02 ***	−3.84 ***
15-min	−4.64 ***	−3.82 ***	−3.67 ***	−3.88 ***	−3.74 ***
20-min	−4.67 ***	−3.96 ***	−3.66 ***	−4.01 ***	−3.93 ***
25-min	−4.54 ***	−4.14 ***	−3.48 ***	−4.07 ***	−4.01 ***
30-min	−4.51 ***	−4.25 ***	−3.60 ***	−4.18 ***	−4.13 ***

*** represents the rejection of the null hypothesis at the 0.01 level, ** represents the rejection of the null hypothesis at the 0.05 level, and * represents the rejection of the null hypothesis at the 0.1 level.

5. Conclusions

This paper studied the passenger flow forecasting and proposed a novel model SSA-AWELM. In the model, the SSA was developed to decompose the original data into the three components of trend, periodicity, and residue; then, the AWELM was developed to forecast each component separately. The three predicted results were summed as the final outcomes. In order to demonstrate the effectiveness of the proposed model, the passenger flow in two transfer stations, which were extracted from an AFC system, were utilized to carry out prediction testing and a comparison analysis. The main conclusions are drawn and listed as follows:

1. The SSA approach can get an insight into the inner characteristics of the passenger flow. The trend represents the overall tendency, the periodicity represents the variants within a day, and the residue represents noise.
2. The AWELM model, which is combined by AdaBoost and WELM, are developed to make a more accurate and faster prediction for each component. Compared to the state-of-the-art model LSTM, the propose model has improved upon the performance by 22% and saved time by 84%, on average.
3. From the results of the evaluation measures and DM statistical test, the proposed model SSA-AWELM can reduce the cumulative errors during the multistep-ahead prediction. These findings have demonstrated that the SSA-AWELM is a robust model for passenger flow forecasting.

The proposed method in this paper still retains two limitations that will be addressed in the future. One is that the testing cases are in only two transfer stations with large travel demands, and the other is that the passenger flows are collected under regular conditions. Thus, in further studies, more cases including regular stations will be tested and discussed. In addition, the passenger flows during some special incidents, such as extreme weather, passenger control, etc., will be focused on to extend the proposed model.

Author Contributions: Conceptualization, W.Z., W.W., and D.Z.; methodology, W.Z.; software, W.Z.; validation, W.Z.; data curation, W.Z. and D.Z.; writing—original draft preparation, W.Z.; writing—review and editing, W.Z., W.W., and D.Z.; visualization, W.Z. and D.Z.; supervision, W.W. and D.Z.; project administration, W.W.; and funding acquisition, W.W. and D.Z. All authors have read and agreed to the published version of the manuscript.

Funding: This research was funded by National Natural Science Foundation of China, grant numbers 51878166 and 71701047.

Acknowledgments: The authors are grateful to Ali Tianchi for opening the AFC datasets.

Conflicts of Interest: The authors declare no conflicts of interest.

Appendix A

Table A1. The determined hyper-parameters of the experimental models.

Model	Component	Hyper-Parameters	Testing Cases			
			Exit Passenger Flow of Q.R. Sta.	Entrance Passenger Flow of Q.R. Sta.	Exit Passenger Flow of J. Sta.	Entrance Passenger Flow of J. Sta.
ARIMA	-	p, d, q	10,0,3	8,0,7	6,0,10	6,0,9
ANN	-	H	24	24	26	34
LSTM	-	H	40	34	28	22
ELM	-	H	50	50	42	42
AWELM	-	H, T	50,13	50,5	50,8	42,16
SSA-AWELM	trend	H, T	24,15	22,20	22,13	24,11
	periodicity	H, T	50,19	50,15	46,15	48,11
	remainder	H, T	46,17	44,9	40,8	48,10

H represents the neuron number of the hidden layer, and T represents the number of base learners.

Table A2. The training time of the experimental models.

Models	Training Time (Seconds)			
	Exit Passenger Flow of Q.R. Sta.	Entrance Passenger Flow of Q.R. Sta.	Exit Passenger Flow of J. Sta.	Entrance Passenger Flow of J. Sta.
ARIMA	~10.1	~8.9	~12.8	~11.0
ANN	~2.9	~1.6	~3.6	~2.4
LSTM	~54.0	~51.8	~51.0	~55.0
ELM	<1	<1	<1	<1
AWELM	~4.6	~1.7	~2.9	~4.1
SSA-AWELM	~9.6	~8.8	~8.2	~7.8

The experiment is conducted in the environment: programming language: *Python*; Main Packages: *Statsmodels, Scikit-learn, Keras,* and *TensorFlow*; OS: Windows 10 (64bit); RAM: 8G; CPU: Intel Core i5-8300H 2.30 GHz; and GPU: NVIDIA GeForce GTX 1650Ti.

References

1. Gallo, M.; De Luca, G.; D'Acierno, L.; Botte, M. Artificial neural networks for forecasting passenger flows on metro lines. *Sensors* **2019**, *19*, 3424. [CrossRef] [PubMed]
2. Chen, Q.; Wen, D.; Li, X.; Chen, D.; Lv, H.; Zhang, J.; Gao, P. Empirical mode decomposition based long short-term memory neural network forecasting model for the short-term metro passenger flow. *PLoS ONE* **2019**, *14*, e0222365. [CrossRef] [PubMed]
3. Lin, P.; Weng, J.; Fu, Y.; Alivanistos, D.; Yin, B. Study on the topology and dynamics of the rail transit network based on automatic fare collection data. *Phys. A Stat. Mech. Appl.* **2020**, *545*. [CrossRef]
4. Zhang, J.; Wang, S.; Zhang, Z.; Zou, K.; Shu, Z. Characteristics on hub networks of urban rail transit networks. *Phys. A Stat. Mech. Appl.* **2016**, *447*, 502–507. [CrossRef]
5. Liu, Z.Q.; Song, R. Reliability analysis of Guangzhou rail transit with complex network theory. *J. Transp. Syst. Eng. Inf. Technol.* **2010**, *10*, 194–200.
6. Du, Z.; Tang, J.; Qi, Y.; Wang, Y.; Han, C.; Yang, Y. Identifying critical nodes in metro network considering topological potential: A case study in Shenzhen city—China. *Phys. A Stat. Mech. Appl.* **2020**, *539*. [CrossRef]
7. Tang, L.; Zhao, Y.; Cabrera, J.; Ma, J.; Tsui, K.L. Forecasting Short-Term Passenger Flow: An Empirical Study on Shenzhen Metro. *IEEE Trans. Intell. Transp. Syst.* **2019**, *20*, 3613–3622. [CrossRef]
8. Danfeng, Y.; Jing, W. Subway Passenger Flow Forecasting with Multi-Station and External Factors. *IEEE Access* **2019**, *7*, 57415–57423. [CrossRef]
9. Ding, X.; Liu, Z.; Xu, H. The passenger flow status identification based on image and WiFi detection for urban rail transit stations. *J. Vis. Commun. Image Represent.* **2019**, *58*, 119–129. [CrossRef]
10. Liu, S.; Yao, E. Holiday passenger flow forecasting based on the modified least-square support vector machine for the metro system. *J. Transp. Eng.* **2017**, *143*. [CrossRef]
11. Jiao, P.; Li, R.; Sun, T.; Hou, Z.; Ibrahim, A. Three Revised Kalman Filtering Models for Short-Term Rail Transit Passenger Flow Prediction. *Math. Probl. Eng.* **2016**, *2016*, 1–10. [CrossRef]
12. Liu, Y.; Liu, Z.; Jia, R.; Deep, P.F. A deep learning based architecture for metro passenger flow prediction. *Transp. Res. Part C Emerg. Technol.* **2019**, *101*, 18–34. [CrossRef]
13. Fu, X.; Gu, Y. Impact of a New Metro Line: Analysis of Metro Passenger Flow and Travel Time Based on Smart Card Data. *J. Adv. Transp.* **2018**, *2018*. [CrossRef]
14. Tavassoli, A.; Mesbah, M.; Shobeirinejad, A. Modelling passenger waiting time using large-scale automatic fare collection data: An Australian case study. *Transp. Res. Part F Traffic Psychol. Behav.* **2018**, *58*, 500–510. [CrossRef]
15. Xu, X.; Xie, L.; Li, H.; Qin, L. Learning the route choice behavior of subway passengers from AFC data. *Expert Syst. Appl.* **2018**, *95*, 324–332. [CrossRef]
16. Hao, S.; Lee, D.H.; Zhao, D. Sequence to sequence learning with attention mechanism for short-term passenger flow prediction in large-scale metro system. *Transp. Res. Part C Emerg. Technol.* **2019**, *107*, 287–300. [CrossRef]
17. Lee, S.; Fambro, D.B. Application of subset autoregressive integrated moving average model for short-term freeway traffic volume forecasting. *Transp. Res. Rec.* **1999**, 179–188. [CrossRef]

18.	Milenković, M.; Švadlenka, L.; Melichar, V.; Bojović, N.; Avramović, Z. SARIMA modelling approach for railway passenger flow forecasting. *Transport* **2018**, *33*, 1113–1120. [CrossRef]

19.	Wang, Y.H.; Jin, J.; Li, M. Forecasting the section passenger flow of the subway based on exponential smoothing. *Appl. Mech. Mat.* **2013**, *409–410*, 1315–1319. [CrossRef]

20.	Yu, B.; Song, X.; Guan, F.; Yang, Z.; Yao, B. K-Nearest Neighbor Model for Multiple-Time-Step Prediction of Short-Term Traffic Condition. *J. Transp. Eng.* **2016**, *142*. [CrossRef]

21.	Cai, P.; Wang, Y.; Lu, G.; Chen, P.; Ding, C.; Sun, J. A spatiotemporal correlative k-nearest neighbor model for short-term traffic multistep forecasting. *Transp. Res. Part C. Emerg. Technol.* **2016**, *62*, 21–34. [CrossRef]

22.	Tsai, T.H.; Lee, C.K.; Wei, C.H. Neural network based temporal feature models for short-term railway passenger demand forecasting. *Expert Syst. Appl.* **2009**, *36*, 3728–3736. [CrossRef]

23.	Zhang, Y.; Zhang, Y.; Haghani, A. A hybrid short-term traffic flow forecasting method based on spectral analysis and statistical volatility model. *Transp. Res. Part C Emerg. Technol.* **2014**, *43*, 65–78. [CrossRef]

24.	Zeng, D.; Xu, J.; Gu, J.; Liu, L.; Xu, G. Short term traffic flow prediction using hybrid ARIMA and ANN models. In Proceedings of the 2008 Workshop on Power Electronics and Intelligent Transportation System (PEITS 2008), Guangzhou, China, 2–3 August 2008; pp. 621–625. [CrossRef]

25.	Sun, Y.; Leng, B.; Guan, W. A novel wavelet-SVM short-time passenger flow prediction in Beijing subway system. *Neurocomputing* **2015**, *166*, 109–121. [CrossRef]

26.	Wei, Y.; Chen, M.C. Forecasting the short-term metro passenger flow with empirical mode decomposition and neural networks. *Transp. Res. Part C Emerg. Technol.* **2012**, *21*, 148–162. [CrossRef]

27.	Yang, D.; Chen, K.; Yang, M.; Zhao, X. Urban rail transit passenger flow forecast based on LSTM with enhanced long-term features. *IET Intell. Transp. Syst.* **2019**, *13*, 1475–1482. [CrossRef]

28.	Bai, Y.; Sun, Z.; Zeng, B.; Deng, J.; Li, C. A multi-pattern deep fusion model for short-term bus passenger flow forecasting. *Appl. Soft Comput. J.* **2017**, *58*, 669–680. [CrossRef]

29.	Liu, L.; Chen, R.C. A novel passenger flow prediction model using deep learning methods. *Transp. Res. Part C Emerg. Technol.* **2017**, *84*, 74–91. [CrossRef]

30.	Ma, X.; Dai, Z.; He, Z.; Ma, J.; Wang, Y.; Wang, Y. Learning traffic as images: A deep convolutional neural network for large-scale transportation network speed prediction. *Sensors* **2017**, *17*, 818. [CrossRef]

31.	Yang, C.; Guo, Z.; Xian, L. Time series data prediction based on sequence to sequence model. *IOP Conf. Ser. Mat. Sci. Eng.* **2019**, *692*. [CrossRef]

32.	Li, W.; Wang, J.; Fan, R.; Zhang, Y.; Guo, Q.; Siddique, C.; Ban, X. (Jeff). Short-term traffic state prediction from latent structures: Accuracy vs. efficiency. *Transp. Res. Part C Emerg. Technol.* **2020**, *111*, 72–90. [CrossRef]

33.	Liu, R.; Wang, Y.; Zhou, H.; Qian, Z. Short-Term Passenger Flow Prediction Based on Wavelet Transform and Kernel Extreme Learning Machine. *IEEE Access* **2019**, *7*, 158025–158034. [CrossRef]

34.	Chen, M.C.; Wei, Y. Exploring time variants for short-term passenger flow. *J. Trans. Geogr.* **2011**, *19*, 488–498. [CrossRef]

35.	Qin, L.; Li, W.; Li, S. Effective passenger flow forecasting using STL and ESN based on two improvement strategies. *Neurocomputing* **2019**, *356*, 244–256. [CrossRef]

36.	Chen, D.; Zhang, J.; Jiang, S. Forecasting the Short-Term Metro Ridership with Seasonal and Trend Decomposition Using Loess and LSTM Neural Networks. *IEEE Access* **2020**, *8*, 91181–91187. [CrossRef]

37.	Mao, X.; Shang, P. Multivariate singular spectrum analysis for traffic time series. *Phys. A Stat. Mech. Appl.* **2019**, *526*, 1–13. [CrossRef]

38.	Shang, Q.; Lin, C.; Yang, Z.; Bing, Q.; Zhou, X. A hybrid short-term traffic flow prediction model based on singular spectrum analysis and kernel extreme learning machine. *PLoS ONE* **2016**, *11*. [CrossRef]

39.	Guo, F.; Krishnan, R.; Polak, J. A computationally efficient two-stage method for short-term traffic prediction on urban roads. *Transp. Plan. Technol.* **2013**, *36*, 62–75. [CrossRef]

40.	Qiu, H.; Zhang, N.; Xu, W.; He, T. Research of Architecture on Rail Transit's AFC System. *Urb. Rapid Rail Transit* **2014**, *27*, 86–89. [CrossRef]

41.	Taieb, S.B. *and Hyndman, R.J. Recursive and Direct Multi-Step Forecasting: The Best of Both Worlds*; Monash Econometrics and Business Statistics Working Papers; Monash University: Melbourne, Australia, 2012.

42.	Bontempi, G.; Ben Taieb, S.; Le Borgne, Y.A. Machine learning strategies for time series forecasting. In *Lecture Notes in Business Information Processing, LNBIP*; Springer: Berlin/Heidelberg, Germany, 2013; Volume 138, pp. 62–77. [CrossRef]

43. Golyandina, N.; Nekrutkin, V.V.; Zhigljavsky, A.A. *Analysis of Time Series Structure: SSA and Related Techniques*; Monographs on Statistics and Applied Probability; Chapman & Hall/CRC: Boca Raton, FL, USA, 2001; Volume 90, ISBN 1584881941.
44. Freund, Y.; Schapire, R.E. Experiments with a New Boosting Algorithm. In Proceedings of the 13th International Conference on Machine Learning, Bari, Italy, 3–6 July 1996; pp. 148–156.
45. Drucker, H. Improving regressors using boosting techniques. In Proceedings of the 14th International Conference on Machine Learning, San Francisco, CA, USA, 8–12 July 1997; pp. 107–115.
46. Solomatine, D.P.; Shrestha, D.L. AdaBoost.RT: A boosting algorithm for regression problems. In Proceedings of the 2004 IEEE International Conference on Neural Networks, Budapest, Hungary, 25–29 July 2004; pp. 1163–1168. [CrossRef]
47. Shrestha, D.L.; Solomatine, D.P. Experiments with AdaBoost.RT, an improved boosting scheme for regression. *Neural Comput.* **2006**, *18*, 1678–1710. [CrossRef]
48. Huang, G.B.; Zhu, Q.Y.; Siew, C.K. Extreme learning machine: A new learning scheme of feedforward neural networks. In Proceedings of the 2004 IEEE International Conference on Neural Networks, Budapest, Hungary, 25–29 July 2004; pp. 985–990. [CrossRef]
49. Tianchi, A. The AI Challenge of Urban Computing. Available online: https://tianchi.aliyun.com/competition/entrance/231712/information (accessed on 20 February 2020).
50. Sun, Y.; Zhang, G.; Yin, H. Passenger flow prediction of subway transfer stations based on nonparametric regression model. *Discret. Dyn. Nat. Soc.* **2014**, *2014*, 1–8. [CrossRef]
51. Harvey, A.C. *Forecasting, Structural Time Series Models and the Kalman Filter*; Cambridge University Press: Cambridge, UK, 1990.
52. Diebold, F.X. Comparing Predictive Accuracy, Twenty Years Later: A Personal Perspective on the Use and Abuse of Diebold-Mariano Tests. *SSRN Electr. J.* **2013**. [CrossRef]
53. Zhang, X.; Wang, J.; Gao, Y. A hybrid short-term electricity price forecasting framework: Cuckoo search-based feature selection with singular spectrum analysis and SVM. *Energy Econ.* **2019**, *81*, 899–913. [CrossRef]

 *sensors*

Letter

A Unified Fourth-Order Tensor-Based Smart Community System

Chang Liu [1], Huaiyu Wu [2], Junyuan Wang [2,3,*] and Mingkai Wang [4]

[1] College of Design and Innovation, Tongji University, 281 Fuxin Road, Shanghai 200092, China; liuc@tongji.edu.cn
[2] Department of Information and Communication Engineering, College of Electronics and Information Engineering, Tongji University, 4800 Cao'an Highway, Shanghai 201804, China; why_not@tongji.edu.cn
[3] Institute for Advanced Study, Tongji University, 1239 Siping Road, Shanghai 200092, China
[4] School of Mathematical Sciences, Tongji University, 1239 Siping Road, Shanghai 200092, China; wmk@tongji.edu.cn
* Correspondence: junyuanwang@tongji.edu.cn

Received: 19 September 2020; Accepted: 20 October 2020; Published: 22 October 2020

Abstract: Empowered by the ubiquitous sensing capabilities of Internet of Things (IoT) technologies, smart communities could benefit our daily life in many aspects. Various smart community studies and practices have been conducted, especially in China thanks to the government's support. However, most intelligent systems are designed and built individually by different manufacturers in diverging platforms with different functionalities. Therefore, multiple individual systems must be deployed in a smart community to have a set of functions, which could lead to hardware waste, high energy consumption and high deployment cost. More importantly, current smart community systems mainly focus on the technologies involved, while the effects of human activity are neglected. In this paper, a fourth-order tensor model representing object, time, location and human activity is proposed for human-centered smart communities, based on which a unified smart community system is designed. Thanks to the powerful data management abilities of a high-order tensor, multiple functions can be integrated into our system. In addition, since the tensor model embeds human activity information, complex functions could be implemented by exploring the effects of human activity. Two exemplary applications are presented to demonstrate the flexibility of the proposed unified fourth-order tensor-based smart community system.

Keywords: tensor modeling; smart community system; Internet of Things

---

1. Introduction

Community, as a social unit with commonality, has been increasingly complex and significant due to the issues and concerns brought by social problems and environmental changes. In China, the term of community (*shequ* in Chinese) is used in an official governmental discourse, which is designated as the basic unit of urban social, political and administrative organization within a certain geographic area and a certain population [1,2]. With the rapid increase of the urban population in China, the concentration of urban functions and resources in the city center has posed plenty of challenges in transportation, medical care, education, housing, etc. This year, the COVID-19 pandemic outbreak brought Chinese bustling cities to an unexpected halt. In order to ease pressure at hospitals and other public service institutes, communities started playing an important role in city governance and management. A typical example is to monitor the body temperatures of the residents and report to the government immediately if a fever is detected, which requires extensive repetitive work. This urges the development of a human-centered smart community system that could continuously monitor and quickly react to abnormal events by taking advantage of Internet of Things (IoT) technologies [3,4].

In China, seven cities/districts that engage different partners and the public in the determination process have been designated as smart communities [5] to implement intelligent applications in city management, public services, medical care and industrial development [6]. By connecting various smart communities together via wireless/wired broadband networks, remote access to many services across the whole country or even worldwide would be possible, e.g., consultation of medical care and education. At the same time, satellite cities and new communities have been built to enhance the smart community pilot program so as to increase efficiency, improve connectedness and centralize information on residents from public finance, taxation, city planning, housing, commerce, education and justice. For example, [7] applied the PROMETHEE-II algorithm in Industrial 4.0 to help with the selection of Intelligent Internet of Things (IIoT) platforms. A distributed demand side management system was proposed in [8] to facilitate smart energy trading in a local community. [9] discussed the vision of a holistic smart city that takes advantage of crowdsourced data to make customizable and scalable decisions.

A common fact underlies the practices: that is, these communities/cities are meeting a growing demand for being interconnected and intelligent and are labeled as "smart". However, in smart communities' research and practices, we see the gap between visions and facts. Existing smart communities are mainly practiced in a new town/city, where few people live before being built. As a result, the core of these smart communities is the intelligent infrastructure, including deploying sensors and connecting them to the Internet. A majority of studies discussed individual diverging smart systems based on different IoT platforms and cloud computing technologies, e.g., medical care, e-governance, job creation, etc. [10–14]. In order to meet the diverse requirements in smart communities, a complex intelligent system that consists of multiple individual subsystems operating independently will need to be deployed, which could lead to significant wasting of digital technological equipment. How to build a unified smart community system to enable flexible add-on functions still remains largely unknown.

On the other hand, a number of scholars raised questions about whether the overuse and overemphasis of digital technologies in shaping communities and cities is beneficial for enabling a livable environment [15,16]. One of the problems is that the current smart communities are designed with the focus on the optimization of technologies, while people's needs from different stakeholders have been rarely taken into account.

In this paper, we consider a human-centered smart community, which serves humans' needs based on systematic information. For this purpose, a four-dimensional model is built, as depicted in Figure 1, which is based on object, time, location and human activity. The relationship between any two of them can be reflected by the data collected in the IoT network. For these four dimensions, *location* means a particular place or position in the community. *Object* refers to things and people that exist independently. *Time* is a series of consecutive discrete timestamps. Furthermore, *human activities* are various actions taken by people for commuting, living, reaction or necessity, such as running, walking, standing, talking, sitting, falling, fighting, abnormal activities, etc. With the help of IoT, data can be collected from wearable sensors, or remotely recorded from video, radar and other wireless sensing methods.

Based on the above four-dimensional model of a human-centered smart community system, a fourth-order tensor model that represents object, time, location and human activity is proposed in this paper to store all the data collected by IoT devices. This is motivated by the appealing features of the tensor to manage heterogeneous and enormous data [17,18], which has also been shown to be a powerful tool for public bicycle rental forecasting [19] and breathing monitoring [20], to name a few. As all the data are now formalized into the proposed fourth-order tensor model, a unified smart community system can be proposed, where tensor data can be truncated and extracted to fit various application purposes. These retrieved data could be then processed in parallel. It can be seen that such a fourth-order tensor-based smart community system provides a systematic solution to various living needs in smart communities, and thus avoids the use of multiple application-driven intelligent systems. Therefore, a couple of benefits can be obtained: (1) resources could be saved as a number of key devices can be shared by multiple application functions, e.g., gateways, which also reduces energy consumption; (2) functions can be easily removed or added, enabling a flexible and extendable smart community system, which is especially crucial for old towns/cities where deployment of new equipment is usually complicated and of high cost. In addition, compared to the traditional schemes based on third-order tensors, extra information could be obtained and thus gains could be achieved by adopting our proposed fourth-order tensor-based smart community system.

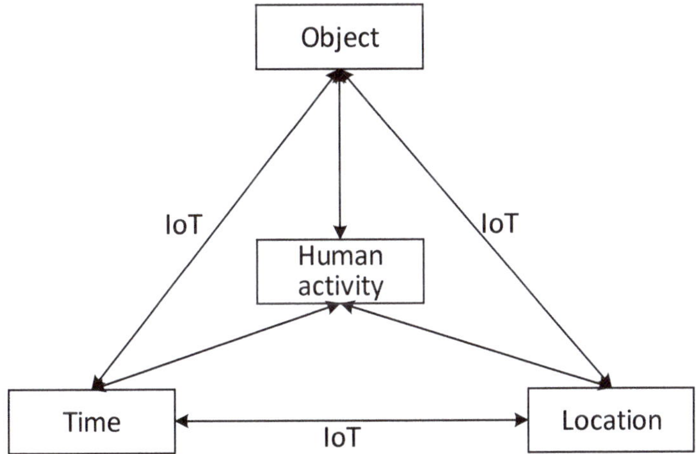

Figure 1. Four-dimensional model of smart community system.

The rest of this paper is organized as follows. Section 2 models the collected data in smart communities as a fourth-order tensor and proposes a unified tensor-based smart community system. Two application examples are then demonstrated in Section 3. Section 4 discusses the key findings, and concluding remarks are summarized in Section 5.

2. Methods

In this section, all the data collected from intelligent sensors are formalized into a fourth-order tensor, based on which a unified smart community system is proposed. Before presenting the fourth-order-based smart community system, let us first introduce some useful definitions regarding tensor [21].

2.1. Basic Definitions

Definition 1 (Tensor). *A tensor is a multidimensional array. The order of a tensor is its number of dimensions, while the dimensions are usually referred to as modes or ways. An Nth-order tensor is denoted as $\mathcal{X} \in \mathbb{R}^{I_1 \times I_2 \times \cdots \times I_N}$, where I_n is the size of mode n. Particularly, a zeroth-order tensor is a scalar, a first-order tensor is a vector, and a second-order tensor is a matrix.*

Definition 2 (Slice). *A two-dimensional section of a tensor, whose all but two indices are fixed, is referred to as a slice. For example, a third-order tensor $\mathcal{X} \in \mathbb{R}^{I_1 \times I_2 \times I_3}$ has the horizontal (mode 1) slices $\mathcal{X}_{i_1 : :}$, the lateral (mode 2) slices $\mathcal{X}_{: i_2 :}$, and the frontal (mode 3) slices $\mathcal{X}_{: : i_3}$, as illustrated in Figure 2.*

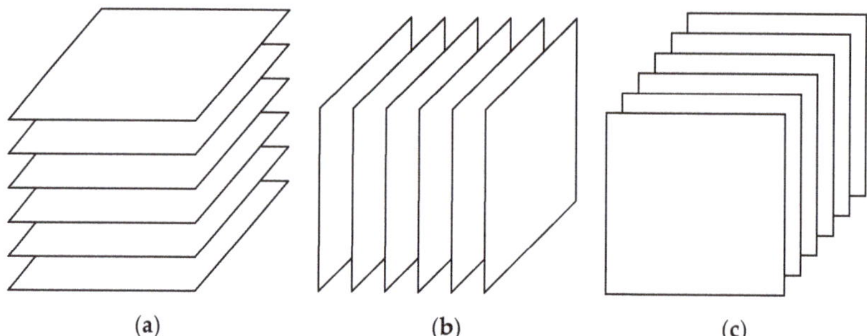

Figure 2. Graphic illustration of slices of a third-order tensor $\mathcal{X}$. (a) Horizontal slices $\mathcal{X}_{i_1 : :}$; (b) Lateral slices $\mathcal{X}_{: i_2 :}$; (c) Frontal slices $\mathcal{X}_{: : i_3}$.

Definition 3 (Mode-*n* Product). *A mode-n product or n-mode product of tensor* $\mathcal{X} \in \mathbb{R}^{I_1 \times I_2 \times \cdots \times I_N}$ *and a matrix* $\mathbf{U} \in \mathbb{R}^{I_n \times J}$ *is denoted by* $\mathcal{Y} = \mathcal{X} \times_n \mathbf{U}$ *with* $\mathcal{Y} \in \mathbb{R}^{I_1 \times \cdots \times I_{n-1} \times J \times I_{n+1} \times \cdots \times I_N}$, *and*

$$\mathcal{Y}_{i_1 \cdots i_{n-1} j i_{n+1} \cdots i_N} = \sum_{i_n=1}^{I_n} x_{i_1 i_2 \cdots i_N} \cdot u_{j i_n}. \tag{1}$$

It is obvious that with the mode-n product, each mode-n fiber (which is defined by fixing every index but one) of the tensor is multiplied by the matrix $\mathbf{U}$.

2.2. Fourth-Order Tensor Modeling of Smart Community Data

In this section, a fourth-order tensor will be constructed to encode all the data in the four dimensions of object, time, location and human activity. As the object, time and location information can be usually obtained directly from intelligent sensors, let us first construct a third-order tensor to represent them and the interplay among them, after which the fourth dimension, human activity, will be added.

Let $\mathcal{X} \in \mathbb{R}^{I \times J \times K}$ denote the third-order tensor of object, time and location, as illustrated in Figure 3. Specifically, the indices of the object dimension (mode 1) are the object indices numbered according to the predefined policy. For example, the objects located at fixed positions, such as surveillance cameras and environmental sensors, are assigned fixed unique indices, while a moving object is assigned a dynamic index once successfully recognized from surveillance video images. This is because visitors and new objects could appear from time to time in smart communities, making it impossible to assign a fixed unique index for every object.

Based on the above four-dimensional model of a human-centered smart community system, a fourth-order tensor model that represents object, time, location and human activity is proposed in this paper to store all the data collected by IoT devices. This is motivated by the appealing features of the tensor to manage heterogeneous and enormous data [17,18], which has also been shown to be a powerful tool for public bicycle rental forecasting [19] and breathing monitoring [20], to name a few. As all the data are now formalized into the proposed fourth-order tensor model, a unified smart community system can be proposed, where tensor data can be truncated and extracted to fit various application purposes. These retrieved data could be then processed in parallel. It can be seen that such a fourth-order tensor-based smart community system provides a systematic solution to various living needs in smart communities, and thus avoids the use of multiple application-driven intelligent systems. Therefore, a couple of benefits can be obtained: (1) resources could be saved as a number of key devices can be shared by multiple application functions, e.g., gateways, which also reduces energy consumption; (2) functions can be easily removed or added, enabling a flexible and extendable smart community system, which is especially crucial for old towns/cities where deployment of new equipment is usually complicated and of high cost. In addition, compared to the traditional schemes based on third-order tensors, extra information could be obtained and thus gains could be achieved by adopting our proposed fourth-order tensor-based smart community system.

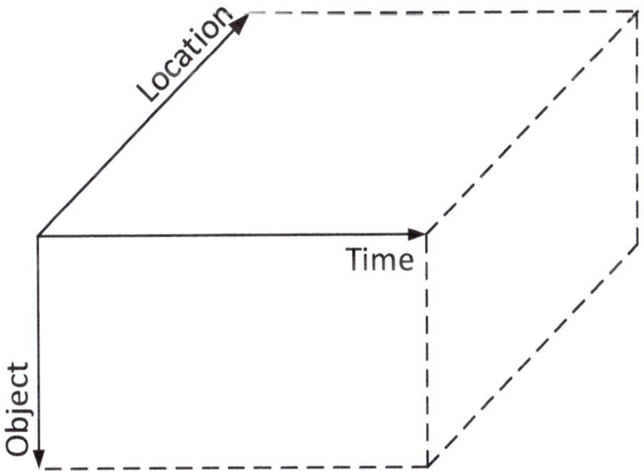

Figure 3. Graphic illustration of the third-order tensor with 3 modes representing object, location and time, respectively.

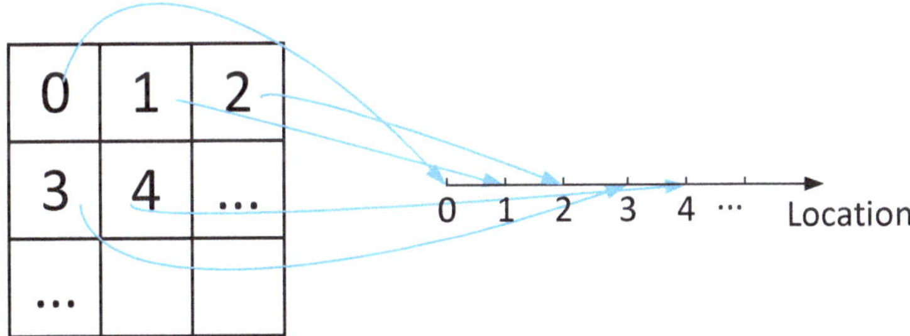

Figure 4. A simple location coding scheme.

An element $\mathcal{X}_{ijk}$ of a third-order tensor $\mathcal{X} \in \mathbb{R}^{I \times J \times K}$, can be seen as a state information matrix that stores the data collected from IoT sensors for object i at timestamp j and location k. The state information matrix should include the data collected and their types, which could be denoted by Unicode for instance. Two examples of element $\mathcal{X}_{ijk}$ are shown in Figure 5 for demonstration purpose, where object i is a man wearing a smart watch that automatically connects to the IoT network in smart communities. The health-related data collected, such as body temperature and heart rate, are then transmitted to the gateway in the IoT network and saved to the corresponding state information matrix. Object i' is a surveillance camera installed, and thus the data collected are video frames, each of which is usually represented by a third-order tensor. It is can be seen that the data in a state information matrix is not limited to a scalar, but also could be a matrix or even a tensor.

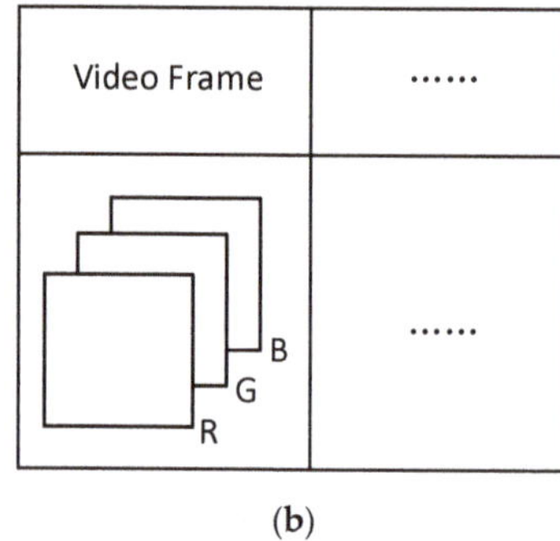

(a)   (b)

Figure 5. Illustrative toy examples of an element of a third-order tensor. (a) X_{ijk} for human object i at timestamp j and location k; (b) $X_{i'j'k'}$ for surveillance camera object i' at timestamp j' and location k'.

By forming such a third-order tensor to store the massive data collected from intelligent sensors in a smart community, the geometry structures of various types of data are preserved, which could be further exploited by tensor-based data processing methods to enhance data mining and analysis and thus assist smart community decision-making in many aspects, such as emergency management, human resource allocation, crime detection and forecasting, etc. For instance, based on the proposed tensor model, people can be identified via face recognition and tracked via object tracking techniques [23], based on which, the trajectory of a person can be obtained and analyzed. Once abnormal behavior is detected, the face image and trajectory information of the suspect will be sent back to the community management team via the IoT network for further investigation and/or possible face-to-face questioning. If a crime happened, the face image of the suspect would be then sent to the police station to help with identity and crime record checking.

In the following, let us take a closer look at the mode-1 slices (time-location), mode-2 slices (object-location), and mode-3 slices (object-time) one by one.

- **Mode-1 (Time-Location) Slices** $X_{i::}$: A time-location object slice $X_{i::}$ is a subarray of tensor $X \in \mathbb{R}^{I \times J \times K}$ maintaining the time and location information for object i, where element X_{ijk}, i.e., the state information matrix at timestamp j and location k for object i introduced above, indicates whether object i is within the kth subarea or not at timestamp j. Specifically, if X_{ijk} is a nonzero matrix, object i is located in the kth subarea; otherwise, it is outside subarea k. The location information can be obtained by using one or multiple localization technologies, e.g., Wi-Fi signal sensing, image recognition, etc. Note that a large object, for example a building, could appear in multiple subareas, resulting in more than one state information matrices in a location fiber. By retrieving a time-location slice from the tensor model, an indicative matrix for object i can be formed, which consists of 1 s and 0 s, as shown in Figure 6, where 1 and 0 denote that the corresponding state information matrix is a nonzero matrix or a zero matrix, respectively. This indicative time-location matrix would be very useful for trajectory analysis and object localization.

- **Mode-2 (Object-Location) Slices** $X_{:j:}$: Similarly, an object-location slice $X_{:j:}$ is a subarray that consists of object and location information collected at timestamp j. By detecting whether an element, i.e., the state information matrix of a mode-2 slice is a zero matrix or not, and using 1 to denote a nonzero matrix and 0 to denote a zero matrix, an indicative object-location matrix can be obtained, which clearly shows the geographical locations of various objects in the whole area at a given timestamp, which could be used to monitor social distancing state in a smart community/city in real time during the current COVID-19 epidemic.

- **Mode-3 (Object-Time) Slices** $X_{::k}$: An object-time slice $X_{::k}$ is a subarray representing the state information of various objects at various timestamps at a given subarea. In other words, it shows what objects appear in a given area at what time and its state evolution with time. Such object-time slices are especially useful for resident-visitor classification. Specifically, a resident appears regularly at the entrance and exit, while a visitor usually appears randomly and might appear only a couple of times. Tracking of visitors can be then conducted in the cloud center in case of any crimes happened.

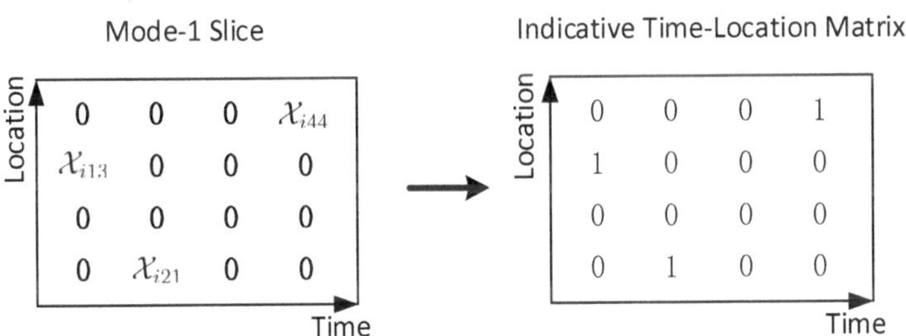

Figure 6. A toy example of mode-1 slice $\mathcal{X}_{i\ :\ :}$ and the corresponding indicative time-location matrix for object i.

The above third-order tensor model can only show the objective components, i.e., object, time and location. However, the data stored in the tensor highly depend on human activity, as our society is naturally human-centered. Therefore, we propose to add a novel fourth dimension that represents human activity, which is referred to as activity in the following for simplicity. As such, a comprehensive fourth-order tensor model is established for smart communities, as depicted in Figure 7. Note that the proposed fourth-order tensor model could better describe real scenarios and thus help with more accurate decision-making and meeting more complex application requirements in smart communities.

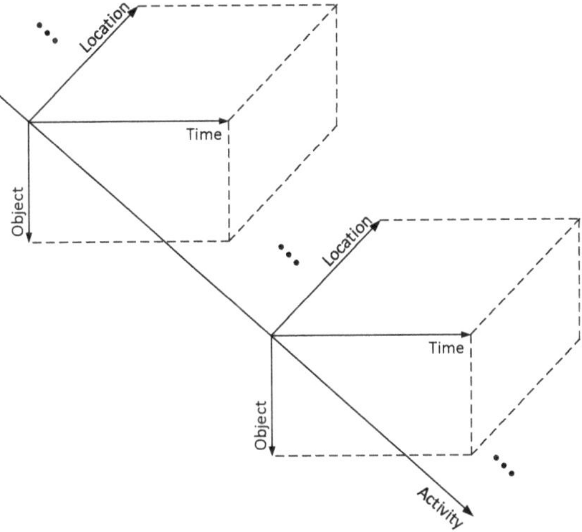

Figure 7. Graphic illustration of the proposed fourth-order tensor model for smart communities.

The activity dimension indices represent various human activities, which, in contrast to the other three dimensions, cannot be obtained directly from intelligent sensors. Instead, preprocessing the object-time-location tensor by using various machine learning algorithms is usually required. A number of common or dangerous human activities can be predefined, such as walking, running, gathering, falling, etc.

2.3. A Unified Fourth-Order Tensor-Based Smart Community System

Based on the fourth-order tensor model that encodes object, time, location and human activity information, a unified smart community system is proposed in this section, as shown in Figure 8. Specifically, data collected from intelligent sensors, surveillance cameras and other devices are first sent to the gateways in the IoT network and then stored and processed in the cloud center. At the cloud side, real-time object detection is conducted based on the video frames fed back from surveillance cameras. By continuing to track an object once it is detected, the data collected for this object are formalized into a third-order tensor with 3 dimensions, being object, time and location, and then saved to the cloud storage. As this third-order tensor builds up, human activity analysis is being done simultaneously for classification. According to the human activity information, the third-order tensor data are reorganized into a fourth-order tensor so as to highlight human activity, which is useful for the applications that aim to investigate its effects and the relationship between human activity and other components.

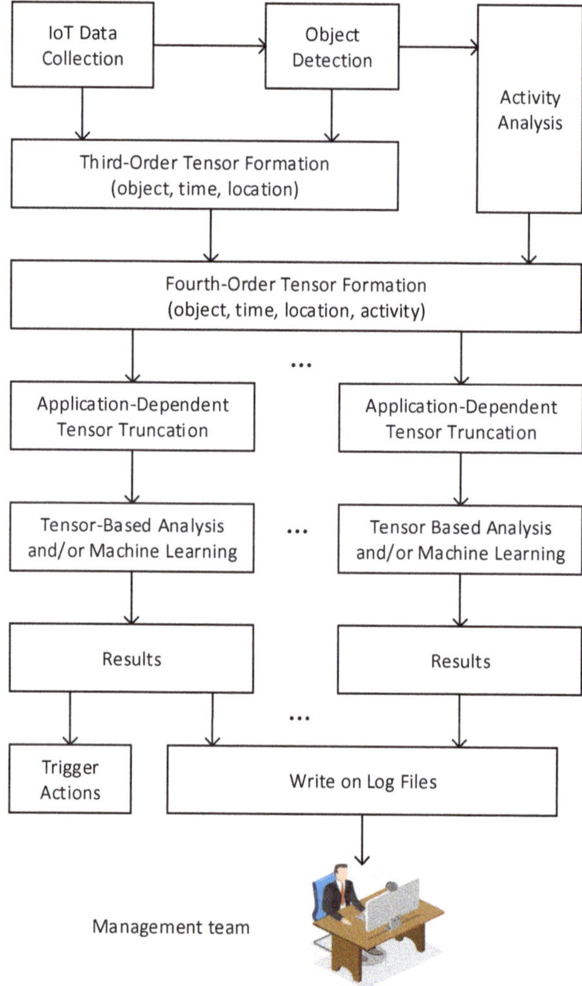

Figure 8. Block diagram of the proposed unified fourth-order tensor-based smart community system.

So far, a fourth-order tensor has been constructed, from which useful data for specific applications with various purposes could be truncated over a certain range of objects, timestamps, locations and activities for further processing. These truncated tensor data can then be carefully analyzed to obtain insightful information or serve as the input of a machine learning framework with a specific purpose. Analytical and/or machine learning results will then be obtained and written onto the log files, which are monitored and reviewed by the management team to assist decision-making and improve the livability of the community. Some analytical and/or machine learning results could also trigger actions directly if required by the corresponding application, e.g., fire alarms will be triggered immediately if the obtained results suggest a suspicious fire. Since the fourth-order tensor stores all the sensed data, applications with new functions can be added to the existing smart community system. In other words, the proposed unified fourth-order tensor-based smart community system embraces flexibility and extendibility.

3. Application Examples

In this section, two exemplary applications will be introduced to demonstrate the flexibility of the proposed unified fourth-order tensor-based smart community system.

3.1. Intelligent Traffic Light Control

Intelligent traffic light control at intersections, which adapts to road traffic including vehicles and pedestrians, can improve everyone's experience. This is one of the benefits we could have in smart communities. To make it happen, accurate traffic forecasting is required, which can be obtained by using our proposed fourth-order tensor-based smart community system.

Specifically, at an intersection, the traffic in a given period of time is related to that in other time periods around it, as people's daily routine usually varies slightly over time. In order to control traffic light intelligently, the comprehensive information road traffic needs to be predicted, i.e., how many vehicles and pedestrians on which road will be turning right, going straight and turning left. In this case, the objects are the drivers and pedestrians; the timestamps are a number of time periods; the locations are the roads; and the activities are drivers' and pedestrians' actions, including turning left, turning right and going straight. Each element of the fourth-order tensor is either 1 or 0, which indicates whether an object at a specific timestamp and a specific location is taking a specific action. As there are so many elements in the fourth-order tensor space, each of which corresponds to one combination of object, timestamp, location and activity, the tensor is usually sparse, implying that the original tensor could be compressed to reduce the size and hence the amount of data so as to speed up further processing. A popular way is to find out a core tensor to represent the original one.

By denoting the original tensor as $\mathcal{X} \in \mathbb{R}^{I_1 \times I_2 \times I_3 \times I_4}$ and applying Tucker decomposition [24], tensor $\mathcal{X}$ is decomposed into a core tensor $\mathcal{G} \in \mathbb{R}^{J_1 \times J_2 \times J_3 \times J_4}$ and four factor matrices $\mathbf{U}_1 \in \mathbb{R}^{I_1 \times J_1}$, $\mathbf{U}_2 \in \mathbb{R}^{I_2 \times J_2}$, $\mathbf{U}_3 \in \mathbb{R}^{I_3 \times J_3}$, $\mathbf{U}_4 \in \mathbb{R}^{I_4 \times J_4}$ with

$$\mathcal{Y} = \mathcal{G} \times_1 \mathbf{U}_1 \times_2 \mathbf{U}_2 \times_3 \mathbf{U}_3 \times_4 \mathbf{U}_4. \tag{2}$$

A typical algorithm to calculate the core tensor and also the factor matrices is the higher-order singular value decomposition (HOSVD) algorithm [25]. After obtaining the historical core tensors of the truncated tensors over the same time period of interest in the past a few months, the historical core tensors can be used as training data to train a machine learning model to predict the core tensor $\mathcal{G}' \in \mathbb{R}^{J_1 \times J_2 \times J_3 \times J_4}$ of road traffic. Based on the core tensor $\mathcal{G}'$, the predicted tensor $\mathcal{Y}' \in \mathbb{R}^{I_1 \times I_2 \times I_3 \times I_4}$, which encodes the predicted road traffic information can be obtained. These predicted road traffic data can then be used for real-time traffic light control.

In practice, four surveillance cameras need to be deployed in the intersection to monitor four crossroad areas. Real-time videos from these four cameras are sent back to the cloud center for object detection and activity analysis, based on which a fourth-order tensor could be formed. The above traffic forecasting algorithm can then be run in the cloud center to predict traffic in the coming minute, i.e., how many vehicles and pedestrians on which road are taking what actions. With the detailed road traffic information, an intelligent traffic control algorithm will be executed to control traffic light.

3.2. Human Behavior Pattern Analysis

Thanks to the multi-dimensional data available in the tensor model, hidden behavior patterns of individuals can be extracted from the huge amount of data collected every day, which could provide statistical support for meeting people's living needs.

In order to analyze individual human behaviors, let us retrieve the third-order tensor $\mathcal{X}_{i:::} \in \mathbb{R}^{I_2 \times I_3 \times I_4}$ for each human object i and construct a corresponding indicative matrix $\mathcal{M}_{i:::} \in \mathbb{R}^{I_2 \times I_3 \times I_4}$, which includes time, location and activity information. By applying the canonical polyadic (CP) decomposition technique [26], which is another popular tensor decomposition method that approximates a tensor by the sum of a number of three-way outer products, we have

$$\mathcal{M}_{i:::} \approx \sum_{r=1}^{R} \mathbf{a}_{ir} \circ \mathbf{b}_{ir} \circ \mathbf{c}_{ir}, \tag{3}$$

where R is the number of decomposed components, which can also be seen as the number of basic behavior patterns. For each basic behavior pattern i, $\mathbf{a}_{ir} \in \mathbb{R}^{I_2}$, $\mathbf{b}_{ir} \in \mathbb{R}^{I_3}$ and $\mathbf{c}_{ir} \in \mathbb{R}^{I_4}$ indicate the distribution of pattern i in the dimensions of time, location and activity, respectively. It should be noted that the number of decomposed basic behavior patterns R is a key parameter that determines the ability of behavior pattern mining, which should be carefully chosen.

Based on the distribution information obtained over time, location and activity, human behavior patterns can be analyzed from different points of view to fit different needs. For example, individuals' behavior patterns regarding activity can be obtained, based on which individuals can be clustered into a number of groups of common interests, which could be utilized to meet people's social needs. For the single inhabitants, a similarity check between one's current behavior pattern and the past ones can be performed to detect abnormal events in case of any emergency needs.

In practice, the time and location information of the residents in a smart community are collected from surveillance camera videos and smart home devices, based on which each resident's activities are detected. The information is then organized into the fourth-order tensor model and stored in the cloud storage. When a public community event is planned, the truncated third-order tensor data for every resident could be analyzed by adopting the above human behavior pattern analysis method to identify interested residents, e.g., promoting interested residents to participate in weekly badminton training. For the purpose of abnormal event detection, the behavior pattern of a resident could be checked against those in the past days. Once a mismatch is detected, the community management team will message or ring the resident for a security check.

4. Discussion

In the last section, two exemplary applications have been presented to demonstrate the operation of our proposed fourth-order tensor-based unified smart community system. It can be seen that our unified smart community system can support various functionalities by retrieving interested data from the fourth-order tensor model of object, time, location and human activity and applying designated data processing methods. Such flexibility makes the smart community system easier to be upgraded with new functions. As a result, no removal of old intelligent systems and redeployment of new ones is needed, which reduces the maintenance cost significantly. In addition, since our proposed smart community system provides a unified platform for various applications with different goals, hardware devices can be shared by multiple applications, which is environmental-friendly in terms of both energy consumption and hardware manufacturing.

Moreover, as our proposed fourth-order tensor model includes more data, particularly human activity data, gains could be achieved from the proposed scheme compared to conventional methods based on third-order tensors. A representative third-order tensor-based scheme was proposed in [19] to forecast the rental data of urban public bicycles to assist redistribution. The three dimensions of the third-order tensor in [19] are time, location and numbers of bicycles rented out and returned back, while the effect of human activity is overlooked. It should be noted that human activity plays a key role in the usage of public bicycles. For example, a large number of university students might cycle together to a place for a big gathering. With such an accidental event, if human activity is not considered, the predicted number of public bicycles required around the gathering place based on previous data could be much smaller than the demand. As a result, a large number of students might not be able to cycle back to the campus after gathering. By adopting our fourth-order tensor model with one dimension representing human activity, student activity can also be predicted, which could be further exploited to optimize public bicycle redistribution to improve user experience.

Apart from the above advantages, the use of the tensor model to store all the collected data can preserve the geometry structures of multiple dimensions, indicating that additional hidden information could be exploited by adopting data mining techniques. Moreover, as the tensor model encodes multidimensional data of various types, in contrast to conventional methods which can usually handle single-parameter tasks, by adopting suitable tensor-based algorithms/methods a set of parameters can be predicated as a whole. Nevertheless, designing tensor-based algorithms to fit specific purposes could be very challenging and needs to be carefully investigated. Although the above benefits can be harvested thanks to the ability of a tensor to embed multi-dimensional data, it also brings challenges. For example, a high-order tensor usually has enormous elements, which require high computing capability to process data. Efficient tensor decomposition algorithms have been recently proposed in [27–29], which can achieve up to 14 times faster tensor decomposition. How to further accelerate tensor decomposition and reduce tensor data size as far as possible is still an open problem, which deserves significant efforts in future work.

5. Conclusions

In this paper, a novel fourth-order tensor model was proposed to formalize the data collected from intelligent sensors in smart communities, based on which a unified smart community system was proposed. By presenting two exemplary applications, it was shown that by truncating the tensor data according to application requirements, various applications with different functionalities can be integrated into the same smart community system. More importantly, new applications with new functionalities can be easily added to the existing system if needed, indicating that our proposed fourth-order tensor-based smart community system is flexible and extendable.

Note that this paper mainly discussed the fourth-order tensor-based theoretical framework to enable a unified smart community system. In the future, the effectiveness and efficiency of our proposed system will be carefully investigated and compared against conventional methods by implementing a number of functions in practice.

Author Contributions: C.L., H.W., J.W., and M.W. proposed the ideas; C.L., H.W. and J.W. designed the unified smart community system framework; C.L., H.W., and J.W. prepared the original draft; C.L., H.W., J.W. and M.W. edited and reviewed the manuscript. All authors have read and agreed to the published version of the manuscript.

Funding: This research was funded by Key-Area Research and Development Program of Guangdong Province (No. 2019B090912002). The work of J.W. was sponsored by Shanghai Sailing Program (No. 20YF1452600) in part.

Conflicts of Interest: The authors declare no conflict of interest.

References

1. Bray, D. Building 'Community': New strategies of governance in urban China. *Econ. Soc.* **2006**, *35*, 530–549. [CrossRef]
2. The Ministry of Civil Affairs Views on Promoting Urban Community Building throughout the Nation. Available online: http://www.cctv.com/news/china/20001212/366.html (accessed on 15 July 2020).
3. Guidance for Community's Precise and Refined Prevention, Control and Service toward COVID-19. Available online: http://www.gov.cn/zhengce/zhengceku/2020-04/16/content_5503261.htm (accessed on 10 August 2020).
4. Notice of the Central Government on Care Measures to Front-Line Community Workers against COVID-19 in Urban and Rural Areas. Available online: http://www.gov.cn/zhengce/content/2020-03/04/content_5486761.htm (accessed on 10 August 2020).
5. Intelligent Community Network of China. Available online: https://www.intelligentcommunity.org/china_communities (accessed on 2 September 2020).
6. Shanghai 13th 5-year Plan of Improving Smart City Construction. Available online: http://fgw.sh.gov.cn/wcm.files/upload/CMSshfgw/201706/201706020412003.pdf (accessed on 3 September 2020).
7. Contreras-Masse, R.; Ochoa-Zezzatti, A.; García, V.; Pérez-Dominguez, L. Implementing a novel use of multicriteria decision analysis to select IIoT platforms for smart manufacturing. *Symmetry* **2020**, *12*, 368. [CrossRef]
8. Afzal, M.; Huang, Q.; Amin, W.; Umer, K.; Raza, A.; Naeem, M. Blockchain enabled distributed demand side management in community energy system with smart homes. *IEEE Access* **2020**, *8*, 37428–37439. [CrossRef]
9. Dixon, B.; Johns, R. Vision for a holistic smart city-HSC: Integrating resiliency framework via crowdsourced community resiliency information system (CRIS). In Proceedings of the 2nd ACM SIGSPATIAL International Workshop on Advances on Resilient and Intelligent Cities (ARIC 2019), Chicago, IL, USA, 5 November 2019; pp. 1–4.
10. Sánchez, L.; Muñoz, L.; Galache, J.A.; Sotres, P.; Santana, J.R.; Gutiérrez, V.; Ramdhany, R.; Gluhak, A.; Krco, S.; Theodoridis, E.; et al. SmartSantander: IoT experimentation over a smart city testbed. *Comput. Netw.* **2014**, *61*, 217–238. [CrossRef]
11. Nam, T.; Theresa, T.A. Conceptualizing smart city with dimensions of technology, people, and institutions. In Proceedings of the 12th annual international digital government research conference: Digital government innovation in challenging times, College Park, MD, USA, 12–15 June 2011; pp. 282–291.
12. Farahani, B.; Firouzi, F.; Chang, V.; Badaroglu, M.; Constant, N.; Mankodiya, K. Towards fog-driven IoT eHealth: Promises and challenges of IoT in medicine and healthcare. *Future Gener. Comput. Syst.* **2018**, *78*, 659–676. [CrossRef]
13. Borgia, E. The Internet of Things vision: Key features, applications and open issues. *Comput. Commun.* **2014**, *54*, 1–31. [CrossRef]
14. Catarinucci, L.; De Donno, D.; Mainetti, L.; Palano, L.; Stefanizzi, M.L.; Tarricone, L. An IoT-aware architecture for smart healthcare systems. *IEEE Internet Things J.* **2015**, *2*, 515–526. [CrossRef]
15. Yigitcanlar, T.; Kamruzzaman, M. Does smart city policy lead to sustainability of cities? *Land Use Policy* **2018**, *73*, 49–58. [CrossRef]
16. Alizadeh, T. An investigation of IBM's smarter cities challenge: What do participating cities want? *Cities* **2017**, *63*, 70–80. [CrossRef]
17. Jindal, A.; Kumar, N.; Singh, M. A unified framework for big data acquisition, storage, and analytics for demand response management in smart cities. *Future Gener. Comput. Syst.* **2020**, *108*, 921–934. [CrossRef]
18. Yin, J.-F.; Kong, X.-H.; Zheng, N. Aitken extrapolation method for computing the largest eigenvalue of nonnegative tensors. *Appl. Math. Comput.* **2015**, *258*, 350–357. [CrossRef]
19. Li, D.; Lin, C.; Gao, W.; Meng, Z.; Song, Q. Short-term rental forecast of urban public bicycle based on the HOSVD-LSTM model in smart city. *Sensors* **2020**, *20*, 3072. [CrossRef] [PubMed]
20. Wang, X.; Yang, C.; Mao, S. TensorBeat: Tensor decomposition for monitoring multiperson breathing beats with commodity WiFi. *ACM Trans. Intell. Syst. Technol.* **2017**, *9*, 8. [CrossRef]
21. Kolda, T.G.; Bader, B.W. Tensor decomposition and applications. *SIAM Rev.* **2009**, *51*, 455–500. [CrossRef]
22. Stocco, L.J.; Schrack, G. On spatial orders and location codes. *IEEE Trans. Comput.* **2009**, *58*, 424–432. [CrossRef]

23. Yang, M.-H.; Kriegman, D.J.; Ahuja, N. Detecting faces in images: A survey. *IEEE PAMI* **2002**, *24*, 34–58. [CrossRef]

24. Tucker, L.R. Some mathematical notes on three-mode factor analysis. *Psychometrika* **1966**, *31*, 279–311. [CrossRef] [PubMed]

25. Lieven, D.L.; Bart, D.M.; Joos, V. A multilinear singular value decomposition. *SIAM J. Matrix Anal. Appl.* **2000**, *21*, 1253–1278.

26. Hitchcock, F.L. The expression of a tensor or a polyadic as a sum of products. *J. Math. Phys.* **1927**, *6*, 164–189. [CrossRef]

27. Oh, S.; Park, N.; Lee, S.; Kang, U. Scalable Tucker factorization for sparse tensors-algorithms and discoveries. In Proceedings of the 2018 IEEE 34th International Conference on Data Engineering (ICDE), Paris, France, 16–19 April 2018; pp. 1120–1131.

28. Papalexakis, E.E.; Faloutsos, C.; Sidiropoulos, N.D. Parcube: Sparse parallelizable tensor decompositions. In *Joint European Conference on Machine Learning and Knowledge Discovery in Databases*; Flach, P.A., De, B.T., Cristianini, N., Eds.; Springer: Berlin/Heidelberg, Germany, 2012; pp. 521–536.

29. Kaya, O.; Uçar, B. Scalable sparse tensor decompositions in distributed memory systems. In Proceedings of the SC'15: Proceedings of the International Conference for High Performance Computing, Networking, Storage and Analysis, Austin, TX, USA, 15–20 November 2015; pp. 1–11.

Publisher's Note: MDPI stays neutral with regard to jurisdictional claims in published maps and institutional affiliations.

MDPI
St. Alban-Anlage 66
4052 Basel
Switzerland
Tel. +41 61 683 77 34
Fax +41 61 302 89 18
www.mdpi.com

Sensors Editorial Office
E-mail: sensors@mdpi.com
www.mdpi.com/journal/sensors

www.ingramcontent.com/pod-product-compliance
Lightning Source LLC
LaVergne TN
LVHW070138170726
843515LV00007B/1826